ECONOMIC MICROBIOLOGY

ECONOMIC MICROBIOLOGY

Edited by

Dr. Esha Rami

Delve
PUBLISHING

www.delvepublishing.com

Economic Microbiology

Dr. Esha Rami

Delve Publishing
4164 Lakeshore Road
Burlington, ON L7L 1A4
Canada
www.delvepublishing.com
Email: orders@arclereducation.com

ISBN: 978-1-77956-364-4 (Hardcover)

ABOUT THE EDITOR

Dr. Esha Rami is presently working as an Assistant Professor, in the Department of Life science, Parul Institute of Applied Science, Parul University, India. She Did her Post graduated from Ganpat University, Ph.D. in biotechnology from Hemchandracharya North Gujarat University, in the year 2015. She has authored a number of national and international publications in reputed journals.

TABLE OF CONTENTS

LIST OF FIGURES

LIST OF TABLES

ABSTRACT

The book "Economic Microbiology" delves into the economic implications of microbial processes, exploring how microorganisms contribute to industrial production. The book comprises ten chapters, each examining a specific product or process with substantial industrial relevance. Chapter 1 introduces the production of primary microbial metabolites, emphasizing their economic roles. Chapters 2 through 4 cover acetone-butanol fermentation, organic acids, and acetic acid, exploring their biochemical pathways, industrial applications, and economic significance. Chapter 5 focuses on nucleotides and their uses in pharmaceuticals and food, while Chapter 6 discusses microbial production of amino acids, essential in various sectors. Chapter 7 investigates lipids and fatty acids, highlighting microbial synthesis and applications. Chapters 8 and 9 explore microbial vitamin production and polysaccharides, showcasing their industrial roles and economic importance. The final chapter examines polyhydroxy alcohols produced by osmotolerant yeasts. This interdisciplinary approach connects microbiology and economic theory, illustrating how microbial processes fulfill economic objectives through resource-efficient, industrial applications.This book serves as a resource for students, researchers, and industry professionals interested in harnessing microbial processes to improve economic productivity and address global challenges sustainably.

PREFACE

Economic decision-making is crucial for achieving the objectives of individuals, private businesses, and public organizations. For individuals, these decisions are primarily driven by personal goals, while for private enterprises and societal organizations, the focus is on broader public goals. Regardless of the specific context, the overarching aim remains the same: to manage or mitigate uncertainty and to allocate and utilize limited resources as efficiently as possible.

Traditionally, classical and neoclassical economic theories have centered on the notion of rational decision-making, where individuals and organizations are assumed to make choices based on logical reasoning and complete information. However, more recent approaches in bio- and neuro-behavioral economics have expanded this view to include factors such as bounded rationality, irrationality, inconsistencies, temporal biases, judgment errors, and cognitive distortions. These contemporary models strive to capture the complexities of a modern society grappling with increasing asymmetries, inequalities, crises, and shocks.

To achieve this, these advanced frameworks integrate insights from various scientific disciplines that were once considered unrelated or isolated. They draw on knowledge from biology, physiology, biochemistry, psychology, and neuroscience. This interdisciplinary approach has helped researchers uncover previously unnoticed or underestimated factors influencing decision-making and the drivers behind economic choices. Moreover, it has provided a more accurate portrayal of real-world economic decision-making, which can be slow, prone to errors, and far from achieving equilibrium consistently. Real-life settings are often characterized by continuous change, high levels of uncertainty, and noise, which makes it challenging to apply theoretical models perfectly.

This book, *Economic Microbiology*, is organized into ten chapters. Chapter 1 covers the production and industrial importance of primary products of microbial metabolism, emphasizing the critical roles that microbes play in producing compounds vital to various industries. It sets the stage for understanding how microbial processes contribute to industrial applications, focusing on their economic and practical significance. Chapter 2 explores acetone-butanol fermentation, discussing its historical development, biochemical pathways, and the solvents' industrial applications, highlighting the economic impact and the technological advancements that have optimized this fermentation process. Chapter 3 examines organic acids, detailing the types of organic acids produced by microorganisms, their biosynthetic pathways, and their extensive applications in food, pharmaceuticals, and other industries, showcasing their versatility and industrial value. Chapter 4 shifts focus to acetic acid and its derivative, vinegar, covering microbial pathways for production, methods of vinegar production, and the economic implications of this industry. This chapter illustrates how microbial processes are integral to producing economically valuable products. Chapter 5 addresses the production of nucleotides by microorganisms, exploring their biosynthetic pathways and applications in pharmaceuticals and food industries, highlighting their economic potential and significance. Chapter 6 focuses on amino acids, essential protein building blocks, discussing microbial production methods, technologies employed, and the economic importance of amino acids in various sectors. Chapter 7 dives into lipids and fatty acids, explaining their microbial production, the processes involved, and their industrial applications, emphasizing their economic and functional relevance. Chapter 8 discusses vitamins, detailing microbial production methods, the significance of vitamins in health and industry, and the economic benefits of microbial synthesis of these essential nutrients. Chapter 9 explores polysaccharides, their microbial production, the biochemical processes involved, and their wide-ranging applications in industries such as food, pharmaceuticals, and biotechnology, underscoring their economic importance. Finally, Chapter 10 examines the production of polyhydroxy alcohols by osmotolerant yeasts, discussing the biochemical pathways, production techniques, and industrial applications of these compounds, highlighting their economic value and the innovative microbial processes that facilitate their production. Each chapter provides a comprehensive look at the intersection of microbiology and industrial processes, showcasing the economic significance of microbial products and the technologies that harness microbial metabolism for industrial benefit.

Chapter **1**

Production and Industrial Importance of Primary Products of Microbial Metabolism

Contents

INTRODUCTION

Primary metabolites, which are vital substances involved in cell growth and function and have important industrial applications, are produced by microbial metabolism. Microorganisms produce amino acids, which are widely utilized in the food, pharmaceutical, and animal feed industries. Examples of these amino acids are glutamate and lysine. Monosodium glutamate (MSG) is used to improve flavor, and lysine is an essential supplement for animal nutrition. Organic acids, which are produced as metabolic byproducts and are widely used in food preservation, flavoring, and the pharmaceutical industry, include citric, lactic, and acetic acids. Lactic acid is essential to dairy products and biodegradable plastics, while citric acid functions as an acidulant and preservative. Microbial fermentation produces alcohols like ethanol, which are essential for use as biofuels, solvents, and in the manufacturing of alcoholic beverages. Furthermore, microbial fermentation produces vitamins like B12 and riboflavin (B2), which are critical for pharmaceuticals and food fortification. These basic metabolites are essential to many different industries; for example, microbial processes play a major role in the production of fermented foods and drinks such as cheese, yogurt, beer, and wine. Microbial fermentation is essential to the pharmaceutical industry's production of vaccines, enzyme supplements, and antibiotics. Additionally, microorganisms are essential to agriculture because they produce biopesticides and biofertilizers, which support environmentally friendly farming methods. The production of biofuels like ethanol and biodiesel, which are made possible by microbial metabolism, provides renewable energy sources that are in line with international sustainability objectives. The potential of microbial metabolism has been further enhanced by developments in genetic engineering and synthetic biology, opening the door to the creation of microbial cell factories that can produce a diverse array of bio-based products. These developments demonstrate the adaptability and significance of microorganisms in industrial settings, offering environmentally friendly substitutes for conventional chemical processes and facilitating the shift to a bio-based economy.

1.1. NATURE OF PRIMARY PRODUCTS OF MICROBIAL METABOLISM

Primary products of microbial metabolism are vital substances generated in the course of microbial cell growth and maintenance. These metabolites are essential for the fundamental operations of microbial cells and their general metabolism.

1.1.1. Growth Phases in Batch Culture

Microbiology students were expected to describe bacterial growth in pure culture more fully and quantitatively after Robert Koch (1843–1910) developed methods for doing so. This breakthrough made the study of microorganisms an infinitely more appealing field of study. To begin with, many of them endeavored to estimate the rate of growth of bacteria which Nägeli (1887), for example, did by determining the amount of acid produced in a culture whilst others (e.g., Buchner et al., 1887) used the plating-out method of Koch quantitatively.

These early investigations demonstrated that bacteria do not start to multiply right away when they are freshly inoculated into a suitable medium; rather, they do so after a brief period. These kinds of observations, which Lane-Claypon (1909) and later Slator (1917) greatly advanced, eventually led to the identification of the growth phases in batch cultures of microorganisms, which is a fundamental concept in the field of microbiology. The credit for defining the lag, exponential, and stationary phases of growth must go to Lane-Claypon (1909). However, Buchanen's (1918) paper is a significant milestone in the description of growth phases in batch culture, even though he was ambitious enough to identify seven distinct phases of growth (Figure 1.1).

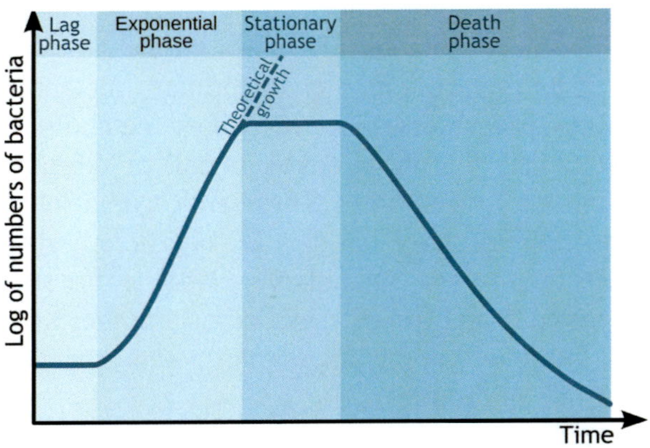

Figure 1.1. *Generalized growth curve for a unicellular micro-organism.*

Source: By M Komorniczak -talk-Illustration by: Michał KomorniczakThis file has been released into the Creative Commons 3.0. Attribution-ShareAlike (CC BY-SA 3.0)If you use on your website or in your publication my images (either original or modified), you are requested to give me details: Michał Komorniczak (Poland) or Michal Komorniczak (Poland).For more information, write to my e-mail address: m.komorniczak.pl@gmail.com – This W3C-unspecified vector image was created with Inkscape., CC BY-SA 3.0, https://commons.wikimedia.org/w/index.php?curid=7731719.

1.1.2. Primary and Secondary Products of Metabolism

A metabolite that is directly involved in normal growth, development, and reproduction is known as a primary metabolite. It typically carries out an organism's physiological function (i.e., an inherent purpose). Most organisms or cells normally contain a primary metabolite. It is also known by the more restrictive term "central metabolite," which is defined as "present in any autonomously growing cell or organism." A few typical instances of primary metabolites are specific amino acids, ethanol, and lactic acid. Because they play a fundamental role in basic cell metabolism, these compounds are found concentrated in seeds and vegetative storage organs of higher plants, where they are essential for physiological development. Primary metabolites that are extracted from higher plants for commercial use are typically bulk chemicals with a high volume and low value. They include goods like vegetable oils, fatty acids (used to make soaps and detergents), and carbohydrates (sucrose, starch, pectin, and cellulose), and are primarily used as industrial raw materials, foods, or food additives. This rule does have some exceptions, though. For instance, because of the difficulties in their extraction, isolation, and purification, myoinositol and ß-carotene are pricey primary metabolites.

Primary metabolites are compounds that a plant makes and uses for growth and metabolism. A primary metabolite is vital to a plant's existence and plays a significant role in its metabolism. Certain primary metabolites function as secondary metabolites' precursors. Since secondary metabolites do not directly contribute to an organism's growth, development, or reproduction, they are not as essential as primary metabolites. These are organic substances that help plants grow and develop normally, but they are not directly responsible for the survival of the plants themselves. Secondary metabolites are substances that are biosynthetically derived from primary metabolites, but they are only found in a specific taxonomic group within the kingdom of plants (species, genus, family, or closely related group of families).

Although secondary compounds are not evident in a plant's primary metabolism, they frequently play an important ecological role. For example, they can be used as chemical defenses against microbes, insects, higher predators, and even other plants (allelochemicals), or they can attract pollinators. Plants usually accumulate secondary metabolites in smaller amounts than primary metabolites. Furthermore, secondary metabolites are challenging. Therefore, in contrast to primary metabolites, secondary metabolites that are used in the pharmaceutical, flavor, fragrance, and pesticide industries as biologically active compounds are typically higher-value, lower-volume products. They are also typically synthesized in specialized cell types and at distinct developmental stages, which makes them extraction and purification metabolites. As a result, many secondary metabolites can be thought of as fine chemicals or specialty materials as opposed to primary metabolites, which are bulk chemicals.

Secondary metabolites are often large organic molecules that require a large number of specific enzymatic steps for production:

- Synthesis of tetracycline requires at least 72 separate enzymatic steps
- Starting materials arise from major biosynthetic pathways

Some examples of commercially valuable secondary metabolites from plants are nicotine, pyrethrins, and rotenone, which are used as pesticides in small amounts, as well as some steroids and alkaloids that are used by the pharmaceutical industry to make drugs. Steroidogenic sapogenins, Digitalis glycosides, belladonna alkaloids (e.g., atropine, hyoscyamine, and scopolamine), anticancer alkaloids from Catharanthus (formerly Vinca), cocaine, colchicine, opium alkaloids (e.g., codeine, morphine, and papaverine), physostigmine, pilocarpine, quinine, quinidine, reserpine, and d-tubocurarine are among the steroids and alkaloids. Limited amounts of other secondary plant metabolites are employed as pharmacological instruments to investigate different biochemical processes. For instance, the latices of several species of Euphorbia, which belong to the spurge family, contain diterpene esters, which include phorbol derivatives. These compounds are effective irritants and cocarcinogens and can be helpful in research on chemical carcinogenesis.

1.1.3. Ethyl Alcohol or Ethanol Production from Molasses by Fermentation

Molasses is produced by the sugar industry through the processing of sugar cane. Molasses, with the chemical formula $C_{12}H_{22}O_{11}$, contain 50–55% sugar in the form of sucrose. This compound's source is utilized to make ethyl alcohol. Molasses can be used to make absolute and rectified spirits, which are two forms of ethanol. An industry needs the following basic raw materials to produce 1 ton of ethyl alcohol: up to 5.6 tons of molasses, 27 kg of sulfuric acid, and 2.5 kg of ammonium sulfate.

The chemical reactions involved in ethanol production from molasses:

- Main reaction:

  $C_{12}H_{22}O_{11} + H_2O \rightarrow 2C_6H_{12}O_6$, with enzyme invertase.

  $C_6H_{12}O_6 \rightarrow 2C_2H_5OH + 2CO_2 – 31.2$ KCal, with enzyme zymase.

- Side reaction:

  $2C_6H_{12}O_6 + H_2O \rightarrow ROH + RCHO$ (high molecular weight alcohols C_2H_5OH)

1.1.3.1. Description of the Ethanol Production Process

Large molasses storage tanks hold fresh molasses from the sugar processing section during the fermentation process and offer a constant supply of molasses. To get a sugar concentration of about 10–15%, water is added to the molasses from the tanks. Because molasses is naturally acidic, which helps yeast grow when sucrose breaks down, acids are added to keep the pH between 4 and 5. This is done by continuous diluter equipment. A yeast culture tank is used to provide nutrients to the yeast, such as ammonium and magnesium phosphate or sulfate. The yeast is more likely to produce the catalytic enzymes zymase and invertase in an acidic environment.

The yeast from storage is fed into the fermentation chamber along with diluted and treated molasses. These days, fermentation tanks are constructed of stainless steel and have jackets or heating coils installed. The heating and cooling system keeps the tanks' temperature between 20 and 30°C. Depending on the temperature, sugar concentration, and yeast count, the fermentation process takes between 30 and 70 hours to complete. After the procedure, a final temperature of 35°C is reached. Carbon dioxide is a byproduct of fermentation produced by microorganism yeast. Following the cycle of the process, distillation is carried out by feeding the product liquid mixture into a beer still. The alcohol and water solution is left behind after the separation of the solid and slurry mass. The liquid mixture's alcohol content would be between 8 and 10%. A range of beer stills is utilized to generate varying-quality beer products. Slops are the slurry-like materials that are extracted from the bottom of beer stills. It is utilized for fertilizer and cattle feed following certain waste treatment procedures. Aldehyde columns are used to eliminate the aldehydes from the solution because they are prohibited in beer that is intended for human consumption. The streams exiting the column at different sections are the bottom stream with water, the middle stream with ethanol mixture and fusel oil, and the top stream with aldehydes.

The middle stream is fed into a rectification column, which yields rectified spirit, a product with 95% ethanol. Benzene is the third component used in anhydrous stills to further convert rectified spirit to absolute alcohol. A common product used as a blending agent in power fuels and as an intermediate for the synthesis of other chemical products is absolute alcohol with 100% ethanol concentration. The production of solvent, acetaldehyde, acetic acid, polyethylene, and synthetic rubber would be the main end uses for the ethanol.

1.1.3.2. Application of Ethanol as Fuel in IC Engines

Methanol or ethanol produced biologically is blended with gasoline. Cars and other gasoline-powered vehicles that are currently driven on public roads use this kind of fuel. According to scientists, utilizing these fuels will lower greenhouse gas emissions. Cars with gasoline engines burn the fuel vapors in the cylinder using spark plug ignition technology. When it comes to octane number, gasoline is not the best fuel. Additives have been added to fuel in recent times to raise its octane value. The environmental effects of gasoline additives are well documented. For instance, tetraethyl lead is a hazardous substance, MTBE contaminates water supplies, and benzene and toluene are the most potent carcinogens. Nowadays, ethanol is used as an additive to replace all other dangerous substances. Its structure contains oxygen, which enhances combustion and reduces the release of carbon monoxide and other hazardous unburned hydrocarbons.

1.2. MICROBIAL METABOLISM AND ITS REGULATION

A microbe's growth and reproduction come from a complex web of intricately ordered reactions. It is convenient for the microbial biochemist to identify three distinct unit processes in microbial growth and reproduction for the sake of simplicity. The first of these is the organism's internalization of chemical compounds from its surroundings or media. Most compounds found in the environment can freely pass through a microorganism's cell wall and extramural layers, and the plasma membrane serves as the primary barrier separating a chemical compound from the metabolic machinery inside the microorganism. Only recently have the mechanisms underlying the passage of compounds into and out of microorganisms started to be understood. This is primarily because the process is vectorial, involving the movement of molecules across space at a distance of approximately 7.5 nm, which makes it less explorable than the scalar or metabolic reactions that make up the other two units processes in microbial growth and reproduction. Many of the foundational texts on microbial physiology provide descriptions of the mechanisms that are at work when compounds pass through the microbial plasma membrane (Rose, 1976; Dawes and Sutherland, 1976). A large number of solute molecules traverse the microbial plasma membrane via the action of transport or carrier mechanisms that entail particular proteins. Like in some sugar-transport processes, the solute may enter the plasma membrane as a phosphate or in a chemically unaltered state. Other solutes that are soluble in membrane lipids may diffuse through the membrane to cross it. Ironically, the process by which ethanol—the oldest and most significant primary product of microbial metabolism and the foundation of a vast array of industries—crosses the yeast plasma membrane remains largely unknown to microbial physiologists.

A nutrient undergoes a series of chemical changes, or metabolization, while it is inside a microorganism. A metabolic pathway is the set of chemical reactions that a compound goes through during its metabolization. A nutrient that enters a pathway is transformed into one or more distinct compounds, referred to as intermediates, each of which is then further metabolized to produce the pathway's final product. A particular enzyme catalyzes each reaction on a metabolic pathway. Precursory is another term that's frequently used when talking about cell metabolism. It refers to any compound that's created inside the cell or supplied in the medium and gets metabolized to produce an end product. A nutrient's molecular complexity either increases or decreases as it moves through a metabolic pathway. To differentiate them from anabolic pathways, which increase the molecular complexity of the compounds being metabolized, those pathways that cause a decrease in molecular complexity are known as catabolic pathways. Low molecular weight (C2, C3, and C4) compounds and energy (ATP) are provided to the cell by catabolic pathways. Some of these compounds then enter anabolic pathways to be utilized in the synthesis of cell components. Additional low molecular weight substances are eliminated as waste from metabolism. Reactions on both catabolic and anabolic pathways are subject to a variety of control or regulatory mechanisms the aim of which is to ensure that just sufficient ATP and low molecular-weight compounds are generated on catabolic pathways, while anabolic pathways do not lead to the overproduction of cell constituents. The production of industrially significant primary products of microbial metabolism takes advantage of these regulatory processes, which are rarely entirely efficient. These inefficiencies are frequently exacerbated by cultivating the microbe in environments that further disrupt cell metabolism.

1.2.1. Primary Metabolic Pathways

1.2.1.1. Catabolic Pathways

Microorganisms derive their energy from three distinct sources: visible radiation, inorganic compounds, and organic compounds. Again, introductory texts on microbial physiology can be consulted for information on all the catabolic pathways that have been identified in microorganisms; this chapter does not attempt to describe them all (Mandelstam and McQuillen, 1973; Rose, 1976). The primary catabolic pathways in organotrophic microbes—microbes that get their energy from organic compounds—are outlined in Figure 1.3. To produce pyruvate, organic substrates—typically sugars—are catabolized via a variety of C_3 and C_2 compounds along one or more of the few known glycolytic pathways (Embden Meyerhof-Parnas, hexose monophosphate, phosphoketolase, and

Entner-Doudoroff). Glycerol and other three-carbon compounds can also enter these glycolytic pathways after converting to an intermediate. ATP is produced by substrate-level phosphorylation during glycolysis. Under aerobic conditions, pyruvate is oxidatively decarboxylated to acetyl-CoA, the acetyl residue which is then completely oxidized in the tricarboxylic acid cycle. An oxidative phosphorylation chain is responsible for the oxidation of the reduced nicotinamide and flavin nucleotides, which are created when the tricarboxylic acid cycle functions. This is the aerobic respiration process, where the electron acceptor is molecular oxygen. When aerobic respiration transitions to anaerobic respiration, certain microorganisms can take over the role of oxygen in accepting electrons, converting it to compounds like nitrate and sulfate. On the other hand, a lot of organotrophic microorganisms perform fermentation, which is an energy-producing process where an organic compound serves as both an electron donor and an electron acceptor when they come into anaerobic environments. While different microorganisms handle pyruvate differently during fermentation, all of them require the regeneration of NAD+ for additional glycolysis to occur. Numerous fermentation byproducts, such as ethanol, acetone, butanol, and lactic acid—all of which are covered in this volume—have real or potential industrial value.

Strictly speaking, a fermentation process must be an anaerobic process, since an organic compound rather than molecular oxygen is acting as the electron acceptor. Even so, industrial microbiologists frequently refer to any medium or large-scale microbial culture as a fermentation and the vessel used to hold it as a fermenter, even though the vessel is frequently being purged with a large amount of air. This usage of the term stems from the early days of industrial microbiology, when acetone and butanol production, as well as the production of beer, wines, and spirits, were all considered authentic fermentation processes. In a basic text written some years ago (Rose, 1961), I made strenuous efforts to use the term fermentation in its correct physiological sense, but my example completely failed to make any impression on industrial microbiologists.

Contrasting with *Escherichia coli*, which efficiently converts its energy sources into carbon dioxide and water through aerobic oxidation, leaving minimal organic waste in the medium, microbes involved in anaerobic fermentations are substantially less efficient due to lower energy returns from substrate catabolism. Numerous microorganisms, particularly filamentous fungi, release modest quantities of organic acids, such as those involved in the tricarboxylic acid cycle, or compounds like itaconic acid, when growing aerobically. The production of citric acid commercially, utilizing *Aspergillus Niger*, takes advantage of this inefficiency, despite altering the fungal metabolism to boost citric acid excretion. A select few microbes, like acetic-acid bacteria, even while growing aerobically, perform incomplete oxidation of their organic substrate, yielding compounds such as acetic acid and oxo gluconic acids.

1.2.1.2. Anabolic Pathways

The nature of intermediates on pathways that lead to the synthesis of cell constituents, namely anabolic pathways, was discovered to a large extent sometime after catabolic pathways had been charted, and while not yet complete this body of information is very nearly so. As information on anabolic pathways grew, the unity that the Dutch microbiologist Albert J Kluyver had stressed early on in the biochemistry of catabolic pathways began to emerge. It was discovered that (Kluyver, 1931; Kluyver and Van Niel, 1956) existed on biosynthetic pathways as well. Given the similarities in the pathways that lead to biosynthetic end products among all microorganisms, it makes sense that an industrial microbiologist would employ a diverse array of microbes to produce commercially valuable compounds on a large scale. However, this is not so, for the majority of industrially useful micro-organisms are organotrophs. The reasons for this are mainly economic. Growing algae on a large scale requires artificial light and higher partial pressures of carbon dioxide, which can be expensive (Vincent, 1971). On the other hand, growing chemolithotrophic bacteria can be challenging and usually results in low cell yields.

- **Nucleic Acids and Proteins:** Cell metabolism is based on the synthesis of proteins and nucleic acids (DNA and RNA), as most of the proteins that are produced—which translate information encoded in DNA—have regulatory or enzymatic functions. The ammonium ion, which is the form in which nitrogen is made available in many fermentation processes, is incorporated into one or more of three amino acids, namely α-oxoglutarate, aspartate, or alanine. Following that, these amino acids are utilized as sources of the amino group for the synthesis of 15 other amino acids. Aspartate and glutamate are amidated to form asparagine and glutamine, respectively. All microorganisms have an intracellular pool of amino acids, and in certain cases, this pool is large enough or out of balance to permit the organism to expel amino acids. On polyribosomes, amino acid residues polymerize into polypeptides in the form of their tRNA derivatives. Phosphoribosyl pyrophosphate and glutamine are precursors of inosinic acid, which is the starting point for the synthesis of the purine ribonucleotides adenylic acid and guanylic acid. A pyrimidine nucleotide called uridylic acid is created via a pathway that starts with aspartate and carbamoyl phosphate. ATP and NH_4^+ are needed for the reaction that converts UTP into CTP. A ribonucleoside diphosphate reductase catalyzes reactions that result in the formation of deoxyribonucleoside diphosphates from their corresponding ribonucleoside diphosphates. Similar to the amino acid pool, the ribonucleoside and deoxyribonucleoside pool found in microbes can grow uncontrollably large and unbalanced, which causes some of the pool's constituents to be excreted. Reactions involving nucleoside triphosphates and the proper polymerases catalyze the synthesis of deoxyribonucleic acid and ribonucleic acid.

- **Polysaccharides:** Microbial walls contain a variety of polysaccharides, often heteropolysaccharides, added to which some microbes can produce extramural layers (microcapsules and slime layers) which with a few exceptions are composed of polysaccharides. In the case that hydrolysis is required, the sugars utilized in the media for microorganisms of commercial importance are transformed into monomers and, typically, into other sugar residues at the level of the nucleoside diphosphate derivative. Many transferases are involved in the polymerization of sugar residues into polysaccharides, and the specificities of these enzymes dictate the structure of the resulting polymer. A lipid intermediate, such as dolichol phosphate or undecaprenyl phosphate, is required for the synthesis of a few wall polysaccharides, though, as it transfers preformed repeating units to the expanding polysaccharide chain. Many microbes, especially those that synthesize extramural layers, tend to overproduce polysaccharides due to the lax regulation of polysaccharide synthesis. This is especially true when the medium is relatively depleted of nitrogen sources and rich in carbon sources. The overproduction of polysaccharides by microbes has been steadily exploited in recent years, which is a development.
- **Lipids:** Polar and neutral lipids are the two types of lipids that microorganisms synthesize. The principal polar lipids are phospholipids and glycolipids, which are located in membranes and because of their amphipathic nature confer on a membrane (in collaboration with membrane proteins) barrier properties. Eukaryotic microbes can also synthesize neutral lipids, which are primarily triacylglycerols. Microbial phospholipid synthesis is strictly regulated, and there are very few, if any, reports of microorganisms producing excessive amounts of phospholipids. Not so as regards triacylglycerols, for these neutral lipids appear to be over synthesized in appreciable quantities by many eukaryotic micro-organisms although the excess lipid remains intracellular.

Nevertheless, a few microorganisms, such as filamentous fungi and yeasts (species of Candida, Hansenula, and Torulopsis), excrete different kinds of lipids (Hunter and Rose, 1971). Sphingolipids, polyol fatty-acyl esters, fatty acid sophorosides, and substituted acids are some of these lipids.

The fatty-acyl residues found in microbial phospholipids and triacylglycerols are of some commercial interest, although this project has not yet been proven to be commercially feasible. Certain yeasts can produce polyols extracellularly, which may have commercial value. Lipid synthesis by microorganisms involves, firstly, the production of the coenzyme-A ester of a fatty acid, followed by esterification of the hydroxyl groups on a glycerol residue in glycerol 3-phosphate. Lipid synthesis in microorganisms is covered in detail in the texts by Weete (1974), Erwin (1973), and Wakil (1970).

- **Tetrapyrroles and Terpenes:** The two main classes of tetrapyrroles that microorganisms produce are chlorophylls and haems. Glycine and succinyl-CoA are their biosynthetic precursors. The pathway is strictly regulated because neither class of end product is overproduced to a significant degree. In contrast, many microbes overproduce the two main terpene classes found in microorganisms: carotenoids and sterols, which are derived from leucine and acetyl-CoA and are still retained within the cell. There has been interest in *Saccharomyces cerevisiae* and other filamentous fungi and yeasts as sources of sterols for use in the production of contraceptive medications. Sacch can be obtained by appropriately modifying the cultural environment ergosterol, most likely in the form of fatty-acyl ester, in up to 10% of the dry weight of cerevisiae (Dulaney et al., 1954). The reason for the ongoing interest in yeast as a sterol source is the potential scarcity of plant sterols needed to make contraceptive medications.
- **Vitamins:** Microorganisms, like all other living cells, need vitamins, primarily those in the B group, primarily for coenzymes. Goodwin's text (1963) provides an account of the various biosynthetic pathways that result in the synthesis of vitamins. The tiny amounts of vitamins that microorganisms synthesize are due to their metabolic role as components of enzymes. In addition, a lot of microbes can excrete sizable amounts of B-group vitamins due to the relatively inefficient metabolic regulation on these pathways. The majority of microorganisms are not greatly harmed by this inefficient regulation of vitamin synthesis because there is obviously little energy waste concerning the amount of vitamin expelled. This isn't the case, though, for some bacteria that are cultivated for their vitamins and whose concentration in culture fluid can reach up to 10 mg/ml.

1.2.2. Regulation of Microbial Metabolism

The control of microbial metabolism is still largely unknown to microbial physiologists, but it is evident that there are three main ways in which this control is carried out: through regulation of genetic information, regulation of enzyme action, and regulation of enzyme synthesis.

1.2.2.1. Genetic Control

The genetic material found in cells, collectively referred to as the genome is composed of DNA that has instructions for the cell's "running" encoded on it. Each gene, or sequence of genetic information consisting of roughly 600 base pairs, controls the synthesis of a single enzyme, or more precisely, a single polypeptide. A microbe is obviously unable to synthesize the corresponding enzyme if the gene for that enzyme is absent from its genome. However, not all genes are structural genes—that is, genes that code for proteins with an enzyme function. Other genes produce proteins that either stimulate or inhibit different metabolic processes. These proteins have a regulatory function.

1.2.2.2. Regulation of Enzyme Action

Regulation of enzyme action is effected in at least four different ways.

- **Access of Substrates to Enzymes:** Limiting the enzyme's ability to access the substrate is a fundamental, but frequently disregarded, method of reducing the activity of an enzyme. This is caused by several factors, including the regulation of the pace at which a substrate is transferred into the cell through the plasma membrane and the development of membrane-bound organelles like mitochondria in eukaryotic microorganisms.
- **Feedback Inhibition of Enzyme Action:** Feedback inhibition is the mechanism by which microbes control the activity of enzymes that have been studied the most. If a metabolic product, £, is created through a sequence of reactions where A is transformed into B, B into C, and so forth, feedback inhibition may function by preventing one or more of the enzymes that catalyze reactions earlier in the sequence from acting on the end product E, thereby limiting the amount of £ that is created. A class of low molecular weight end products known as effectors interact with enzymes by changing the conformation of the enzyme. They accomplish this by joining forces at the allosteric site—a location distinct from the active site—of the enzyme. Allosteric enzymes are those that interact in this way with effectors. Numerous nuances on the fundamental concept

of feedback inhibition of enzyme activity have been discovered in microorganisms. It became evident that branched metabolic pathways that produce two or more end products but share certain reactions would require a different kind of mechanism as soon as the phenomenon was demonstrated to function on linear metabolic pathways. Even though the cell might not have produced enough of the other end product to meet its immediate metabolic demands, it is obvious that overproduction of one of the end products could limit the synthesis of the other end product by feedback inhibition of the first enzyme on the pathway. One solution to this issue is to have a branch's final product on a pathway only inhibit the activity of the enzyme at the branch's base. One of several isoenzymes—proteins with the same enzymatic activity—that catalyze the initial reaction on a pathway can also inhibit the end product, slowing down the flow through the pathway.

The complexity of the operation of feedback mechanisms is exemplified by two other phenomena. One has been termed concerted feedback which operates, for example, on a pathway in bacteria that leads to the synthesis of lysine, methionine and threonine. The first step on this pathway is catalyzed by aspartate kinase (Rose, 1976), but there are only one enzyme species produced even though this is a branched pathway. This enzyme is inhibited little or not at all by individual amino-acid end products, but strongly by two of the amino-acid end products.

Because the product of glutamine synthetase's activity is involved in numerous anabolic reactions, it is a crucial enzyme in the regulation of metabolism. Its activity is controlled by the inhibition of cumulative feedback. The total residual activity of glutamine synthetase in reaction mixtures containing multiple feedback inhibitors is equal to the product of the fractional activities manifest when each is present separately in the reaction mixture, according to in vitro experiments using extracts of prokaryotic and eukaryotic microbes. Moreover, feedback inhibition affects catabolic pathways, which yield ATP and reduced nicotinamide nucleotides as their principal byproducts. While the effects of cellular concentrations of AMP, ADP, and ATP on enzyme activity are somewhat understood, the mechanism by which reduced nicotinamide nucleotides regulate enzyme activity remains poorly understood. The relative concentrations of these nucleotides in a cell can be used to calculate the energy charge or adenylate charge of the cells, using the equation:

$$\text{Energy charge} = \frac{0.5\{[\text{ADP}] + 2[\text{ATP}]\}}{[\text{AMP}] + [\text{ADP}] + [\text{ATP}]}$$

According to Chapman and Atkinson's 1977 study, the energy charge of a microorganism can be calculated using the following equation: 0–1.0. In vitro studies have demonstrated that increasing the energy charge in the reaction mixture to a high value (more than roughly 0.8) inhibits the activity of several enzymes linked to the synthesis of ATP, including isocitrate dehydrogenase. On the other hand, high energy charges in the reaction mixture speed up the activity of enzymes that catalyze reactions on anabolic pathways and consume ATP, such as aspartate kinase.

Additionally, feedback inhibition affects transport proteins. For instance, the apparent Km value for *Saccharomyces cerevisiae* uptake of glucose cerevisiae is higher in cells cultured aerobically than in anaerobic conditions. In aerobically grown yeast, the glucose-transport proteins' reduced affinity for solute seems to partially account for Pasteur's observation that, when Sacch. After being moved from an anaerobic to an aerobic environment, cerevisiae grows faster and absorbs less sugar. This is known as the Pasteur effect and was among the first observations on the regulation of microbial metabolism.

- **Modification to Enzyme Structure:** Microbes can undergo reversible chemical or physical modifications that can inhibit the activity of specific enzymes. Glycogen synthetase in *Saccharomyces cerevisiae*, for instance. When it is phosphorylated, it can change *Neurospora crassa*. Adenylylation specifically modifies and renders inactive other enzymes. When an enzyme is physically modified, it usually results from the protein joining forces with cations, which alters the protein's conformation and makes it inactive.
- **Degradation of Individual Enzymes:** Probably the most complex mechanism that has been demonstrated to date for the regulation of enzyme activity in microorganisms is the breakdown of individual enzymes that are no longer needed for metabolism by highly specific proteolytic enzymes. Once again demonstrating the microbe's popularity in fundamental cell research on *Saccharomyces cerevisiae* has produced some of the best-documented examples. When this yeast reaches the stationary phase of growth—that is, when the environment becomes such that tryptophan synthesis is no longer required—the enzyme tryptophan synthase is specifically inactivated in the cells.

1.2.2.3. Regulation of Enzyme Synthesis

The microbe can stop the overproduction of low molecular weight end products quickly and sensitively by inhibiting the action of enzymes in a variety of ways.

However, it is essentially inefficient because once enzymes are synthesized, they are either blocked from working or even destroyed. Other, more effective mechanisms that microbes use to conserve cellular protein include regulating the synthesis of enzymes, which prevents the microbe from synthesizing unnecessary enzymes.

- **Induction of Enzyme Synthesis:** A microorganism can synthesize certain enzymes regardless of the nutrients in the medium it grows in. To differentiate them from induced enzymes, which are only produced in response to the presence of an inducer in the medium—typically the enzyme's substrate or another structurally related compound—these are known as constitutive enzymes. Sequential induction is the process by which an enzyme that has been induced to form a product on the inducer can often induce the synthesis of another enzyme, and so on. The most extensively researched instance of microbial enzyme induction is unquestionably the induction of j3-galactosidase in *Escherichia coli*, an enzyme that catalyzes the hydrolysis of lactose to glucose and galactose.
- **Repression of Enzyme Synthesis:** Microbes can halt the synthesis of other enzymes when the end product of the pathway on which it functions is no longer needed, just as they can conserve cellular protein by synthesizing certain enzymes only when the substrates for these enzymes are available. The term "repression of enzyme synthesis" refers to the latter process. In contrast to feedback inhibition of enzyme activity, end-product repression of enzyme synthesis on a pathway typically affects every enzyme on the pathway. In certain metabolic pathways, like the one that produces histidine in *Salmonella typhimurium*, the final product represses the synthesis of every enzyme to the same degree. This is referred to as coordinate repression.

The processes that result in the synthesis of the branched-chain amino acids leucine, valine, and isoleucine are the subject of an intriguing example of enzyme repression. A homologous series of reactions are involved in these two relatively distinct pathways, and enzymes catalyze reactions on both pathways. Research involving multiple bacteria has demonstrated that the production of common enzymes is not inhibited until excessive concentrations of valine, leucine, and isoleucine are present concurrently. The term "multivalent repression" refers to this kind of coordinated action.

The coordinated action of the different mechanisms that control enzyme synthesis and those that modify enzyme action results in the regulation of microbial metabolism. The role that various mechanisms play in overall metabolic regulation for some pathways has been thoroughly studied. The ß-keto-adipate pathway for the catabolism of aromatic compounds in pseudomonads is one of the most well-studied (Stanier and Ornston, 1973). Numerous bacterial pathways involved in the synthesis of amino acids have also been thoroughly investigated.

- **Molecular Mechanisms in the Regulation of Enzyme Synthesis:** The identification of the molecular mechanisms governing the regulation of gene expression, or the on/off switching of genes, during the last 15 years has been one of the greatest advances in molecular biology. Both positive and negative control of protein synthesis are involved in these mechanisms. The concept of negative control of protein synthesis was first introduced in the seminal work of the late Jacques Monod and Francois Jacob (1961). They postulated that structural genes, which are grouped on the genome, regulate the synthesis of enzymic proteins, also known as polypeptides. They also proposed that distinct regulator genes are in charge of generating compounds, or proteins, that function by regulating the expression of a collection of structural genes. These proteins are referred to as repressor proteins, and they regulate the expression of mRNA synthesis on a distinct region of the genome rather than on individual structural genes. We refer to this location as the operator gene. Another genetic component is the promotor region, which is the location on the genome where RNA polymerase attaches itself. Its existence was not suggested by Jacob and Monod, but subsequent experiments revealed its function. An operon is a collection of structural genes, promotor region, and related regulator and operator genes. It is important to keep in mind that, despite the overwhelming evidence of its operation during gene expression in prokaryotic microbes, the Jacob and Monod (1961) mechanism's presence in the eukaryotic genome is still only partially supported.

According to Jacob and Monod's 1961 proposal, protein synthesis can only be induced when the operator gene and repressor protein are not in contact. When the repressor protein is in contact with the operator gene and there is no inducer present, either in the medium or produced by the cell, the expression of the related structural genes is inhibited. It is believed that the repressor protein functions by partially obstructing the promotor region, thereby preventing RNA polymerase from coming into contact with it. The idea behind the repression of enzyme synthesis was that the repressor protein does not have any affinity for the operator gene when the effector molecule is not present, but that it develops an affinity for the operator gene after changing its conformation through interaction with the effector. These mechanisms are all characterized by the fact that gene expression is contingent upon the operator gene's freedom; consequently, the mechanisms are classified as negative.

Less research has been done on positive control of protein synthesis, which requires molecules to be present at specific loci on the genome. It does, however, function when *E. coli* uses arabinose (Rose (1976). One gene (*ara* C) on this operon produces a protein that joins forces with the inducer, arabinose, to create an activator molecule that functions at an initiator site within the operon. Thus, positive regulation of the expression of the *ara* structural genes is attributed to this activator molecule. During catabolite repression, positive control of protein synthesis is also in effect. Cyclic adenosine monophosphate (cAMP) relieves catabolite repression, which is the start of gene expression. It does this by forming a complex with a protein called the catabolite gene activator, which makes it easier for RNA polymerase to bind to the promotor region of the genome. Catabolites function by regulating the amount of cAMP inside cells.

1.3. BIOCHEMICAL BASIS FOR INDUSTRIAL PRODUCTION OF PRIMARY PRODUCTS OF METABOLISM

The principal metabolic pathways of microbes and the mechanisms governing their activity. Emphasis was placed on how inefficient the various pathways' operations are, as demonstrated by the microbe's excretion of an end product of primary metabolism or of an intermediate on a pathway leading to the synthesis of one of these products, or by overproducing these products. The general goal of an industrial microbiologist working on the synthesis of primary products of metabolism is to choose the right strains of the microbe and modify the medium and surrounding environment to optimize the production of the primary product. Compared to the production of secondary products of metabolism, which are typically excreted to a much greater extent than primary products, the task is far more difficult. Understanding the metabolic pathway that the primary product is synthesized along greatly aids the industrial microbiologist in choosing the best strain and culture conditions, or in controlling the microbe's metabolism. However, as many of the chapters in this volume point out, understanding the biosynthetic pathway is not necessary because a production process can be significantly improved with a purely practical approach.

1.3.1. Manipulation of Microbial Metabolism

The ability of certain chemical compounds to depress or inhibit microbial growth has been known for many years. It is possible to say that these substances can disturb microbial metabolism to the point where growth is impacted. However, the industrial microbiologist is more interested in a slightly different kind of manipulation wherein a microbe's metabolism is changed to produce a different end product or excrete more compounds, all without having a material impact on the organism's growth. Numerous instances of this kind of manipulation are reported in the literature, and they can be divided into two categories: those where metabolism is changed by changing the medium's composition, and those where the genome's composition is changed.

When the Embden-Meyerhof-Parnas pathway—the first metabolic pathway—was being mapped out in 1918, the first documented instance of modifying microbial metabolism was being reported. In 1918, Neuberg and Reinfurth found that when sulfite was added to a medium containing *Saccharomyces cerevisiae*. When cerevisiae fermented sugar, glycerol rather than a combination of ethanol and carbon dioxide was the primary byproduct. This was referred to as Neuberg's "second form of fermentation," the first being the process that yields carbon dioxide and ethanol. The second type of fermentation relies on sulfite's capacity to bind acetaldehyde and obstruct the processes that regenerate NAD+. Rather, during the reduction of dihydroxyacetone phosphate to glycerol 3-phosphate, which is then hydrolyzed to produce glycerol, the oxidized carrier is renewed (Sols et al., 1971). A brief commercial application of Neuberg's second fermentation process produced glycerol a third method of *Saccharomyces cerevisiae* fermentation. Subsequently, it was reported that the yeast medium's pH was changed to an alkaline level, leading to the growth of *Saccharomyces cerevisiae*. Since these initial reports, a great deal more environmental perturbations of microbial metabolism have been reported; these are too many to list here, but they result in either increased excretion of a product or excretion of a completely different compound. In many of these, a microbe is stressed by being deprived of an adequate amount of a necessary nutrient.

George Beadle and associates developed methods in the early 1940s for selecting mutant strains of microorganisms. Their *Neurospora crassa* mutants demonstrated the enormous value of mutants in microbial physiology studies and contributed significantly to the mapping of the first biosynthetic pathways. Thus, it can be said that having mutant strains of a microbe is preferable to not having them for a microbial physiologist. It was not long before methods for choosing mutant strains of bacteria and fungi were created. Following the publication of the Jacob and Monod (1961) theory of regulation of gene expression, this research was evidently moving in a more positive direction as it became possible to examine the molecular underpinnings of a mutant's behavior (Drake, 1970). These days, studying the physiology of microorganisms relies heavily on microbial mutants.

1.3.2. Application of Metabolic Manipulations in Industrial Fermentations

1.3.2.1. Environmentally-Controlled Perturbations

Industrial microbiologists immediately launch research programs to find the best environmental conditions for maximum excretion (or sometimes optimum excretion, taking cost considerations into account) of a specific chemical compound that is excreted by, or found in the cell mass of, a microbe. This is because the chemical compound can be produced more cheaply through a microbiological process than through chemical synthesis. Generally speaking, this is a practical exercise, but when known, information about the nature of precursors and intermediates on metabolic pathways is used to guide this experimental strategy. Optimizing the cultural conditions, including temperature, oxygen transfer rate, and pH level, can typically result in a significant increase in productivity at the start of this exercise. One instance is the synthesis of vitamin B12 by Propionibacterium species was significantly increased by first growing the bacteria anaerobically and then growing it aerobically.

Anaerobic conditions are imposed on *Saccharomyces cerevisiae*, which results in the fermentation end products of ethanol and carbon dioxide, could be regarded as a type of metabolic modification. Adding a product's precursor to the medium increases the production of numerous secondary products of metabolism. Fewer instances of this kind of "directed fermentation" (Perlman, 1973) are found in the synthesis of primary metabolic products, though *Propionibacterium* spp. uses cobalt salts to boost their production of vitamin B12 and species of Streptomyces. There are numerous other instances of the manipulation of a microbe's metabolism to boost productivity through the use of a metabolic inhibitor or poison in the large-scale production of primary products of metabolism. It is suggested that adding ferrocyanide to the beet-molasses medium, which *Aspergullus Niger* uses to produce citric acid, significantly increases productivity because ferrocyanide sequesters cations. A similar explanation has been adduced to explain the stimulatory effects of methanol and other short-chain aliphatic alcohols on citric acid production by *A. Niger* in a blackstrap-molasses medium. Riboflavin production by clostridia is stimulated by:-dipyridyl, although no satisfactory explanation of the effect of this inhibitor has been forthcoming (Perlman, 1973).

Amino acid fermentations exhibit some of the most intriguing effects of inhibitors on the synthesis of primary products of metabolism. For instance, when the permeability of the cell envelope increases, mutant strains of bacteria that are auxotrophic for biotin, oleic acid, or glycerol can excrete glutamate more readily. This phenomenon can also be caused by the action of penicillin. The industrial production process makes use of the final one of these compounds.

1.3.2.2. Strain Selection

A logical first step in process development is to investigate additional strains of the microbe once it has been demonstrated that a specific strain or species of microbe is capable of producing a commercially useful compound on a scale that can make the microbiological process competitive with the chemical synthesis of the compound. This was discovered in one of the oldest fermentation processes, the clostridial production of butanol and acetone, when strains that produced more butanol and less acetone were isolated, with the former compound being the desired product. In recent times, strain selection—which typically entails the generation of mutants—has evolved into a crucial step in the process of developing any industrial microbiological procedure. Utilizing mutants can be very fruitful if combined with a thorough understanding of the relevant biosynthesis pathway. The commercial production of primary products of metabolism through the use of mutant strains of microbes is perhaps best exemplified by the amino-acid fermentation process. Reading through the proceedings of the first two International Symposia on the Genetics of Industrial Microorganisms will give you a good idea of the impact that genetics is making in industrial microbiological processes.

1.4. INDUSTRIAL IMPORTANCE OF PRIMARY PRODUCTS OF MICROBIAL METABOLISM

Many different primary products come from microbial metabolism. The ones that are produced commercially are chemical compounds that are important to industry and are more expensive to synthesize chemically than they are when produced through a microbiological process. Perlman (1973b) has updated and expanded his useful inventory of the principal industrial fermentations that were in use at the time (Perlman, 1977). Additionally, Brown (1976) has put together a review of the global production of fermentation products.

1.4.1. Fermentation Products

There are very few commercially produced true fermentation products—that is, compounds that are byproducts of catabolism in which organic compounds serve as both the electron acceptor and the donor—that are truly fermentation products. The most significant of these historically and in terms of output is ethanol, which is made by distilling the fermented medium after sugars are fermented by *Saccharomyces cerevisiae* aromyces cerevisiae strains. These steps are extremely similar to those in this series for producing neutral, drinkable spirit. Before the 1930s, when cheap petroleum became available, ethanol was produced largely through fermentation. However, this significant industrial solvent could now be produced much more cheaply using only chemical methods. Certain ethanol consumers, such as the pharmaceutical sector and less developed nations lacking access to plentiful petroleum resources, persisted in using fermentation ethanol to sustain the fermentation industry. The past three years have seen an explosive rise in petroleum prices worldwide, which has led many industrial microbiological companies to reevaluate whether it would be feasible to increase fermentation ethanol production. Numerous organizations have already carried this out, typically with molasses serving as the substrate (D. Rose 1976).

In the past, clostridia was used to produce butanol and acetone, two more significant industrial solvents, on a substantial scale. With its association with Chaim Weizmann, this fermentation bears significant political implications (Bunker, 1957; Hastings, 1971). However, the advent of inexpensive petroleum also caused a hindrance to this process. Despite this, the production of these solvents through fermentation was still quite active in less developed nations, such as Argentina and Spain (World Wide Survey of Fermentation Industries, 1971). Given the recent rise in petroleum prices, this industry may see a resurgence. It is possible to create lactic acid synthetically or through fermentation. Acetic acid is another organic acid that is a useful industrial chemical. This is also made entirely by chemical synthesis, though microbes produce acetic acid during the acetification process, which is the method used to manufacture vinegar from fermented worts or juices. Synthetic vinegar production is either a small industry or is prohibited by law in many nations.

During World War II, the scarcity of natural rubber led to the development of 2,3-butanediol, an industrial chemical whose microbial synthesis. During that war, the United States of America increased its production of synthetic rubber from 5,000 to nearly a million tons annually, using 1,3-butadiene and chloroprene as raw materials. While 2,3-butanediol can be chemically converted into 1,3-butadiene, microorganisms do not produce butadiene in large quantities. The primary organisms investigated as a source of 2,3-butanediol were strains of Aerobacter, Bacillus, and Serratia. The groundbreaking research was primarily conducted at the Ottawa laboratories of the National Research Council of Canada (Haskins, 1972) and in the U.S. In Peoria, Illinois, the Department of Agriculture Laboratories (Ward, 1970). The current microbiological production of 2,3-butanediol is not discussed in this volume and is quite small (Perlman, 1977 lists only one producer). But given the cost of the alternative source of synthetic rubber raw materials, which is petroleum, it might become more significant once more. Long and Patrick (1963) examined this fermentation's post-war advancements. The energy of a substrate is partitioned into the various fermentation products in microbial fermentations, as Kluyver (1931) noted many years ago. The global concern over the scarcity of energy has led microbiologists to explore the possibility of using additional fermentation products as energy sources. In a thoughtful paper, Hungate (1974) examines the relative roles of methane and molecular hydrogen and concludes that methane could very well be produced on a much larger scale as an energy source. He believes that hydrogen is less appealing than methane because it has less energy per unit volume. Many sewage treatment plants currently produce methane as a fuel source, and cattle are mobile, self-sufficient methanogenic fermenters. The storage of methane produced by microbes is one issue that comes with using this energy source more frequently. The process of turning it into methanol could help with this.

1.4.2. Metabolic Intermediates

Many of the organic acids that are produced microbiologically and serve as intermediates on the tricarboxylic acid cycle are also significant industrial chemicals. Production of this food acidulent from lemons was drastically reduced as a result of the large-scale production of citric acid by molds, which was invented in the United States of America in the 1930s. The nation most impacted by this development was Italy, even though that nation still produces a sizable amount of citric acid from fruits. According to Perlman (1977), fermentation also produces six other organic acids, which are erythorbic acid, gluconic acid, itaconic acid, 2-oxogluconic acid, a-oxoglutaric acid, and malic acid. Microbiological production of the nutritionally necessary amino acids came later, and then mostly from Japanese laboratories. Glutamic acid was produced microbiologically since the early 1950s and sold as a condiment or flavoring agent under the name monosodium glutamate. These are some of the most advanced industrial microbiological procedures, especially when it comes to using bacterial mutant strains. The steps involved in microorganisms producing nucleotides exhibit a similar degree of complexity.

The first vitamins made microbiologically were acquired in the late 1930s when a few American companies set up a plant to recover riboflavin from the leftovers of the fermentation of acetone-butanol. Despite fierce competition from chemical synthesis, some riboflavin is still produced microbiologically (Perlman, 1977). Vitamin B12 is the other vitamin that is commercially produced using microorganisms (Perlman, 1977).

1.4.3. Cell Components

Protein is without a doubt the most significant component of a microbial cell from a commercial standpoint. Large-scale microbes—often referred to as "single-cell protein"—are cultivated as a source of protein for human consumption. This series contains descriptions of these procedures. The current volume's three chapters cover the manufacturing of crucial cell components for commerce. Since the use of dextrans made by microbes as blood-plasma extenders more than 25 years ago, there has been a mild but persistent interest in the large-scale production of microbial poly *Saccharomyces cerevisiae* arides (Grönwall and Ingelman, 1944). Over the last five years, there has been a renewed focus on the large-scale production of microbial poly *Saccharomyces cerevisiae* arides, partly due to the unpredictability of plant and seaweed poly *Saccharomyces cerevisiae* aride sources. The industrial production of fatty acids using microorganisms has long been of interest, but it has never been demonstrated to be economically feasible. While the internal accumulation and secretion of polyols by osmotolerant yeasts have long been recognized, it has only been recently that Brown's clarification of their function as compatible solutes has revealed the physiological significance of the process.

1.5. MICROBIAL PRODUCTS IN THE HEALTH INDUSTRY

In the medical field, microbial products are essential because they support a wide range of therapeutic, diagnostic, and preventive uses. One of the most important microbial products is penicillin, which is produced by the Penicillium species. Penicillin was the first widely used antibiotic and helped to develop many others, including tetracycline and streptomycin. Antibiotics have revolutionized medicine by effectively treating bacterial infections. Probiotics, or live beneficial bacteria, are frequently found in supplements and fermented foods and are used to support gut health and strengthen the immune system. Important vitamins like riboflavin and B12, which are utilized in dietary supplements and fortified foods, are also produced by microbial fermentation. Microorganism-derived enzymes play a crucial role in the treatment of diabetes. For example, recombinant DNA technology in *Escherichia coli* produces insulin, which is essential for the treatment of diabetes. Other microbial enzymes aid in wound healing and digestion. Moreover, microbes are used in the manufacturing of vaccines. For example, yeast cells modified to express the viral protein are used in the production of the hepatitis B vaccine. Microbial products are frequently used in diagnostic tools, such as diagnostic kits and monoclonal antibodies, to detect diseases. Microbial products, in general, are essential to contemporary healthcare because they provide creative approaches to diagnosis, treatment, and prevention, improving patient outcomes and quality of life.

1.5.1. Industrial Production of Antibiotics

Industrial production of antibiotics involves the fermentation process, in which the source microorganism is cultivated in large (100,000–150,000 liters or more) containers with a liquid growth medium. The ideal conditions for oxygen concentration, temperature, pH, and nutrient levels must be maintained and promptly monitored, as needed. Antibiotics are secondary metabolites, so it's important to carefully regulate the population size to get the maximum yield before the cells die. After everything is finished, the antibiotic needs to be extracted and refined into a crystalline form. If the antibiotic is soluble in an organic solvent, this is easier to accomplish. If not, it needs to be eliminated by chemical precipitation, adsorption, or ion exchange.

Fermentation microorganisms are rarely exactly the same as their wild counterparts. This is so that they can produce the greatest amount of antibiotics possible through genetic modification of the species. The introduction of mutagens, such as UV light, X-rays, or specific chemicals, promotes and frequently causes mutation. Yields can be increased by 20 times or more by selecting and reproducing the strains that yield more over a number of generations. Gene amplification is another method for increasing yields; using vectors like plasmids, copies of the genes encoding the enzymes involved in the synthesis of antibiotics can be reinserted into the cell. The production and efficacy of antibiotics must be closely correlated with this process (Figure 1.2).

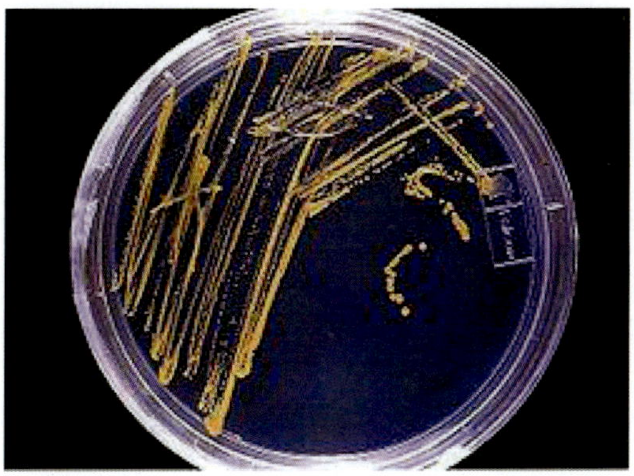

Figure 1.2. *An agar plate with microorganisms.*

Source: https://bio.libretexts.org/Bookshelves/Microbiology/Microbiology_(Boundless)/17%3A_Industrial_Microbiology/17.02%3A_Microbial_Products_in_the_Health_Industry/17.2A%3A_Industrial_Production_of_Antibiotics.

Even though there are many different types of antibiotics, less than 1% of antimicrobial agents are useful for medicine or industry. For instance, many antibiotics do not generally affect human cells, but penicillin has a high therapeutic index because of this. Other antibiotics just don't work any better than those that are already on the market or have any other useful uses. Screening procedures are frequently used to find useful antibiotics.

To perform this kind of screen, isolates from a wide variety of microorganisms are cultivated and examined for the ability to produce diffusible products that impede the growth of the test organisms. Since the majority of antibiotics found by such a screen are already well-known, they must be ignored. The remaining candidates need to be evaluated for their therapeutic potential and selectivity, and the most promising ones can then be further investigated and potentially refined. A rational design program is a more recent iteration of this strategy. Rather than using tests to demonstrate general inhibition of culture, this entails screening aimed at discovering novel natural products that inhibit a specific target, such as an enzyme present only in the target pathogen.

1.5.2. Amino Acids and Vitamins

The functional groups of amines ($–NH_2$) and carboxylic acids (–COOH), as well as a side chain unique to each amino acid, combine to form amino acids, which are biologically significant organic compounds. An amino acid is primarily composed of nitrogen, oxygen, hydrogen, and carbon. There are about 500 known amino acids, which can be categorized in numerous ways. Numerous rare amino acids can be synthesized by plants and microorganisms. For instance, certain microorganisms produce lanthionine, a sulfide-bridged derivative of alanine, and 2-aminoisobutyric acid. Alamethicin and other peptidic lantibiotics contain both of these amino acids. As a crucial intermediate in the synthesis of the plant hormone ethylene, 1-aminocyclopropane-1-carboxylic acid is a tiny, disubstituted cyclic amino acid found in plants.

A class of fat-soluble vitamins with structural similarities, vitamin K is essential for the posttranslational modification of some proteins needed for blood coagulation as well as for metabolic pathways in bone and other tissues. They are derivatives of 2-methyl-1,4-naphthoquinone (3-). Two naturally occurring vitamins are included in this category: vitamin K1 and vitamin K2.

Since vitamin K1 is directly involved in photosynthesis, it is also referred to as phylloquinone, phytomenadione, or phytonadione. Plants produce vitamin K1, which is primarily found in green leafy vegetables. One of the various subtypes of vitamin K2 is connected to bone metabolism. The quantity of isoprenoid residues in the side chain of vitamin K2 homologs, or menaquinones, is what distinguishes them.

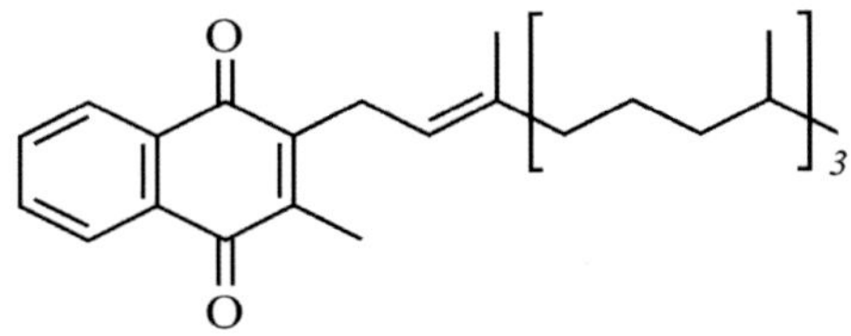

Figure 1.3. *Vitamin K1 (phylloquinone): This is vitamin K1 (phylloquinone). Both forms of the vitamin contain a functional naphthoquinone ring and an aliphatic side-chain. Phylloquinone has a phytyl side-chain.*

Source: https://bio.libretexts.org/Bookshelves/Microbiology/Microbiology_(Boundless)/17%3A_Industrial_Microbiology/17.02%3A_Microbial_Products_in_the_Health_Industry/17.2B%3A_Vitamins_and_Amino_Acids.

The acronym MK-n stands for menaquinones, where n is the number of side chain residues that are isoprenoid. For instance, the side chain of menaquinone-4, also known as MK-4, contains four isoprene residues. Large intestine (colonial) bacteria are capable of converting K1 into K2 (MK-7 homolog) and producing a variety of vitamin K2 forms. Vitamins K1 and K2 have no known toxicity. It is known that there are three synthetic forms of vitamin K: K3, K4, and K5. The artificial form K3, or menadione, has demonstrated toxicity despite the fact that the natural forms K1 and K2 are nontoxic.

1.5.3. Steroids

A steroid is a class of organic compounds characterized by a unique configuration of four joined cycloalkane rings. The sex hormones testosterone and estradiol, the dietary fat cholesterol, and the anti-inflammatory medication dexamethasone are a few examples of steroids. 20 carbon atoms bound together to form four fused rings—three cyclohexane rings (shown as rings A, B, and C in the figure to the right) and one cyclopentane ring (the D ring)—make up the core of steroids. The oxidation state of the rings and the functional groups affixed to this four-ring core distinguish the different steroids. With a hydroxyl group at position 3 and a skeleton derived from cholestane, steroids are unique types of steroids.

It is possible to find hundreds of different steroids in fungi, animals, and plants. All steroids are produced in cells, either from lanosterol (found in fungi and animals) or cycloartenol (found in plants). The triterpene squalene undergoes cyclization to produce lanosterol and cycloartenol (Figure 1.4).

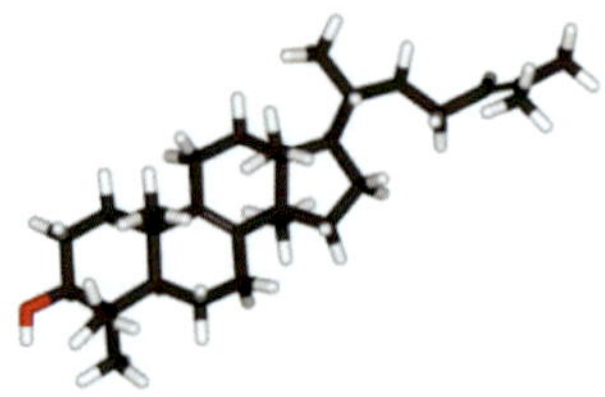

Figure 1.4. *Steroid lanosterol.*

Source: https://bio.libretexts.org/Bookshelves/Microbiology/Microbiology_(Boundless)/17%3A_Industrial_Microbiology/17.02%3A_Microbial_Products_in_the_Health_Industry/17.2C%3A_Steroids.

Anabolic metabolic pathways like steroid biosynthesis convert basic precursors into steroids. Animals follow a distinct biosynthesis pathway from many other organisms, which makes the pathway a popular target for antibiotics and other anti-infective medications. Acetyl-CoA is used as building blocks in the mevalonate pathway, which is how steroids are biosynthesized in humans and other animals, to create dimethylallyl pyrophosphate (DMAPP) and isopentenyl pyrophosphate (IPP). Pyruvate and glyceraldehyde 3-phosphate are the substrates used by the non-mevalonate pathway in bacteria and plants (Figure 1.5).

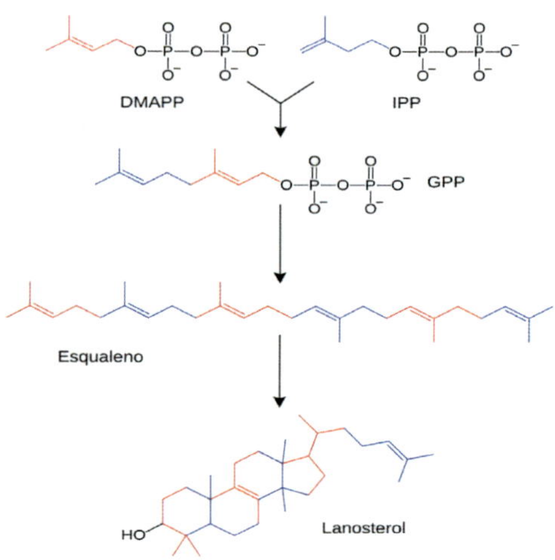

Figure 1.5. *Steroid synthesis.*

Source: https://bio.libretexts.org/Bookshelves/Microbiology/Microbiology_(Boundless)/17%3A_Industrial_Microbiology/17.02%3A_Microbial_Products_in_the_Health_Industry/17.2C%3A_Steroids.

The alternative metabolic pathway that produces isopentenyl pyrophosphate (IPP) and dimethylallyl pyrophosphate (DMAPP) during isoprenoid biosynthesis is known as the non-mevalonate pathway, or 2-C-methyl-D-erythritol 4-phosphate/1-deoxy-D-xylulose 5-phosphate pathway (MEP/DOXP pathway). All higher eukaryotes and many bacteria contain the classical mevalonate pathway, also known as the HMG-CoA reductase pathway, which is a crucial cellular metabolic pathway. It is necessary for the synthesis of IPP and DMAPP, which are the building blocks for the biosynthesis of chemicals utilized in a variety of processes, including hormone production, protein anchoring, protein prenylation, cell membrane maintenance, and N-glycosylation. Plants and apicomplexan protozoa, including malaria parasites, can synthesize their isoprenoids (terpenoids) through a different pathway called the non-mevalonate pathway, which occurs in their plastids, as opposed to the traditional mevalonate pathway of isoprenoid biosynthesis. Furthermore, the non-mevalonate pathway is used by the majority of bacteria, including significant pathogens like *Mycobacterium tuberculosis*, to synthesis IPP and DMAPP.

1.5.4. Enzymes Used in Industry

Biological molecules called enzymes catalyze chemical reactions, meaning they speed them up. The molecules at the start of the process, known as substrates, are changed into other molecules, known as products, in enzymatic reactions. Enzymes are required for almost all chemical reactions in biological cells to proceed at rates high enough to support life. The set of enzymes produced in a cell dictates which metabolic pathways take place in that cell because enzymes are specific to their substrates and only accelerate a small number of reactions out of many potential ones (Figure 1.6).

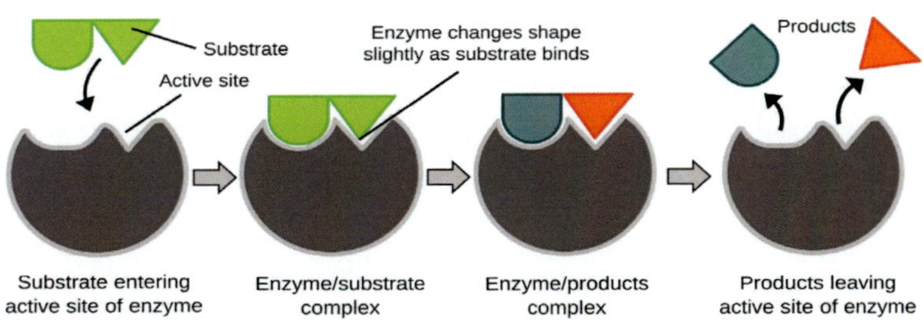

Figure 1.6. *Enzymatic catalysis.*

Source: https://bio.libretexts.org/Bookshelves/Microbiology/Microbiology_(Boundless)/17%3A_Industrial_Microbiology/17.02%3A_Microbial_Products_in_the_Health_Industry/17.2D%3A_Enzymes_Used_in_Industry.

Enzymes function similarly to other catalysts in that they speed up reactions significantly by reducing the activation energy of a reaction. As a result, reactions reach their equilibrium state faster and products are formed more quickly. Comparable uncatalyzed reaction rates are typically millions of times slower than those of the majority of enzyme reactions. Enzymes, like all other catalysts, neither change the equilibrium of the reactions they catalyze nor are consumed by them. Enzymes are very specific to their substrates, which sets them apart from the majority of other catalysts. Certain regions of the ribosome are notable examples of how certain RNA molecules known as ribozymes can catalyze reactions. Artificial enzymes, which are synthetic molecules, exhibit catalysis resembling that of enzymes.

Other molecules can have an impact on the activity of enzymes. Enzyme activity can be increased by activators and decreased by inhibitors. Enzyme inhibitors are found in many poisons and drugs. In addition, factors like temperature, pressure, the chemical environment (e.g., pH), and the amount of substrate. Certain enzymes are employed in the pharmaceutical industry, such as when creating antibiotics. Enzymes are also used in certain household products to accelerate biochemical reactions (e.g., Enzymes in meat tenderizers break down proteins into smaller molecules, making the meat easier to chew; enzymes in biological washing powders break down protein or fat stains on clothing).

When very specific catalysts are needed, the chemical industry and other industrial applications use enzymes. Nevertheless, the range of reactions that enzymes can catalyze is restricted due to their instability in organic solvents and high temperatures. As a result, efforts are being made in the field of protein engineering research to develop new enzymes with unique characteristics, either by in vitro evolution or rational design. Some enzymes have now been created "from scratch" to catalyze reactions that do not happen in nature as a result of these efforts, which are starting to show results.

The enzymes used in food processing include plant and fungal amylases. These enzymes are employed in the synthesis of starch-based sugars, including high-fructose corn syrup. They catalyze the conversion of flour's starch to sugar during baking. The carbon dioxide that raises the dough is produced when sugar is fermented by yeast. Biscuit makers use proteases to reduce the protein content of flour. Baby food is predigested using trypsin. Fruit juices are processed using cellulases and pectinases to make them more lucid. Meat is tenderized for cooking by using papain. Rennin, which is obtained from the stomachs of young ruminant animals (such as lambs and calves) and used to hydrolyze protein, is used in the dairy industry to manufacture cheese. Lipases are used in the process of making Roquefort cheese to improve the blue-mold cheese's ripening. Lactose is broken down into glucose and galactose by enzymes called lactases.

Enzymes from barley are released in the brewing process during the mashing phase of making beer. To create simple sugar, amino acids, and peptides that yeast uses for fermentation, they break down proteins and starch. The natural enzymes found in barley are frequently replaced in the brewing process with commercially produced enzymes. The poly*Saccharomyces cerevisiae* arides and proteins in the malt are broken down by amylase, glucanases, and proteases. Utilizing arabinoxylanases and beta-glucanases enhances the wort and beer's filtration properties. Pullulanases and amyloglucosidase are used to adjust fermentability and make low-calorie beer. Proteases are used to eliminate cloudiness that develops while storing beer.

Amylases, amyloglucosideases, and glucoamylases are enzymes found in the starch industry that break down starch into glucose and different syrups. For the purpose of creating high-fructose syrups from starchy materials, glucose isomerase converts glucose into fructose. Amylases, xylanases, cellulases, and ligninases are used in the paper industry to break down starch in order to reduce viscosity, which helps with coating and sizing paper. Cellulases are used in the biofuel industry to convert cellulose into fermentable sugars. Proteases, which are extracellularly produced by bacteria, are utilized in pre-soak conditions and direct liquid applications to aid in the removal of protein stains from clothing in the manufacturing of biological detergents. Restrictions enzymes, DNA ligases, and polymerases are crucial for restriction digestion and the polymerase chain reaction in molecular biology. They are also used in genetic engineering and pharmacology, agriculture, and medicine. A significant area of study in forensic science is molecular biology (Bartlett and Mercer, 1974).

REFERENCES

1. Bartlett, K., & Mercer, E. I. (1974). *Phyto-Chemistry*, *13*(1115).
2. Brown, A. D. (1978). In A. H. Rose & J. G. Morris (Eds.), *Advances in Microbial Physiology* (Vol. 17). Academic Press.
3. Brown, W. K. (1976). *Die Branntweinwirtschaft*, *116*, 216.
4. Buchanan, R. E. (1918). *Journal of Infectious Diseases*, *23*, 109.
5. Büchner, H., Longard, K., & Riedlin, G. (1887). *Centralblatt für Bakteriologie, Parasitenkunde, Infektionskrankheiten und Hygiene*, *2*, 1.
6. Bu'Lock, J. D. (1961). In W. W. Umbreit (Ed.), *Advances in Applied Microbiology* (Vol. 3, p. 293). Academic Press.
7. Bu'Lock, J. D. (1965). *The Biosynthesis of Natural Products: An Introduction to Secondary Metabolism*. McGraw-Hill.
8. Bu'Lock, J. D. (1967). In *Essays in Biosynthesis and Microbial Development* (p. 71). John Wiley & Sons.
9. Bu'Lock, J. D. (1975). In J. E. Smith & D. R. Berry (Eds.), *The Filamentous Fungi* (Vol. 1, pp. 33–58). Edward Arnold.
10. Chapman, A. G., & Atkinson, D. E. (1977). In A. H. Rose & D. W. Tempest (Eds.), *Advances in Microbial Physiology* (Vol. 16, p. 253). Academic Press.
11. Dagley, S., & Nicholson, D. E. (1970). *An Introduction to Metabolic Pathways* (343 pp.). Blackwell Scientific Publications.
12. Dawes, I. W., & Sutherland, I. W. (1976). *Microbial Physiology* (185 pp.). Blackwell Scientific Publications.
13. Drake, J. W. (1970). *Molecular Basis of Mutation* (273 pp.). Holden & Day.
14. Dulaney, E. L., Stapley, E. O., & Simpf, K. (1954). *Applied Microbiology*, *2*, 371.
15. Erwin, J. A. (1973). *Lipids and Biomembranes of Eukaryotic Microorganisms* (354 pp.). Academic Press.
16. Goodwin, T. W. (1963). *Biosynthesis of Vitamins and Related Compounds* (366 pp.). Academic Press.
17. Grönwall, A., & Ingelman, B. G. A. (1944). *Acta Physiologica Scandinavica*, *7*, 97.
18. Haskins, R. H. (1972). In D. Perlman (Ed.), *Advances in Applied Microbiology* (Vol. 15, p. 415). Academic Press.
19. Hossack, J. A., Belk, D. M., & Rose, A. H. (1977). *Archives of Microbiology*, *114*, 137.

20. Hungate, R. E. (1974). *American Society for Microbiology News, 40*, 833.
21. Hunter, K., & Rose, A. H. (1971). In A. H. Rose & J. S. Harrison (Eds.), *The Yeasts* (Vol. 2, pp. 211–270). Academic Press.
22. Jackson, E. A. (1976). *Process Biochemistry, 11*(5), 29.
23. Jacob, F., & Monod, J. (1961). *Journal of Molecular Biology, 3*, 318.
24. Kluyver, A. J. (1931). *The Chemical Activities of Micro-Organisms* (109 pp.). London University Press.
25. Kluyver, A. J., & Van Niel, C. B. (1956). *The Microbe's Contribution to Biology* (182 pp.). Harvard University Press.
26. Long, S. K., & Patrick, R. (1963). In W. W. Umbreit (Ed.), *Advances in Applied Microbiology* (Vol. 5, pp. 135). Academic Press.
27. Mac Donald, K. D. (1976). *Second International Symposium on the Genetics of Industrial Micro-Organisms* (630 pp.). Academic Press.
28. Mandelstam, J., & McQuillan, K. (1973). *Biochemistry of Bacterial Growth* (2nd ed., 582 pp.). Blackwell Scientific Publications.
29. Nägeli, K. W. (1877). *Das Mikroscop* (2nd ed., 461 pp.). Leipzig.
30. Perlman, D. (1973a). *Process Biochemistry, 2*(7), 18.
31. Perlman, D. (1973b). *American Society for Microbiology News, 39*, 648.
32. Rose, A. H. (1961). *Industrial Microbiology* (286 pp.). Butterworths.
33. Rose, A. H. (1976). *Chemical Microbiology* (3rd ed., 469 pp.). Butterworths.
34. Simpson, A. C. (1977). In A. H. Rose (Ed.), *Economic Microbiology* (pp. 537–593). Academic Press.
35. Sols, A., Gancedo, C., & Gancedo, G. (1971). In A. H. Rose & J. S. Harrison (Eds.), *The Yeasts* (Vol. 2, pp. 271–307). Academic Press.
36. Stanier, R. Y., & Ornston, L. N. (1973). In A. H. Rose & D. W. Tempest (Eds.), *Advances in Microbial Physiology* (Vol. 9, p. 89). Academic Press.
37. Vanek, Z., Hoslalek, Z., & Cudlin, J. (1973). *Genetics of Industrial Micro-Organisms* (Vol. 1, 496 pp.; Vol. 2, 510 pp.). Elsevier.
38. Vincent, W. A. (1971). *Symposium of the Society for General Microbiology, 21*, 47.
39. Weete, J. K. (1974). *Fungal Lipid Biochemistry* (393 pp.). Plenum Press.
40. World Wide Survey of Fermentation Industries. (1971). *International Union of Pure and Applied Chemistry: Information Bulletin. Technical Report No. 3*. International Union of Pure and Applied Chemistry.

Chapter 2

Acetone-Butanol Fermentation

Contents

INTRODUCTION

One of the first large-scale industrial fermentation processes to be developed was the production of acetone and butanol using solvent-producing strains of *Clostridium* spp. During the first part of this century, it was second in importance only to ethanol fermentation. The fermentation process's incapacity to financially rival the chemical synthesis of solvents was the primary cause of the fermentation's near-complete collapse in the early 1960s. However, there is now renewed interest in the fermentation route as a potential source of solvent production due to recent advancements in biotechnology and the use of renewable resources as feedstocks for chemical production. In order to create a more effective and competitive fermentation process, there has been a surge in research over the past seven years with the goal of better understanding this intriguing and complex fermentation.

Several authors have documented different facets of the evolution and history of acetone-butanol (AB) fermentation; however, the data is fragmented and lacking, and no all-encompassing history of AB fermentation exists. We have examined the history and evolution of traditional industrial AB fermentation around the globe and talked about the state of biochemical, genetic, and process engineering research today about the challenges and opportunities of reviving an industrial AB fermentation process.

2.1. HISTORY

2.1.1. Origin of AB Fermentation

Pasteur published the first account of butanol production in microbial fermentation in 1861. Many researchers looked into the process by which anaerobic bacteria produced butanol in the later half of the 1800s. Schrodinger did not, however, report the fermentation-based acetone production until 1905.

A scarcity of natural rubber at the turn of the century sparked curiosity about the viability of synthesizing rubber. One of the chemists working on the rubber synthesis problem was Chaim Weizmann, who came to Manchester in 1904 at the age of 30 from Berlin. He was hired by Manchester University to work under Professor Perkins. 10 years ago, in 1910, Strange and Graham Ltd. started a project to create synthetic rubber in England. They first enlisted the help of Perkins and Weizmann from the Institute Pasteur, and then Fernbach and Schoen. It was determined that using butanol or isoamyl alcohol was the most effective way to produce butadiene or isoprene. This sparked an inquiry into the viability of using microbial fermentation to produce the compounds. Fernbach isolated a culture in 1911 that could ferment potato starch to produce butanol, but not maize starch. Weizmann broke off his relationship with Strange and Graham Ltd. in 1912 but carried on with his studies at Manchester University. Despite not being a microbiologist, he decided to train himself to become one after concluding that the synthetic rubber process needed to successfully produce butanol or isoamyl alcohol through fermentation. He isolated and examined several cultures between 1912 and 1914; he named one of them BY, which was subsequently identified as *Clostridium acetobutylicum*. This organism was different from Fernbach's original culture in several ways, such as its capacity to use a wide range of starchy substances and its significantly higher yields of butanol and acetone. In the interim, Strange and Graham Ltd. proceeded with their business and submitted an English patent application for a method utilizing Fernbach's bacillus. They started producing acetone and butanol from potatoes at a plant in Rainham around the middle of 1913. After a year, the Rainham plant was shut down, and operations were moved to a new location in King's Lynn.

The outbreak of the First World War in August 1914 had a profound impact on the future development of AB fermentation. Large quantities of smokeless powder, or cordite, were needed by the British army to make munitions. Cordite was made by utilizing acetone as the colloidal solvent for nitrocellulose. Before the war, Germany, Austria, and the United States were the main importers of small amounts of calcium acetate, which was used to make acetone. The majority of supplies were cut off with the start of the war, and the little that was supplied by the US was insufficient. The British War Office contacted Strange and Graham Ltd. in the event of an emergency. They were given a contract in 1915 to provide the British government with acetone. The King's Lynn plant produced some acetone using potatoes as a substrate, but the average production was only 970 pounds (approximately).

At the University of Manchester, Weizmann carried on with his independent research during this time, with plans to publish his results in a scientific journal. This was changed, though, when war broke out and it became clear that his work could help the British war effort. He began bringing his work to the attention of the authorities in response to a request for cooperation from the scientific community, and a private demonstration was set up for the head of Nobel's Explosive Company's Chemical Department. The benefits of the Weizmann process impressed the head of the Chemical Department, who suggested that Weizmann file for a patent, which was granted in March 1915. Nevertheless, any additional advancements were halted by an explosion at the company's facility. After a year, Weizmann received a summons from the British Admiralty requesting him to work on a project involving the production of acetone. He consented, and pilot-scale research was initiated at the London-based Nicholson Distillery. Weizmann was granted permission to assemble a group of young scientists to work on the project after meeting with Churchill, the First Lord of the Admiralty. Midway through 1916, Weizmann left his position at Manchester University, and the group was granted access to facilities in the south of England. He oversaw larger pilot-scale development work at distilleries in London and Greenock, as well as additional research at the Lister Institute in London. The Admiralty chose to convert six distilleries in Great Britain to produce acetone using the Weizmann process and to build a plant for acetone production at the Royal Naval Cordite Factory in Poole, Dorset, as a result of the extremely encouraging results.

In 1916 the Government also took possession of the Strange and Graham Ltd. plant at King's Lynn under "The Defense of the Realm Act" and requested a change to the Weizmann process with maize as a substrate. As a result, production was increased to over 2,000 pounds of acetone per week.

However the German blockade had an impact on grain supplies, and it became clear that the fermentation process could not continue in England on the planned scale. Rationing was instituted due to the severe shortage of grain and other foods, and the government was unable to release starch for the production of solvents. Children across Britain were asked to gather horse chestnuts as a substrate for the fermentation in a last-ditch effort to save the process. Even though a laboratory experiment produced solvents from horse chestnuts, attempts to run the fermentation in England were abandoned due to the plant's difficult-to-control foaming characteristics and limited availability. Transferring the fermentation process to a region of the British Empire where a suitable substrate would be easily accessible was the obvious solution, and in 1916 the fermentation was moved to Canada. Herbert Speakman led the recruitment of several young Canadian scientists while Weizmann stayed in London. August 1916 saw the opening of the Weizmann process-adapted Gooderham and Worts Distillery in Toronto. Up until the armistice in 1918, it was in operation and produced 6,000 tons of butanol and 3,000 tons of acetone. There was also some acetone produced in France, and another plant was constructed in India, but it wasn't finished until the end of the war, at which point it was sold to the government of Bombay to be converted into a distillery.

After the United States entered the war in March 1917, the Midwest corn belt of the United States was chosen as the site of a proposed acetone production project. Two distilleries in Terre Haute, Indiana, were jointly purchased by the US government and the British War Mission. A group of scientists was dispatched from Canada, and Weizmann persisted in serving as the main advisor. The factories produced acetone for the manufacture of cordite and airplane dope from May to November 1918. Large vats were used to store the butanol that was produced, though some of it was converted to methyl ethyl ketone in Canada.

Since acetone was no longer needed, all plants closed in November 1918, when the armistice was declared.

The British government was keen to pay tribute to Weizmann for his contributions as soon as peace was declared. He made it plain that his one wish was to see a home established for the Jews in Palestine, but he declined all personal honors and rewards. Weizmann rose to prominence as the Zionist Organization's leader following the Balfour Declaration, and he eventually became Israel's first president.

2.1.2. Development of AB Fermentation

Because acetone was needed during the war, butanol accumulated as an undesirable fermentation byproduct. The butanol was kept in storage, and attempts were made to salvage it following the war. The automotive industry was growing quickly at the time, and quick-drying lacquer was needed to give car bodies a nice finish. E. I. du Pont de Nemours & Co. was at the forefront of the development of nitrocellulose lacquer, and the production of low-viscosity nitrocellulose was made possible by the process's development. It was discovered that butanol and its ester, butyl acetate, were the perfect solvents for these lacquers in response to the need for a suitable solvent.

A group of American businessmen founded the Commercial Solvents Corporation of Maryland at the end of the war because they saw the potential for industry. At an auction in late 1919, they purchased the plant in Terre Haute from the Allied War Board. Additionally, they were granted an exclusive license under the 1919 US Weizmann patent. The Terre Haute plant began producing solvents again in 1920. They began operations with a US license but subsequently obtained global patent rights to the Weizmann process and an Eloi Ricard process that was similar. The plant had to close for nine months in 1920 due to a downturn in business, but by 1923, there was a greater need for butanol, and the Terre Haute facility had grown from 40 to 52 fermentors. However, in 1923, a significant fermentation issue arose (which was later diagnosed as a bacteriophage infection). As a result, for a year, the fermentation yield was reduced by half. At the Terre Haute facility, a sizable research department was established, and Weizmann was kept on as advisor and consultant. Additionally, the company provided funding for universities to research AB fermentation, and Fred, Peterson, and others started a research program at the University of Wisconsin. While the phage infection issue at the Terre Haute plant was being looked into, the decision was made to construct a completely new plant in another state due to the growing demand for butanol. In Peoria, Illinois, the Majestic Distillery was purchased by the Commercial Solvents Corp. in 1923. When the plant first opened in December 1923, it had 32 fermentors with a capacity of 50,000 gal (189,250 liters). The plants expanded quickly between 1924 and 1927 as a result of the increasing demand for butanol. In the years 1924–1925, the Peoria plant added 50 fermentors, and between August 1925 and August 1926, output doubled. 60 tons of the 100 tons of solvents produced daily by the two plants were used to make lacquer. In 1926, the Peoria plant underwent yet another expansion, and new laboratories were constructed. Additional expansions were made in 1927, bringing the total number of fermentors operating at both plants to 148. By the end of that year, 96 fermentors were operating at Peoria. Studies were conducted on the utilization of by-products, specifically the use of the gases for the production of methanol and ammonia.

Strange and Graham Ltd 1923 took back control of their King's Lynn factory and started producing butanol and acetone again using the Weizmann culture. But Commercial Solvents Corp. began a lawsuit for infringement of British patent 4845 of 1915 at the High Court of Justice in London, in which the Weizmann patent had become vested. In the first patent case involving a biochemical process in Great Britain, Commercial Solvents Corp. prevailed in court. British patent 4845 was found to be legitimate and infringed upon in a historic ruling. An explosion at the King's Lynn plant shortly after it began full-scale operations severely damaged the facility. The factory was not restarted, and Strange and Graham Ltd. entered liquidation. In 1931, the US patent was also contested in the US, and US patent 1,315,585 was found to be legitimate and infringed upon.

Research initially focused on isolating strains capable of fermenting higher concentrations of starch because it was already recognized that the development of synthetic acetone from petroleum posed a threat. Not much progress was made, despite the isolation of numerous novel strains. An abundance of molasses was available at a price that was very competitive with starch at the start of the 1930s, prompting research into the isolation of organisms capable of fermenting sugars. Along with several other technical benefits, it was also believed that the more fluid sugar medium might allow the fermentation of a higher concentration of carbohydrates. In addition to adding soybean meal and other protein sources, attempts were made to replace a portion of the grain mash with molasses. As a result, fermentation times were extremely long, and the outcome was not very successful. After screening the current culture collections, several strains that would ferment molasses sugars to differing degrees were discovered. One of the cultures was able to ferment in media with 6% sugar, and it developed into the well-known CSC no. 8 strain—the first in a long line of improved strains. The cost of distillation was cut in half and a far less expensive substrate could be used thanks to the fermentation of 6.5% sugars to produce nearly 2 percent solvents.

In the late 1920s, the Distillers Co. in England reevaluated the process on a pilot plant scale and came to the conclusion that it was not economically viable in the conditions that prevailed at the time. But by 1935, the availability of high test molasses, low cost of the substrate, and the development of cultures capable of fermenting up to 6.5 percent sugars in molasses had changed the game. After carefully examining the circumstances, the Commercial Solvents Corp. and the Distiller's Co. decided to build a plant in England at Bromborough, which is located across the Mersey River from Liverpool. The nearby docks of Port Sunlight offered harbor facilities for the biggest tankers of the era, and shipping molasses by water was less expensive than the lengthy land route to the two major US plants in Indiana and Illinois. The factory was fully staffed by British-trained employees under the direction of J. Hastings, and it was intended to produce industrial alcohol as well as acetone and butanol using separate fermentation plants and distillation units for the two processes. The solvent process, which used a phage-immunized strain of *C. aroacetobutylicum* created in the Terre Haute research laboratories, went into operation at the end of 1935 following some minor teething issues. Edible oils were hydrogenated using the hydrogen produced during the AB fermentation.

In 1936 the Weizmann patent expired and new AB fermentation plants were built in Philadelphia, Pa., Baltimore, Md., and Puerto Rico. From 1936 onwards the production of acetone and butanol from the fermentation of molasses progressed rapidly, with the isolation of numerous new strains capable of carrying out the fermentation. Each company possessed many strains with different properties, and between 1935 and 1941 at least 18 patents were issued covering different strains which were used by many of the large producers, including Commercial Solvents Corp., Publiker Industrial Inc., U.S. Industrial Chemicals Co., and the Western Condensing Co.

When the Second World War broke out, molasses-based ethanol production was all but abandoned in the UK, but AB fermentation was given top priority due to the need for acetone in the manufacture of munitions. The need for acetone increased significantly as the war went on, necessitating an expansion of the Bromborough fermentation facility. The appropriate alcohol fermentors were sequestered to further increase the fermentation capacity. As a result, production levels were far higher than what the current batch distillation plant could handle, and attempts to use multiple-column continuous distillation units—which are used to produce industrial alcohol—were successful. Continuous distillation took over as the preferred technique after that. Furthermore, effective semicontinuous fermentation techniques were developed, reducing the fermentation cycle to 30 to 32 hours. A chemical plant was built to produce calcium carbide, which could be transformed into acetylene and ultimately acetone, as a result of the difficulties encountered in bringing molasses into Britain during the war. However, no acetone was produced by this method due to the high demand for acetylene for welding. Large volumes of acetone were imported from the United States, where plants were operating at full capacity to meet the needs of the US and its allies, to meet the demand for acetone. However, due to a lack of molasses, American plants were forced to resume using maize mash as a substrate.

In addition, plants were constructed in several other nations after 1936, including South Africa, Australia, India, and Japan. While research on the process was conducted for several years, the process was still in use in Japan during the Second World War and only stopped producing solvents through fermentation in the early 1960s.

In 1937, a plant was established in the South African maize-growing region of Germiston, which produced acetone and butanol from maize mash. The plant was converted to using molasses as the raw material during the later stages of World War II, and it ran on this basis until 1983.

Additionally, small plants were set up in several less-developed nations where molasses was readily available. After World War II, a fermentation plant in Formosa, Taiwan, produced butanol, acetone, and isopropanol from molasses by employing *C. toanum.* This plant was in operation for several years. It also appears that blackstrap molasses was used on a small scale in Brazil and Egypt to produce butanol and acetone.

Fermentation was used to produce solvents in China and some other Eastern Block countries, as well as acetone and butanol on a fairly large scale in the USSR. Numerous studies on the continuous fermentation method of producing acetone and butanol were published in the 1950s and 1960s, and in 1960 the acetone plant in Dokshukino, USSR, started producing acetone on a large scale.

The raw materials for AB fermentation in Russia were maize, wheat, and rye. However, at the end of the 1950s, attempts were made to replace these substrates with other ones, such as combinations of molasses and grain and corn cob hydrolysates, as a result of a decision made by the 20th Congress of the Communist Party of the Soviet Union.

2.1.3. Decline of AB Fermentation

Two-thirds of butanol and one-tenth of acetone produced in the US were still made through fermentation in 1945 when the war came to an end. But in the 1950s, fermentation's share of the overall output fell off quickly, and by 1960, fermentation-based production of these solvents had all but stopped in the US and Great Britain. After the war, there were two reasons why AB fermentation decreased. First, the petrochemical sector expanded at a never-before-seen pace, and by the late 1950s, there was fierce rivalry between the chemical and fermentation processes. Second, molasses started to be used in significant quantities in cattle feed, especially in the United States. Molasses became much more expensive as a fermentation material due to rapid price increases brought on by competitive buying in the open market. The farm vote was encouraged by US government administrations, and farmers were able to purchase molasses at a lower cost than distillers could match thanks to subsidies. When supply and demand were equal, molasses saw its first seller's market.

Following the war, the Bromborough plant in England made an effort to resume regular business operations. But in addition to having to pay a premium for molasses, it discovered that chemically produced solvents posed a significant threat. Higher fermentation yields were obtained when the fermentation was shifted to using homemade beet molasses. But eventually, the plant was forced to close in 1957 because the competition was too fierce.

Due to an abundance of inexpensive coal and molasses as well as the lack of an easily accessible source of cheap petroleum, the AB Fermentation Plant in South Africa was able to maintain its economic viability long after the process had stopped working in other Western nations. However, a severe molasses shortage brought on by southern Africa's devastating droughts in 1981 forced the closure of the AB fermentation process. The plant was temporarily reopened in 1982, but it was eventually closed due to a number of issues with fermentation, molasses availability, and cost.

2.2. AB FERMENTATION

2.2.1. Microorganisms

Currently, clostridia that produce butanol are classified into several species, primarily distinguished by the kind and proportion of the solvents they produce. While *C. aurantibutyricum* produces both acetone and isopropanol in addition to butanol, *C. beijerinckii* (*C. butylicum*) produces solvents in a ratio that is roughly the same as *C. acetobutylicum*. A recently discovered species called *C. tetanomorphum* yields ethanol and butanol in nearly equal proportions but no other solvents. Sadly, not much taxonomic research has been done on the solvent-producing clostridia, and as of right now, there are no accepted guidelines for classifying these organisms.

Since solvent-producing clostridial strains are both sporeformers and obligate anaerobes, which have comparatively simple growth requirements, and because the methods used for isolation have been well documented, the process of isolating them has proven to be relatively simple. Rather than decomposing plant matter or soil, these bacteria are most frequently linked to living plant material. It has been noted that potatoes, the roots of legumes that fix nitrogen, and other root crops are good sources of material for isolating these bacteria. Furthermore, it has been documented that fruit, including gooseberries, cereal crops, and agricultural soil are effective sources of these bacteria.

The choice of strains for industrial fermentation was contingent upon the type of raw material employed, the necessary ratio of end products, the requirement for supplementary nutrients, and the presence of phage resistance. Most of the time, strains with the desired traits were isolated as new cultures, and it doesn't seem like much work has been done to genetically modify strains to make them better.

2.2.2. Industrial Process

Industrial solvent production was done in a batch process with fermentors that held between 50,000 and 200,000 gallons of capacity and no mechanical agitation system. 8–10% maize mash was used in the original industrial process, which involved cooking it for 60 to 90 minutes at 130 to 133°C. For the majority of strains, no additional nutritional additions were required. Because molasses was a more advantageous fermentation substrate than maize mash, it replaced maize mash in most industrial processes starting in the mid-1930s. To achieve a concentration of fermentable sugars of roughly 6.5% (5.0 to 7.5%, wt/vol), blackstrap, invert (high-test), or beet molasses were diluted. It was customary to add extra sources of organic and inorganic nitrogen, phosphorus, and a buffering agent to the molasses after it had been cooked and sterilized for 15 to 60 minutes at 107 to 120°C. It was also common practice to replace up to 33% of the makeup water with distillation slops when using both molasses and maize mash.

Under a layer of carbon dioxide, the fermentors were filled to 90–95% capacity. To aid in mixing, sterile carbon dioxide was frequently bubbled through both before and after inoculation. Typically, cultures were maintained as spores in sterile soil or sand. After two to four buildup stages, the cells were inoculated into the fermentor, either during or shortly after filling, at a concentration of two to four percent. The inocula were prepared by heat-activating spores at 65 to 100°C for one to three minutes.

Utilizing maize mash, fermentations were conducted at 34 to 39°C for 40 to 60 hours, yielding approximately 25 to 26% dry-weight corn equivalents. The final concentration of the solvents produced ranged from 12 to 20 g/liter and was typically lower than those obtained with molasses. The strain and the conditions of the fermentation affected the solvent ratios, although a ratio of 6:3:1 (butanol, acetone, and ethanol) was typical for the Weizmann fermentation.

Lower temperatures (29 to 35°C) were used for molasses substrate fermentations; for many strains, 37 to 32°C was the ideal range. In practice, lower concentrations were frequently obtained. Solvent yields based on the fermentable sugars were typically between 29 and 33%, and cell metabolism was inhibited when the concentration of solvents reached 18 to 22 g/liter.

The hydrogen and carbon dioxide created during fermentation were extracted, separated, and used in several ways in many plants. Following fermentation, the distillate was fractionally distilled to yield pure acetone and butanol as well as an additional fraction of mixed solvents. The solvents were then extracted from the liquor using primary batch or continuous distillation. Following distillation, the liquid effluent's total solids content was between 4 and 4.5% (wt/vol). With significant amounts of group B vitamins and between 28 and 30% bacterial protein, the solids had a relatively high nutritional value. There was widespread use of the dried solids from the effluent in animal feed.

2.2.3. Limitations of the Conventional AB Batch Fermentation Process

The traditional batch AB fermentation method had several significant shortcomings.

- The process depended on using standard renewable carbohydrate substrates like molasses and corn. With rising raw material costs, especially after World War II, the fermentation route could no longer compete with the synthetic route, which relied on petrochemical feedstocks.
- Because butanol is extremely toxic to biological systems even at very low concentrations, the final fermentation broth could contain no more than 2% of solvents. Given the high energy needed, recovering these low solvent concentrations through distillation was costly.
- Intrinsic limitations in the fermentation process itself led to comparatively low solvent yields and the production of solvent ratios that were not always desirable.
- The fermentation process required to be carried out in a sterile environment due to its complexity. Phage infections in particular led to contamination, which created issues.
- Large volumes of effluent were produced during the fermentation process, necessitating the development of specialized handling, treatment, and processing procedures.

The costs of raw materials, the production and recovery costs of solvents, and the plant's capital cost were the main determinants of AB fermentation's economic feasibility. The cost of the substrate, which accounted for nearly 60% of the total cost, was the most significant economic factor among them. Transporting the heavy material to the plant increased the cost in addition to the raw material's actual cost. The economic viability of the process depended on the plentiful supply of inexpensive substrate. Regretfully, when these plants were converted to using molasses, the location of many of the original plants—which were situated in regions with an abundance of maize, like the highveld of South Africa and the Midwest of the United States—led to a significant increase in transportation costs. According to recent estimates, the cost of the raw materials for solvent production via fermentation on conventional carbohydrate substrates ranges from 57 to 116% of the solvents' selling price. Coal (or another energy source) for steam generation accounted for 15–20% of the total cost, making it the second largest expense. The remaining steam was used for cooking and sterilizing, with about 65% of it going toward distillation. The process's economic feasibility was largely dependent on the availability of a cheap supply of coal.

The cost of the fermentation plant's capital was typically much lower than that of the plant needed for the chemical synthesis of solvents.

Research on industrial AB fermentation declined in tandem with the industrial AB fermentation's decline in importance following World War II. However, there was a resurgence of interest in fermentation processes as a potential substitute for producing liquid fuels and chemicals following the 1973–1974 oil crisis. Many of the fundamental aspects of the physiology, biochemistry, and molecular biology of *C. acetobutylicum* and related solvent-producing species have been illuminated by the recent growth in AB fermentation research. It is our hope that as we gain a deeper understanding of this fermentation, the process itself will be improved.

2.3. ALTERNATIVE FERMENTATION SUBSTRATES

One major factor affecting the economic viability of the AB fermentation has been identified as the high cost of conventional starch (maize, wheat, millet, etc.) or sugar (molasses) substrates. This, along with the fact that *Saccharomyces cerevisiae* and arolytic clostridia can use a wide variety of carbohydrates, has prompted investigation into the use of substitute, less expensive substrates. A preliminary investigation into the fermentation of different carbohydrates to produce acetone and butanol revealed that the following carbohydrates were entirely consumed: glucose, fructose, mannose, sucrose, lactose, starch, and dextrin; galactose, xylose, arabinose, raffinose, melezitose, inulin, and mannitol; trehalose, rhamnose, melibiose, and glycerol. Other researchers, however, demonstrated that the majority of AB bacteria were capable of fully fermenting xylose and arabinose. It has been demonstrated that several clostridia strains that produce acetone and butanol can ferment the carbohydrates found in wood and dairy waste.

2.3.1. Noncellulosic Substrates

With varied degrees of success, starches from a variety of other sources, such as potatoes, rice, and tapioca, have also been used as a substrate in addition to starches from maize, wheat, rye, and millet. Investigations into cassava and Jerusalem artichokes as possible substrates for AB fermentation have also been conducted. According to Banzon, while cassava by itself was not a good substrate, it could be used in conjunction with other sources of organic nitrogen or 20% corn meal.

2.3.1.1. Jerusalem Artichokes

A crop that has a lot of promise for use as a carbohydrate substrate in AB fermentation is the Jerusalem artichoke. The majority of the tuber's carbohydrates are found as short oligomeric fructans, which have an inulinic structure and need to be hydrolyzed by inulase or acid to be fermented. It was suggested that the hydrolysate needed to be further supplemented with soy or maize meal. But more recently, as part of a French research program on the synthesis of fuel extenders from biomass, the use of Jerusalem artichokes as a fermentation substrate has been studied. Inulinase was added to produce hydrolysates; aside from ammonia, no additional nutritional supplementation was required; after 36 hours of batch fermentation under ideal circumstances with sufficient pH control, 23 to 24 g of solvents per liter were obtained. A pilot plant has been used to test this substrate.

2.3.1.2. Cheese Whey

Cheese whey's lactose content, disposal issue, and international availability have made it a popular candidate for use as an alternate substrate for AB fermentation. Whey filtrate, which is left over after casein is precipitated and removed, has a low sugar content (4–5% lactose) and is not suitable for most fermentations that don't first concentrate. However, it works well for AB fermentations, where the amount of sugar that can be consumed is limited due to product toxicity. The use of whey and lactose in AB fermentation has been studied by a number of researchers; however, when total reactor productivities in batch fermentations are taken into consideration (solvents, 5 to 15 g/liter; productivity, 0.1 g/liter per h; yield, 0.23 to 0.41 g/g), whey permeate has proven to be a relatively poor substrate when compared with starch and molasses substrates. Additionally, incomplete utilization of the lactose is a significant issue. Product inhibition does not seem to be a factor in the inadequate acetone and butanol production from whey, as the concentrations of total solvents produced are lower than those produced on conventional substrates. It has been demonstrated that the butanol/acetone ratios produced from the fermentation of whey by *C. acetobutylicum* differ from those produced from glucose. Linden and associates found that whey had butanol/acetone ratios as high as 12:1 to 20:1, whereas glucose had a ratio of 3:1. Although the cause of the change is unknown, several factors appear to be at play. The production of acetone and butanol from whey is also influenced by growth factors and temperature.

2.3.1.3. Apple Pomace

Apple pomace is a solid agricultural waste that has a weight-per-weight percentage of carbohydrates of about 10% (fructose: 67%, glucose: 23%, and sucrose: 10%). Voget and associates looked into producing butanol from apple pomace, and results showed that butanol yields from fresh apple pomace ranged from 1 to 2 percent. 80 to 84% of the sugars were consumed, and the new residue left over after butanol separation might make a fantastic source of animal feed.

2.3.1.4. Algal Biomass

It is thought that marine algal biomass is a fermentation substrate that offers some benefits for the use and bioconversion of a sizable renewable resource. According to research on the suitability of the halophilic microalgae Dunaliella, *C. pasteurianum* can produce about 16 g of solvents per liter from an algal biomass mixture that has been supplemented with 4 percent glycerol. Unlike *C. acetobutylicum* fermentations that employ molasses or starch, a solvent mixture rich in butanol and 1,3-propanediol was produced instead of acetone. Nakas et al. did not observe any inhibition of solvent production by salt. With algal concentrates, salt removal or the isolation of solvent-producing, salt-tolerant clostridia may be necessary for the success of large-scale algal cultivation connected to bacterial fermentations.

2.3.2. Substrates Derived from Lignocellulose

The most plentiful renewable resource, lignocellulose, is known to have enormous potential as a fermentation substrate—that is, if the cellulose and hemicellulose components can be effectively broken down and used. Hemicellulose makes up 20 to 40% of cellulosic biomass, with D-xylose being its main component. All of the common sugars found in wood hemicellulose and cellulose hydrolysates can be utilized by *C. acetobutylicum*. The ideal fermentation conditions have been determined, resulting in final solvent levels of 8 to 17 g/liter for both individual and combinations of sugars.

2.3.2.1. Pentose Sugars

Since pentose sugars, and specifically xylose, are important constituents of hemicellulose, several researchers have looked into the fermentation of xylose by *C. acetobutylicum*. *C. acetobutylicum* ferments pentoses, although the yields from the solvent are typically lower. Ounine along with others revealed a limited xylose consumption of 45 g/liter compared to a 62-g/liter glucose conversion, but with a lower fermentation rate and a comparatively high conversion yield of 28% xylose into solvents—close to the maximum value of 32% for glucose. Recent studies have compared the fermentation kinetics of *C. acetobutylicum* in fed-batch and batch cultures in cells grown on xylose, glucose, and mixtures of the both sugars. At glucose concentrations greater than 15 g/liter, xylose utilization is both induced and inhibited. The highest amount of fermented sugars, up to 68 g/liter, were produced in batch cultures when glucose and xylose were mixed together. This was suggested to be because xylose strongly reabsorbed acid and glucose was used preferentially and quickly. Fond et al. investigated the kinetics of AB fermentation at low concentrations of glucose, xylose, and mixtures of the two sugars using fed-batch cultures. The only times results were close to batch cultures were when feeding rates were relatively high (above 18 g/liter per day). During the first part of the fermentation, glucose and xylose were taken up at the same rate at low feeding rates, with the glucose concentration below 15 g/liter. When fermentation was stopped, an accumulation of xylose indicated that xylose utilization was suppressed when the cell's catabolic flux of glucose alone was sufficient to meet its metabolic needs. These fed-batch and kinetic batch fermentations clarified a number of crucial elements pertaining to the control of *C. acetobutylicum* metabolism (Fond et al., 1985).

2.3.2.2. Sulfite Waste Liquors

Glucose, xylose, and arabinose are found in sulfite waste liquors from the pulp and paper industry, which have been studied as potential substitute substrates for AB fermentation. Nonetheless, recent advancements in pulp and paper technology have altered the character of effluent in an effort to lessen issues with it. Wiley along with others demonstrated that when the excess calcium, lignin, and sulfur dioxide were eliminated from the sulfite liquor, good solvent yields were obtained. In a recent study, Wayman and Yu looked into the fermentation of a sugar mixture created to mimic sulfite waste liquor. They found that 96% of the sugar was consumed and that there were 0.36 g of solvents for every gram of sugar.

2.3.2.3. Hydrolysates

Early studies indicated that a range of cellulosic materials could be potentially useful as a substrate for AB fermentation through acid hydrolysis; more recently, Saddler and colleagues have looked into the preparation and application of wood cellulose and hemicellulose hydrolysates. After steam-exploded wood chips were extracted with water, it was found that the water-soluble fraction contained more than 75% of the hemicellulose, leaving behind a cellulose-rich, water-insoluble substrate that was easily hydrolyzed to glucose. The water-soluble, hemicellulose-rich substrate was further hydrolyzed by acid or enzymes to liberate sugars that *C. acetobutylicum* could use. This hydrolysate lacked any inhibitory compounds that had been observed in other hydrolysates.

Yu and colleagues studied the direct use of cellulose and hemicellulose in acid-hydrolyzed, steam-exploded wood in order to prevent the separation of the hemicellulose component. With almost theoretical product yields of 0.26 g of butanol per gram of sugar consumed, or 9 g of butanol produced per liter, it is possible to bioconvert combined wood cellulose and hemicellulose carbohydrates.

2.3.3. Direct Utilization of Lignocellulose

The Institut Frangois du Petrole and Technip recently constructed a sizable pilot-scale process at Soutons in southwest France that uses a steam cracking technique to split lignocellulosic material (corn stover) into hemicellulose, lignin, and cellulose. This procedure separates the constituent parts and uses enzymatic hydrolysis to *Saccharomyces cerevisiae* arify the cellulose. The sugars obtained from cellulose and hemicellulose will be used as substrates for fermentation to produce butanol and acetone.

2.3.3.1. Coculture Systems

The direct conversion of cellulose using single or mixed cultures of microorganisms with enzymes capable of hydrolyzing cellulose and hemicellulose is an alternate strategy for using cellulosic biomass.

In addition to hexose sugars, bagasse and rice straw are good substrates for the synthesis of solvents because the hydrolysates contain cellobiose, cellodextrins, and pentoses, all of which can be used by clostridia that produce solvents. Fermentable sugars have been extracted from bagasse and rice straw using mixed-culture filtrates from the cellulolytic fungi *Aspergillus wentii* and *Trichoderma reesei*. Further, *C. aroperbutylacetonicum* produced 16 g of butanol per liter from the hydrolysate following treatment to remove unwanted impurities. Utilizing *C. acetobutylicum* in a fermentation medium enhanced with a *T. reesei* cellulase preparation, the fermentation of alkali-pretreated wheat straw resulted in solvent concentrations of 17.3 g/liter and solvent yields of 18.3% relative to pretreated wheat straw. These outcomes, which show an enhanced performance over the independent hydrolysis and fermentation operation, were attained after 36 hours.

Coculture systems have the advantage of eliminating the enzymatic hydrolysis stage, which requires the costly steps of separate enzyme production and handling. The direct conversion of cellulose by a mesophilic coculture of a cellulolytic *C. cellulolyticum* H.10 strain with *C. acetobutylicum* has been demonstrated. The mixed culture on cellulose produced mainly butyric acid with small amounts of acetic acid, ethanol, and butanol. The accumulation of butyric acid seemed to be due to the low level of cellulose hydrolysis and the concomitant lack of glucose which is required for solvent production during the solventogenic phase. Research is being carried out to obtain a mutant of *C. cellulolyticum* H.10 with increased cellulase activity, to shift the mixed culture towards the production of solvents.

Yu and co. examined the use of a sequential coculture method to turn solka floc or a combination of Solka floc and aspenwood Xylan into solvents. Because *C. thermocellum* produces the enzymes cellulase and xylanase, it was selected for coculture with *C. acetobutylicum*. When grown on lignocellulose, it can convert cellulose directly to ethanol and accumulates glucose and pentose sugars in the culture filtrate. All hydrolysis products from the lignocellulose substrates were efficiently utilized in cocultures with these two anaerobic bacteria. The bulk of the fermentation products were acids, and the outcomes resembled those of Fond et al.'s experiments with *C. cellulolyticum* H.10 and *C. acetobutylicum*. Yu along with others raised the glucose levels in the cocultures of *C. thermocellum* and *C. acetobutylicum*, but little to no solvent was found and acids continued to be produced. This implied that the production of solvents was not solely regulated by glucose levels. Butyric acid was added to the coculture system, which caused the solventogenic phase to be induced, as research had shown that levels of this acid may be crucial in initiating the production of solvents.

The use of cocultures to boost *Clostridium* spp.'s butanol production has been studied. Since butyric acid is reduced to produce butanol, it might be able to supply more butyric acid by using a precursor organism to raise the levels of butanol. High concentrations of butyric acid are produced from glucose by *C.*

butyricum and *C. pasteurianum* cocultures. In pure culture, *C. pasteurianum* and *C. beijerinckii* yielded butanol concentrations that were 20% higher than those of *C. beijerinckii.* Butanol concentrations were not increased by coculturing *C. butyricum* or *C. pasteurianum* with *C. acetobutylicum* or *C. butyricum* with *C. beijerinckii.*

2.3.3.2. Cellulolytic and Xylanolytic Activities

An inducible carboxymethyl cellulase and a cellobiase were produced by the industrial *C. acetobutylicum* P270 strain, according to Allcock and Woods. Although the amounts of cellulase activity that were obtained were low, genetic modification and mutation might be able to increase the levels of cellulase activity. Screening and isolating additional solvent-producing *C. acetobutylicum* strains with strong cellulolytic and xylanolytic activity is an alternate strategy. Lee and associates examined 20 Clostridium species that produced solvents for their ability to break down cell walls and Xylan. The activities of extracellular and cell bound endoglucanase, cellobiase, and cellulolytic activity were only seen in two strains of *Clostridium acetobutylicum*, NRRL B527 and ATCC 824. These two strains showed extracellular xylanase, xylopyranosidase, and arabinofuranosidase activities in addition to hydrolyzing Xylan. Xylanolytic activity was found in 17 out of the 20 solvent-producing Clostridium strains that were studied, suggesting that this activity was widespread. Potential strains for the direct production of solvents from cellulosic biomass include NRRL B527, ATCC 824, and the cellulolytic industrial *C. acetobutylicum* strains P262 and P270. However, extensive basic biochemical and genetic research on these bacteria is required before an effective industrial process can be considered.

2.4. BIOCHEMISTRY AND PHYSIOLOGY

During the first growth phase (acidogenic phase) of a typical batch culture, solvent-producing *Clostridium* species produce hydrogen, carbon dioxide, acetate, and butyrate, which lowers the pH of the culture medium. The cells' metabolism changes to produce solvents as the culture moves into the stationary growth phase (solventogenic phase). The reassimilation of acids, which happens concurrently with the continuous consumption of carbohydrates, typically causes the pH of the culture medium to rise during the second phase of fermentation. Early on in the development of the industrial fermentation process, it was determined that there is a relationship between the fermentation's pH breakpoint and the start of solvent production, which happens at the start of the second phase of the fermentation. Doelle and Gottschalk described the primary processes by which *C. acetobutylicum* converts carbohydrates into acids, end products, and solvents. Thauer et al. have also reviewed the general mechanisms of energy metabolism that occur in anaerobic chemolithotrophic bacteria, including the clostridia. More recently, Haggstrom reviewed the energy metabolism of *Clostridium acetobutylicum*, and Rodgers reviewed the relationship between the biochemical pathways and fermentation strategies that occur among the clostridia about the development of fermentative processes (Haggstrom and Molin. 1980).

It is well known which biochemical pathways *C. acetobutylicum* is used to convert carbohydrates to fatty acids, hydrogen, carbon dioxide, and solvents.

The Embden-Meyerhof pathway is used to metabolize hexose sugars, which include mono-, di-, tri-, and polysaccharides. This process converts 1 mol of hexose to 2 mol of pyruvate, resulting in the net production of 2 mol of ATP and 2 mol of reduced nicotinamide adenine dinucleotide (NADH). Pentose sugars are metabolized by the solvent-producing clostridia via the pentose phosphate pathway. Fermenting pentoses yield pentose 5-phosphate, which is then dissimilated via the transketolase-transaldolase sequence to produce fructose 6-phosphate and glyceraldehyde 3-phosphate, which are then released into the glycolytic pathway. 5 mol of ATP and 5 mol of NADH are produced during the fermentation of 3 mol of pentose.

In the presence of coenzyme A (CoA), pyruvate ferredoxin oxidoreductase cleaves the pyruvate produced by glycolysis to produce reduced ferredoxin, acetyl-CoA, and carbon dioxide. The key intermediary in the branched fermentation pathways that produce both acid and solvent is acetyl-CoA, which is generated by the phosphorolytic cleavage.

2.4.1. Electron Flow

Ferredoxin appears to be present in every one of the clostridia. This protein has a low molecular weight and contains iron and sulfur. It can accept and donate electrons at a very low potential that is similar to the hydrogen electrode (E_0' = –410 mV). As an electron carrier in the cell's electron distribution, ferredoxin is essential. When the right circumstances are met, the reduced ferredoxin can transfer electrons to an iron-containing hydrogenase, allowing protons to be used as a last electron acceptor and producing molecular hydrogen. In this stage, the cell releases hydrogen gas and ferredoxin is reoxidized.

NADH ferredoxin oxidoreductase is another important enzyme in the electron distribution system. It works by equilibrating the electrons between NAD and ferredoxin, thereby causing either oxidation or reduction of NAD.

A fast flow of electrons is produced during acid-producing metabolism, both from the phosphorolytic cleavage of pyruvate and from NADH via ferredoxin to produce molecular hydrogen. It appears that a different enzyme called NADPH ferredoxin oxidoreductase is involved in the regulated synthesis of NADPH from reduced ferredoxin, which is necessary for biosynthesis. Since most clostridia don't seem to have the enzymes needed to oxidize glucose 6-phosphate and produce NADPH, this seems to be the only way that NADPH can be produced. Although it has been reported that sporulating cultures of *C. thermosaccharolyticum* can induce pentose 5-phosphate dehydrogenase and 6-phosphogluconate dehydrogenase, these enzymes do not seem to be present in sporulating cells of *C. acetobutylicum*.

Many clostridia, including *C. acetobutylicum*, also contain rubredoxin, a second low-molecular-weight nonsulfur protein, in addition to ferredoxin. However, the concentration of rubredoxin exhibits notable fluctuations in response to the pH and composition of the culture medium, as well as the growth phase of the cells, unlike ferredoxin, which is constitutively produced. The highest levels of rubredoxin were found in *C. acetobutylicum* ATCC 824 experiments when the cells grown in a chemically defined medium without pH control were nearing the end of the acid producing phase. The concentration of ferredoxin, which makes up between 0.8 and 1.6 percent of the soluble protein in the cell, was roughly three times that of rubredoxin under these circumstances. While rubredoxin is widely expressed in solvent-producing clostridia, not all strains of the bacterium have been shown to express NADH rubredoxin oxidoreductase. The synthesis of this enzyme varies when it is present as well, depending on the growth phase and type of medium. It has been shown that NADH rubredoxin oxidoreductase mediates the transfer of electrons to rubredoxin; however, when NADPH was used in place of NADH, the reaction did not take place.

There is no known specific electron acceptor for reduced rubredoxin, in contrast to ferredoxin, whose function in electron transfer has been well-established. It has been shown that rubredoxin can only, and only much less frequently, replace ferredoxin in a small number of oxidation-reduction reactions. This electron carrier is thought to play a significant role in the clostridia due to its abundance and wide distribution. These proteins may be involved in an electron transport system that uses NADH as an electron donor and becomes active in low pH and high acid concentration cultures because both rubredoxin and NADH rubredoxin oxidoreductase appear to be induced under these circumstances.

2.4.2. Acid-Producing Pathways

Under specific circumstances, *C. acetobutylicum* can also convert pyruvate to lactate in addition to the phosporoclastic cleavage of pyruvate to acetyl-CoA.

The lactic acid pathway is not active in normal circumstances, and it only seems to function as a less effective substitute to permit the production of energy and the oxidation of NADH if the pathways that dispose of protons and electrons through the creation of molecular hydrogen are obstructed. Reduced levels of ferredoxin and hydrogenase in cells depleted of iron have been reported to cause lactate production, as has carbon monoxide inhibition of hydrogenase activity.

The four enzymes that are part of the metabolic pathway that converts acetylCoA into butyryl-CoA are crotonase, 3-hydroxybutyryl-CoA dehydrogenase, butyryl-CoA dehydrogenase, and thiolase. These enzymes are similar to the ones that work in this pathway in *C. kluyveri*, and their presence in *C. acetobutylicum* has been shown. A CoA transferase of this kind has not been found in *C. acetobutylicum* or *C. beijerinckii* and does not seem to be involved in butyrate synthesis, despite reports that it is involved in butyrate formation in *C. kluyveri* and *C. aurobutyricum*.

2.4.3. Solvent-Producing Pathways

The carbon flow shifts from the pathways that produce acid to those that produce solvents when solvent production begins. Acetyl- and butyryl-CoA are essential intermediates in the solvent production process that are necessary for the synthesis of butanol and ethanol. These pathways result in the intermediates acetylaldehyde and butyraldehyde, respectively. To complete the reductions required to produce ethanol and butanol, the pathway depends on the activity of two sets of dehydrogenases.

However, both butyraldehyde and butanol dehydrogenase are involved in the reduction of butyryl-CoA to butanol. It has been reported that butanol dehydrogenase activity is NADPH dependent rather than NADH dependent in both *C. acetobutylicum* and *Clostridium beijerinck.* While ethanol dehydrogenase and acetaldehyde dehydrogenase have not been identified as distinct enzymes from those involved in butanol production, this seems likely given that *C. acetobutylicum* can produce ethanol independently from acetone and butanol under specific culture conditions.

Early researchers noticed that the acid end products created during the first fermentation phase were reassimilated after the transition to solvent production had occurred. However, the uptake of butyrate and acetate was limited to the case where sugars were metabolized simultaneously. According to some theories, the uptake of acids during solvent production—which is typically accompanied by a pH increase—acts as a detoxification process that is triggered in response to the buildup of acid end products, which create an environment that is unsuitable for growth.

It was also observed by early researchers that the yield of acetone and butanol was increased by the addition of either butyrate or acetate. Numerous more recent studies have supported these observations. Research carried out by Wood et al. employing isotopically labeled acetate and butyrate, showed that the acetone contained 15 and 2%, respectively, of the 14C label recovered from the two acids, while the butanol contained 55 and 85%, respectively, of the same label.

The uptake and activation of acids have been linked to several different mechanisms. Valentine and Wolfe postulated that during solvent production, the pathway that generated acetate and butyrate might reverse. Additional reassimilation pathways include the uptake of acids via ATP-dependent acetyl-CoA and butyryl-CoA synthetase or acetate and butyrate kinase reversal, which is followed by the direct reduction of the acyl phosphates to the corresponding aldehydes. However, recent research has shown that acetoacetyl-CoA: acetate/butyrate: CoA transferase is directly responsible for the reassimilation of acetate and butyrate and the subsequent production of acetone. Doelle and Andersch et al. proposed that this enzyme plays a role in the uptake of acids during the production of acetone.

It has been demonstrated that acetoacetyl-CoA transferase can use butyrate or acetate as the CoA acceptor when converting acetoacetyl-CoA to acetoacetate, which preserves the CoA unit. The resultant acetoacetate is subsequently irreversibly converted to acetone by decarboxylation. It has been proposed that the transferase reaction is steered toward the formation of acetoacetate by this crucial decarboxylation step. Andersch and associates showed that acetoacetyl decarboxylase had to be induced before acetate and butyrate could be absorbed. Acetoacetate techniques showed that the cell could absorb butyrate and acetate and convert them straight into solvents without accumulating intermediates. These same authors demonstrated that when the shift to solvent production occurred, the activities of phosphate acetyltransferase, phosphate butyryl-transferase, and acetate kinase rapidly decreased, suggesting that the uptake of acids does not occur through a reversal of the acid-forming pathways. Furthermore, no butyryl phosphate-reducing or short-chain ATP-dependent acetyl-CoA synthetase activities were found.

2.4.4. Stoichiometry

Only precise quantitative measurements of the substrate used and products formed during a fermentation process allow for the determination of the stoichiometry of substrate conversion to products. By using these data, a balance can be created to ascertain how the final products that are produced from a given quantity of substrate relate to one another. Wood created a fermentation balance sheet for several clostridia, such as *C. beijerinckii* and *C. acetobutylicum*. But in figuring out these balances, only the fermentation products that remained at the end of the fermentation process were considered. If the fermentation products produced during the acid-producing and solvent-producing phases are presented as separate fermentation balance sheets, a more meaningful stoichiometry for the fermentation can be obtained. These results demonstrate that the balance obtained during the fermentation's initial phase is strikingly similar to that obtained during a typical butyric fermentation. The amount of acetate produced during this stage of fermentation balances the additional hydrogen produced. It has been calculated that during this phase, ATP is produced at a rate of roughly 3.25 mol/mol of glucose consumed. The yield of molecular hydrogen falls to less than 2 mol during the solvent-producing phase, while the yield of carbon dioxide rises in direct proportion to the amount of acetone generated. Because of reassimilation during solvent production, the yield of acetate and butyrate is significantly lower when a combined fermentation balance is prepared.

If a maximum theoretical yield could be determined, that information would be helpful in evaluating the fermentation process' effectiveness. However, it is impossible to determine an exact value for the maximum theoretical yield due to the branching pathways involved, the variability of the individual solvent ratios, the ratio of acids to solvents, the ratio of hydrogen to reduced end products, and the ratio of biomass to end products. However, a maximum theoretical solvent yield of roughly 0.39 g/liter for *C. acetobutylicum* has been calculated, assuming a fixed biomass ratio and a fixed solvent ratio. The theoretical yield of butanol formed from 1 mol of glucose would be 1 mol (0.41 g/g) of butanol if the fermentation could be controlled to produce only butanol and CO_2.

In an attempt to determine the interrelationships between biomass production and the various end products formed during AB fermentation, a more detailed equation has been derived by Papoutsakis, based on an assumed ATP yield, "two biological regularities," and the biochemistry of product formation.

2.5. TRIGGERING OF SOLVENTOGENESIS

Understanding how the synthesis of solvents is started and maintained requires an understanding of the metabolic transition factors and the physiological state connected to the change from the acidogenic to the solventogenic phase.

Research into the elements that led to the shift to solvent production has been made much easier by the solvent-producing species of *Clostridium's* capacity to grow and produce solvents on a range of chemically defined minimal media. The effects of nutrient limitation and end product accumulation on the beginning and maintenance of solvent production have been studied in a number of recent studies conducted in batch and continuous cultures. Many experiments have produced seemingly contradictory results, despite the fact that these studies have produced a wealth of new information.

2.5.1. Role of External pH

It has long been acknowledged that pH plays a significant role in how AB fermentation turns out. In fact, many of the earliest reports on the industrial synthesis of solvents pointed out that solvent production didn't start until the mash's pH dropped to between 4.5 and 5.0. Numerous more recent studies that have reported that cultures maintained at high pH produce primarily acids, while cultures maintained at low pH typically produce more solvents, have confirmed these observations. However, depending on the specific strain and the culture conditions employed, it appears that there is a significant variation in the pH range over which solvent formation may occur. It has been reported that several strains of *C. acetobutylicum* DSM can only produce solvents with a pH of less than 5.0. These strains have been found to require a pH of approximately 4.3, and they can still produce solvents at a pH of 3.8. It has been reported that the ATCC 824 strain of *C. acetobutylicum* type culture produces good levels of solvents between pH 5.5 and 4.3. According to Kim and Zeikus, good solvent levels were obtained at pH 4.5, while no solvents were produced at pH 5.8 in controlled batch fermentations. For *C. acetobutylicum* P262 and related strains, which were used for the industrial production of solvents, the ideal pH range for solvent production is significantly higher.

The pH of the fermentation medium was adjusted for the industrial fermentation process to be roughly 6.0. During the first stage of the fermentation, the pH dropped to roughly 5.2 at the breakpoint, then increased to roughly 5.8 at the end of the fermentation. Studies conducted in the lab using these strains have demonstrated that the pH range of 5.0 to 6.5 can produce solvents at good levels. The growth and metabolism of the cells were inhibited in poorly buffered media during the early stages of fermentation when the pH was allowed to drop below 4.5, and no solvents were produced. An Egyptian isolate of *C. acetobutylicum* has also been reported to produce solvents in cultures kept at pH 6.4. George and Chen reported on the generation of solvents by a strain of *C. beijerinckii* kept at pH 6.8. It has been reported that the amounts of solvents generated during fermentation at a pH close to neutral were nearly identical to those found in uncontrolled cultures where the pH was permitted to fall below 5.0.

While it has been proposed that reaching an appropriately low pH in the medium could initiate the production of solvents, it has been shown that while a drop in pH is crucial for enabling the transition to solvent production, pH is not the initiator.

At the conclusion of the first stage of fermentation, a number of other noteworthy changes take place in addition to the pH dropping. Reduced production of hydrogen and the specific growth rate are linked to low pH, along with an increase in cell count and an increase in the concentration of acid end-products.

2.5.2. Role of Acid End Products

Since reduced energy-rich organic acids are byproducts of metabolism and are toxic to cells by nature, fermentation requires the use of organic compounds as terminal electron acceptors in order to produce ATP. Organic acids that aren't dissociated, like acetate and butyrate, can partition within the cell membrane and act as uncouplers, letting protons from the medium enter the cell. The undissociated acid concentration causes the pH gradient across the membrane to collapse and the NTP/NDP ratio to drop quickly, both of which lead to the complete inhibition of all metabolic processes in the cell. At lower concentrations, substrate utilization and cell metabolism continue, but the accumulation of acid end products and the resulting drop in pH cause a progressive decrease in the specific growth rate that eventually stops cell growth. According to some theories, *C. acetobutylicum* and related species' adaptation to the inhibitory effects of acid end products is reflected in their shift to solvent production. It seems that the switch to solvent synthesis functions as a detoxification mechanism, sparing the cell from the inhibitory effects that arise from the accumulation of acid end products to hazardous proportions.

When acid end products build up in the medium, the pH of the medium typically drops, signaling the start of solvent production. More acids will be present in the undissociated form in these circumstances. The amount of butyric acid that is in the undissociated form at pH 6.0 is only 6%, while at pH 4.5, 66% of the total is.

According to Gottschal and Morris, batch cultures of *C. acetobutylicum* maintained at pH 5.0 showed a rapid induction of solventogenesis upon the addition of acetate and butyrate (10 mM each). This was accompanied by a decrease in the specific growth rate and the rate of H_2 production. Comparable outcomes have been observed in alternative experimental frameworks involving *C. beijerinckii* and *C. acetobutylicum*. Holt et al. showed that adding a high concentration of acetate and butyrate to the medium led to the induction of solvent production in *C. acetobutylicum* cultures kept at pH 7.0. The addition of butyrate also caused a shift in solvent production in continuous cultures kept at a low enough pH. It has also been noted that in batch cultures of *C. acetobutylicum* with uncontrolled pH, the addition of acetoacetate triggers the production of solvent.

Monot et al. reported that, when the concentration of undissociated butyric acid reached a level of 0.5 to 0.8 g/liter, growth was inhibited and the induction of solvents occurred when the concentration of undissociated butyric acid reached a level of 1.5 to 1.9 g/liter. However, these workers have also reported that growth and acid production are inhibited at undissociated butyric acid concentrations of between 0.2 and 0.4 g/liter, with solvent production being initiated at concentrations of 0.5 to 1.5 g/liter. The excretion of butyric acid ceased when the concentration of undissociated butyric acid reached 1.7 to 1.9 g/liter.

Nevertheless, no other experimental system has shown a similar correlation between the induction of solvent production and the concentration of undissociated butyric acid in the external medium. Monot et al employed a particular inhibitor of membrane-bound adenosine triphosphatase (ATPase) to lower the cells' internal pH and raise the amount of undissociated acids they contained. As a result, the maximum biomass concentration was lowered and solvent production was increased. Through the use of fed-batch fermentation, which allowed for the adjustment of sugar addition to achieve both fast and slow catabolic rates, the impact of acetate and butyrate on the induction of solvents has also been studied. The addition of acetate, butyrate, or both accelerated the transition to solvent production by a factor of 10 to 20 at slow catabolic rates, but it necessitated significantly higher acid concentrations than the culture with fast catabolic rates. Acids were reassimilated at lower concentrations in fed-batch cultures with high catabolic rates, and it was proposed that cultures with a rapid carbohydrate catabolism accumulate a higher intracellular concentration of acids.

2.5.3. Role of Internal pH

While some bacteria are known to keep their internal pH levels above the external medium at a relatively constant level, it has been noted that bacteria that produce weak acids, like butyric and acetic acids, are not able to do so. Instead, when the pH of the external medium drops, these bacteria keep a small but roughly constant pH gradient across the membrane. The proton-translocating ATPase system in fermentative bacteria allows protons to be extruded across the membrane, maintaining the pH gradient (pH). The cell uses ATP metabolic energy to create a proton motive force, which is necessary for growth and metabolism.

Riebeling and associates stated that *C. pasteurianum* demonstrated a proportionate drop in the internal pH of the cell (from 7.5 to 5.9) as the external medium's pH (from 7.1 to 5.1) decreased. This suggests that the cells were able to sustain a roughly constant ΔpH of 0.4 to 0.8 pH units. Proton conductors and

ATPase inhibitors eliminated the ΔpH and were shown to inhibit cell growth when present in low concentrations. Furthermore, in cells that had run out of an energy substrate, no ΔpH was found. These findings implied that cell growth and metabolism depend on the preservation of an ATPase-driven pH gradient.

2.5.4. Role of Nutrient Limitation

Many researchers have looked at how nutrient limitation affects batch- and continuous-culture systems' ability to initiate and sustain solvent production. Only when the carbon source's concentration was restricted did acids arise in batch, fed-batch, and continuous cultures. No shift to solvent production was observed in fed-batch cultures with a feed rate of less than 4 g/liter per day, or in batch cultures where glucose was present at less than 7 g/liter. Similar outcomes were seen in continuous culture, and it is now widely acknowledged that, in the presence of a limited carbon supply, the quantity of acid end products that can be produced is insufficient to raise the concentration threshold needed to trigger the production of solvents. It is the feed rate that matters in continuous culture, not the concentration itself. Solvent production has been achieved with feed rates as low as 5 g/liter per day. Therefore, an excess of sugar is necessary for the start and continuation of solvent production. Catabolite repression of solvent-producing enzymes does not occur because the shift to solvent production can occur in cultures that may have a significant excess of sugar present.

It is less certain how nitrogen limitation affects the synthesis of solvents. Long et al. researched the impact of ammonia limitation in batch culture. They found that the amount of glucose used decreased and residual glucose remained at the end of fermentation when the concentration of ammonia in the medium was lowered to a level that led to a decrease in biomass. Both the amount of biomass produced and the amount of glucose consumed decreased as the ammonia concentration dropped further. Less than one-third of the glucose in the medium was consumed at low ammonia concentrations., and the cells did not shift to producing solvents. Under these circumstances, the inability to produce solvents seemed to be caused by the inability to produce acid end products at threshold concentrations. Research conducted in chemostat cultures with limited ammonium supply by Gottschal, Morris, and Andersch et al. Both were unable to achieve appreciable solvent levels. In a nitrogen-limited chemostat kept at a low pH and operated at a very low dilution rate (0.038/h), Monot and Engasser showed that solvent production could be achieved. The ammonia concentrations used in these investigations, however, were noticeably higher than those found to have an impact on batch culture's biomass production and glucose consumption. Furthermore, in these experiments, the anion of the nitrogen source was acetate, which was fed at concentrations that were getting close to the threshold level noted for acetate accumulation in other studies. Jobses and Roels also reported on the solvents produced by *C. beijerinckii* in a nitrogen-limited chemostat. Roos and associates were also successful in obtaining extremely low solvent concentrations in cultures with nitrogen limitation and pH levels between 3.65 and 4.5.

Bahl et al. observed solvent production in both batch and continuous cultures under phosphate limitation (0.74 mM), and a fruitful two-stage continuous procedure was created based on this discovery. Acids were produced during the first stage of the continuous process, which was operated at a comparatively high dilution rate (0.125/h). These acids were then used to feed the second stage of fermentation, which was operated at a much lower dilution rate (0.04/h). The second stage of the fermentation produced high yields of butanol and acetone, and the process ran continuously for a year without causing any changes in cell activity (Bahl et al., 1982).

2.5.5. Role of Temperature and Oxygen

- **Temperature**: McCutchan and Hickey demonstrated that solvent yields stayed relatively constant at about 31% at 30 and 33°C but dropped to 23 to 25% at 37°C in experiments on three distinct strains that produced solvents. A more recent study using *C. acetobutylicum* NCIB 8052 produced similar results, showing that while the fermentation time decreased with increasing temperature, solvent yields decreased, going from 29% at 25°C to 24% at 40°C. According to these authors, butanol yield remained unchanged while the decline in solvent yield seemed to be related to a decline in acetone production. This is in contrast to previous research that showed that lowering the fermentation temperature from 30 to 24°C after 16 hours increased the butanol ratio.
- **Oxygen**: O'Brien and Mortis have looked into how oxygen affects the growth and metabolism of acetobutylicum. While short-term exposure to oxygen did not prove fatal, cells exposed to high enough concentrations of oxygen stopped growing and stopped synthesizing proteins, deoxyribonucleic acid (DNA), and ribonucleic acid. The production of butyrate, but not acetate, ceased in aerobic conditions, and the amount of intracellular ATP significantly decreased. It appeared that the cells had run out of reducing power. When cells were put back under anaerobic conditions, growth and metabolism resumed. The effects of oxygen inhibition were quickly reversible. Additionally, these researchers showed that the increase in the culture's Eh was not the cause of the oxygen's effect. Spore

formation increased in cultures that were exposed to oxygen. The length of time the cells were exposed to oxygen, the age of the cells, and the culture conditions all affected how much the spore formation was enhanced. It has been observed that cultures can produce 3–9% more butanol when exposed to brief aeration bursts lasting two to three minutes every hour.

2.6. EVENTS ASSOCIATED WITH SOLVENTOGENESIS

2.6.1. Solventogenesis and Enzyme Activity

It has been demonstrated that, as would be predicted, the enzymes involved in the pathways that produce acid and solvent also change in activity in tandem with the change in metabolic activity that occurs when cells transition from the acidogenic to the solventogenic phase. Andersch et al revealed that extracts from acid-producing cells had two–six times higher activity levels of all four terminal enzymes in the acetate and butyrate pathways than extracts from solvent-producing cells. Hartmanis and Gatenbeck reported similar findings, noting that in solvent-producing cells, phosphate butyltransferase activity vanished and phosphate acetyltransferase and acetate kinase activities were significantly decreased. Butyrate kinase, they discovered, maintained its full activity throughout solventogenesis. The activity of the four enzymes that are involved in the conversion of acetyl-CoA to butyryl-CoA was also ascertained by these workers. The fourth enzyme in this pathway, butyryl-CoA dehydrogenase, was only slightly active; the first three enzymes—thiolase, 3-hydroxybutyryl dehydrogenase, and crotonase—appears to be co-ordinatedly expressed and showed maximal activity in solvent-producing cells after growth had stopped.

The rates of product formation in vitro were used to calculate the activity of each of these enzymes. Based on these findings, it was determined that each enzyme's activity was 10 to a thousand times higher than what would be required for the rate at which these metabolic pathways would be processed in vivo. This implies that the regulation of the branched acid-producing pathways may involve the allosteric modification of some or all of these enzymes by metabolic intermediates.

It has been reported that the activities of butanol dehydrogenase and butyraldehyde dehydrogenase are only slightly detectable in solvent-producing cells in *C. beijerinckii* and *C. acetobutylicum*. However, Rogers found that the specific activities of both of these enzymes in *C. acetobutylicum* were 70–90 times higher in extracts from cells that produced solvent. In batch cultures, butanol production was followed by a decrease in butanol dehydrogenase and butyraldehyde dehydrogenase activity as biomass production declined. It is unknown what caused these enzymes' activity to decline at the conclusion of the fermentation.

The uptake of acetate and butyrate during the solvent producing phase has been shown to be directly coupled to the production of acetone via acetoacetyl-CoA: acetate/ butyrate: CoA transferase. This enzyme has also been shown to exhibit a high level of activity in solvent-producing cells in both batch and continuous culture but was hardly detectable in acid-producing cells. The final step leading to the production of acetone occurs via the action of acetoacetate decarboxylase, and again the specific activity of this enzyme was found to be extremely low in acid producing cells but was increased by about 40-fold in solvent-producing cells.

These findings suggest that before the transition to solvent production can take place, the enzymes required in the last reactions of solvent production need to be synthesized or activated. The addition of rifampin and chloramphenicol inhibits the increase in activity of the enzyme, suggesting that new protein synthesis is necessary for the increase in activity of butyraldehyde dehydrogenase and acetoacetate decarboxylase. The simultaneous rise in activity of the different sets of enzymes raises the possibility that their induction is caused by a shared regulatory signal.

It is critical to distinguish between the mechanisms that could be responsible for the regulation of enzyme activity and those that might be involved in the induction of enzyme biosynthesis. Ballongue conducted research on the different factors that contribute to the induction of acetoacetate decarboxylase. They found that while linear acids from C_5 to C_7 could not act as inducers of enzyme biosynthesis, the linear fatty acids from C_1 to C_4 (formate, acetate, propionate, and butyrate) could. When acid concentrations were similar to those that happened during fermentation and matched the pKa of the acids, induction peaked at pH 4.8. The most efficient inducers were butyric, acetic, and acetoacetic acids; propionic and formic acids also worked well. Dimethylsulfoxide did not produce any induction, indicating that the anion itself plays a significant role in the induction mechanisms. However, it is not clear from these results whether the acids exert their effect directly or whether the effect occurs via some other intermediate.

When cells were growing, acetoacetate decarboxylase was inhibited and didn't become active until the growth stopped. This shows that there may be a connection between the induction of this enzyme and the suppression of cell growth or division.

2.6.2. Solventogenesis and Acid Reutilization

The intracellular concentrations of acetate, butyrate, and their metabolic intermediates as well as the amounts of NADH and NADPH in the cell would seem to decrease with the switch to solvent production. The medium's low pH would have less of an inhibitory effect if acetate and butyrate were taken up and used again.

Since the energy in the thioester bond is conserved and transferred to the acids without the need for ATP utilization, the mechanism of acid utilization by means of the CoA transferase reaction would be energetically favorable. Because of the wide carboxylic acid specificity of the acetoacetyl-CoA: acetate/ butyrate: CoA transferase, it is unclear what controls the ratio of acetate to butyrate consumed during reassimilation. Butyrate is typically consumed more quickly than acetate in batch AB fermentation.

One mole of acetoacetic acid is produced for every mole of acetate or butyrate produced by the CoA transferase reaction. But unlike acetyl-CoA produced by glycolysis, acetyl-CoA or butyryl-CoA produced by the cotransferase reaction is not linked to the creation of NADH. However, the cell needs to use two mol of acetyl-CoA in order to produce acetoacetate for every mole of acetyl-CoA or butyryl-CoA produced by this mechanism.

In order to achieve equilibrium between the flow of electrons and carbon during acid reutilization, it seems that the cell needs to metabolize at least two mol of glucose for every mole of acid that is consumed. To maintain the proper electron balance under these circumstances, about half of the reducing equivalents from reduced ferredoxin would need to be redirected from hydrogen formation to the synthesis of reduced pyrimidine nucleotides. The same number of reducing equivalents would be used if 1 mol of either butyrate or acetate were to be recycled. However, since butanol dehydrogenase is an enzyme that needs NADPH, the amount of NADH/NADPH used in the process of producing butanol would vary depending on whether butyrate or acetate was used.

To meet the requirements for butanol formation, all of the reducing equivalents from reduced ferredoxin would need to be transferred from the production of hydrogen to the production of reduced pyridine nucleotides if no acids were consumed during solvent production.

This theoretical analysis suggests that when solvent production is coupled to acid reassimilation, the production of hydrogen would be reduced by approximately half, and the majority of the reduced pyridine nucleotide generated from reduced ferredoxin would need to be in the form of NADPH in order to satisfy the electron balance. Therefore, the ratio of acetate to butyrate consumed may be influenced by the NADH/NADPH ratio.

2.6.3. Solventogenesis and ATP Generation

The anaerobic ability to grow is thought to be restricted by the rate of the energy-producing reaction because of the limited amount of ATP molecules produced during fermentation. The produced ATP molecules can be used in two different ways: first, in biosynthesis that promotes cell growth; and second, in membrane energization that produces an electrical and pH gradient across the cell membrane that is necessary for metabolic activity. This is accomplished by the proton-translocating ATPase. During metabolism, adenosine diphosphate (ADP), Pi, and ATP must all be continuously recycled; therefore, a decrease in the quantity of ATP available would probably have an impact on growth and biosynthesis.

A decrease in a rate-limiting step in ATP production, such as a decrease in flux through the glycolytic pathway or a decrease in sugar uptake, may result in a reduction in the amount of ATP available for biosynthesis. Secondly, an increase in membrane permeability or ApH may cause more ATP to be diverted towards membrane energization, which would explain a decrease in the amount of ATP available for biosynthesis. The cell would seem to be forced to use more energy for membrane energization as the pH drops toward the end of the acid-producing phase in order to counteract the effect of increased membrane permeability and to raise the ApH in order to keep the internal pH of the cell above the threshold level. Furthermore, the accumulation of a high concentration of acid end products within the cell appears to reduce the flux through the branched pathways under these conditions, which would further reduce the generation of ATP.

Reduced net ATP generation from roughly 3.3 mol of ATP per mol of glucose to 2 mol of ATP per mol of glucose is a significant effect of the switch from acid to solvent production. The cell's inability to sustain regular vegetative growth appears to be a direct result of the reduction in ATP available to the cell.

2.6.4. Solventogenesis and Growth

The majority of batch culture studies show a clear correlation between the end of exponential growth and the transition to solvent production, which was previously linked to a decrease in growth during the traditional AB batch fermentation process. If cell number is used as a growth indicator instead of the culture's optical density, the correlation between the onset of solvent formation and the end of exponential growth becomes even clearer. Cell division appears to have stopped around the time of the switch to solvent production, though the optical density of the culture may have continued to rise for several hours after. Growth is measured by counting both the total number of cells and the number of viable cells. The accumulation of storage products inside the cell, changes in the optical characteristics of the cell, or the continuation of cell growth after cell division has stopped are some of the possible causes of the optical density's ongoing increase. When batch culture conditions are met, the rapid onset of sporulation may occur after the cessation of cell division. In these cultures, more than 90% of the cells may develop forespore septa during the solvent-producing phase. For the purpose of producing solvents, nongrowing immobilized cells and solvent-producing cells that have been resuspended in buffer have been employed. It is clear that nongrowing cells can produce solvents, and some researchers have proposed a mechanistic connection between the beginning of solvent production and the growth and metabolism slowdown that happens at the end of the acid-producing phase.

It is less evident how cell growth affects the start and continuation of solvent production in continuous culture systems. The kinetics of product formation, however, follow a similar pattern regardless of the limiting factors used, and the actual values obtained for the specific rate of product formation and substrate consumption agree in the various reported chemostat studies. These investigations clearly show that the dilution rate influences solvent production in a significant way. Acid production dominates at high dilution rates, and as the dilution rate rises, so does the specific rate of acid production. Extrapolations show that at low growth rates, there is no production. With an increase in dilution rate, the specific rate of glucose consumption likewise rises, and an extrapolation to zero suggests the need for maintenance. Both acids and solvents can form at intermediate dilution rates; at low dilution rates, solvent production is preferred and the highest solvent yields and concentrations are achieved. In contrast to the pattern of acid production, the specific rate of solvent production increases as the dilution rate increases up to a certain point before rapidly decreasing as the dilution rate continues to rise. Solvent formation at zero growth rate is shown by extrapolation. The best circumstances for producing solvents in continuous culture have been low dilution rates in the presence of excess glucose, low pH, and phosphate or sulfate limitation.

There seems to be some overlap between the variables influencing solvent production in chemostat cultures and turbidostat cultures. Gottschal and Morris discovered that only acid production happened at low cell densities and high dilution rates, while solvent production predominated at high cell densities and low dilution rates. Both the pathways that produce solvent and acid were operational in steady-state turbidostat cultures with high cell densities.

It follows that under seemingly steady-state conditions, acid production and solvent production can both happen simultaneously in chemostat and turbidostat cultures. In continuous culture, it is typically assumed that all cells will be in the same physiological state under steady-state conditions. This would necessitate an ineffective cycle of acid production and reutilization, though, since it would imply that both pathways could function concurrently in the same cell. It is highly unlikely that both pathways could function simultaneously in the same cell, according to a large body of evidence. Instead, it appears more likely that under these circumstances, distinct cell populations exist in equilibrium within the culture. Actively growing acid-producing cells predominate at high dilution rates, while solvent-producing cells predominate at low dilution rates. However, the fact that solvent-producing cells are still producing acetone suggests that acid production and assimilation must go on in these cultures, suggesting that under these circumstances, acids are produced and reassimilated at nearly the same rate on a molar basis. This implies that cells that produce both acid and solvent are present in continuous cultures that are diluted at low rates. The average cell doubling times at the low dilution rates that have been used to produce good solvent yields range from 6 to 30 hours. Similar to batch culture, solvent-producing cells may experience decreased or inhibited growth and cell division. The loss of these cells due to washout may be offset by the transformation of acid-producing cells into solvent-producing cells.

2.6.5. Solventogenesis and Cell Differentiation

The selection and maintenance of high solvent-producing strains involved cycles of sporulation, heat exposure, germination, and outgrowth. Sporulation was necessary for the maintenance of good solvent production. These data imply that the induction of cell differentiation that results in endospore formation and the induction of solvent production may be related. It seems that both occurrences are connected to the suppression of normal cell division and vegetative growth.

During fermentation, some *C. acetobutylicum* strains, such as the commercial strain P262, display distinct morphological phases. Highly mobile dividing rods are typically present during the first acidogenic phase of fermentation. But a few hours before the pH breakpoint is reached, the cell starts to accumulate granulose, a glycogen-like storage product, and motility starts to decline. The vegetative rods change into swollen phase-bright clostrial forms during the shift to solvent production, which is also accompanied by the formation of an extracellular capsule, or slime layer. The beginning of endospore formation, which entails the creation of a forespore septum at one or both poles of the cell, typically occurs after the development of the clostridial stage. However, mature endospores were not typically generated in industrial fermentation, presumably as a result of toxic butanol concentrations accumulating and inhibiting further development. By the end of the solvent-producing phase, endospores can be produced in more than 90% of the cells in suitable culture media. Endospore production goes through morphological stages that are comparable to those seen in other spore forming bacteria.

2.6.6. Solventogenesis and Culture Stability

More than 90 years have passed since the literature first documented the propensity of solvent-producing *Saccharomyces cerevisiae* arolytic clostridia to experience degenerative changes. Although the mechanism underlying this process is still not well understood, references to this phenomenon have persisted in the literature. Numerous reports show that most strains lose their capacity to produce solvents when they are repeatedly subcultured, and this can also lead to the creation of cultures with altered colonial morphology. Due to the difficulties encountered in serial subculturing, solvent-producing clostridia are now routinely maintained as spores that can be heat-activated and germinate when an inoculum is needed. A recent investigation by Hartmanis et al. proved that the *C. acetobutylicum* ATCC 824 strain could be serially transferred and grown for extended periods of time without losing its capacity to produce solvents under the right circumstances. For 218 days, these workers subcultured the ATCC 824 strain at 24-hour intervals. The strain was still able to generate solvents during this time, but the cultures gradually became more acidogenic. The inoculum size increased from 3.3% to 6.7% after 200 days of subculturing. As a result, the production of butanol increased and acetone, which had been absent for more than 50 days, started to appear again. It seems that the beginning of degenerative alterations is a complicated phenomenon rather than a natural outcome of ongoing subculture. The specific strain and culture conditions employed seem to have an impact on the chance that a culture will experience degenerative changes during serial transfer. The age and condition of the cells at transfer, the time between transfers, the inoculum size, the pH and buffering capacity of the culture medium, and the nutrients it contains all seem to have an impact on when degeneration starts.

The long-term stability of cells grown in continuous culture is significantly impacted by the capacity of cells to degenerate and stop producing solvent after extended growth. Research on the stability of cells in continuous culture has yielded inconsistent findings. According to Finn and Nowrey, serially transferred cells exhibited more degeneration than cells maintained in exponential growth in continuous culture. But in a continuous culture, solvent production was not stable and eventually decreased. Jobses and Roels also reported that although high solvent concentrations could be achieved for varying amounts of time, solvent-producing activity was not stable and decreased over time, and under all conditions used, butanol-producing activity was eventually lost in a study on solvent production by *C. beijerinckii* in continuous culture. It is possible to restore butanol formation in acid-producing cultures grown at pH 6.3 by lowering the pH to 4.5. They did come to the conclusion that long-term, continuous butanol production was not feasible, though. Stephens et al.

reported similar observations in investigations using continuous-culture *C. acetobutylicum* NCIB 8052. In chemostats limited by ammonia, magnesium, or phosphate, they found that solvent production was always temporary and that acid production took over after four to 16 volume changes. Although it was improved, cell stability in a glucose-limited chemostat and a pH auxostat was not permanent. They also noted that no steady state could be reached in any of the continuous-flow systems. To address the issue of cultural deterioration, Afschar et al. used cell recycling to enhance solvent production and stability, as well as to reach higher dilution rates. Additionally, these researchers have documented the use of sheer activation—which is accomplished by pumping cells through capillaries—to increase the rates of stability, cell growth, solvent productivity, and glucose consumption in continuous culture. Lately, Fick and colleagues. reported that using complex media containing 40 g of glucose per liter run at a dilution rate of 0.06/h, they were able to obtain a stable continuous culture of *C. acetobutylicum*. Cultures were kept alive for two months in these circumstances without experiencing a decrease in butanol production. Cultures lost their ability to produce solvent and became unstable when the concentration of glucose increased (Afschar et al., 1985).

2.7. SOLVENT TOXICITY

Cell metabolism typically continues during the solvent-producing phase until the solvent concentration reaches inhibitory levels of about 20 g/liter, at which point it stops. One of the main obstacles to using fermentation for industrial solvent production is the comparatively low solvent concentrations that are produced during the process. Butanol is the most harmful of the solvents generated during fermentation, and it is the only one that is produced in quantities that are inhibitory. When butanol concentration in the industrial fermentation process reaches approximately 13 g/liter, solvent production stops.

Numerous laboratory investigations have verified these findings, demonstrating that adding 7 to 13 g of butanol per liter to cultures cultivating on hexose sugars led to a 50% growth inhibition. There appeared to be a threshold concentration of 4 to 4.8 g of butanol per liter below which no discernible decrease in growth occurred. Growth was completely inhibited at a concentration of 12 to 16 g/liter. It has been observed that butanol's inhibitory effect is amplified in cells cultured on xylose, and that the addition of 8 g of butanol per liter completely inhibited growth. Conversely, it doesn't seem that the fermentation will be inhibited by the concentrations of acetone and ethanol. Growth was inhibited by approximately 50% when acetone and ethanol were added, starting at a concentration of 40 g/liter. At a concentration of 70 g of acetone and 50–60 g of ethanol per liter, growth was completely stopped.

2.7.1. Mechanism of Butanol Toxicity

The primary effect of butanol and other long-chain aliphatic alcohols appears to be the disruption of the phospholipid component of the cell membrane, which is linked to the hydrophobic nature of the compound and the mechanism of its toxicity. While butanol and other longer-chain aliphatic alcohols have the opposite effect and increase membrane fluidity, short-chain aliphatic alcohols like ethanol have the opposite effect. As the chain length increases, so does the alcohol's solubility in the membrane and its impact on membrane fluidity. Low butanol concentrations (<5 g/liter) do not seem to affect the fluidity of *C. acetobutylicum* extracted membranes. However, higher but subinhibitory butanol concentrations (10 g/liter) increased the fluidity of lipid dispersal by 20 to 30%, as measured by electron spin resonance, spin label analysis.

Membranes from stationary phase solvent-producing cells as well as vegetative cells grown in the presence of butanol (0.5 to 1.0%, vol/vol) showed an increase in the ratio of saturated to unsaturated fatty acids. The cell's physiological reaction to counteract the effect of increased membrane fluidity appears to be the increase in the ratio of saturated fatty acids in the membrane. The way that cells grown in butanol change their cell membrane composition is comparable to how cells grown at higher temperatures react, which also causes an increase in membrane fluidity.

As would be predicted, butanol causes an increase in membrane fluidity, which leads to membrane destabilization and the disruption of membrane-linked functions. The complex effects of butanol include the inhibition or disruption of multiple interrelated membrane processes, though it is unknown how these events relate to one another.

Moreira and associates revealed that the addition of aliphatic alcohols caused membrane-bound ATPase activity to be instantly inhibited. Bowles and Ellefson also noted that butanol partially inhibited ATPase activity. It has also been discovered that the addition of butanol eliminates the membrane pH gradient and inhibits the cell's capacity to maintain its internal pH. According to Gottwald and Gottschalk, in *C. acetobutylicum*, adding 7 g of butanol per liter reduced the ApH from 1.2 to slightly below 1.0, and adding 10 g of butanol per liter eliminated the ApH (Bowles et la., 1985).

It was also discovered that the addition of butanol reduced the intracellular level of ATP; however, these experiments revealed that the ATP level decrease happened independently of the pH gradient collapsing.

However, the preservation of the pH gradient across the membrane necessitates both ATP and ATPase activity. An additional effect of butanol toxicity on membranes is the inhibition of sugar and amino acid uptake.

Moreira along with others demonstrated a 50% reduction in the uptake of the nonmetabolizable glucose analog (3-0-methyl glucose) in the presence of 7.4 grams of butanol per liter. The authors proposed that the disruption of a transport system that requires energy was the cause of the inhibitory effect.

The autolytic breakdown of solvent-producing cells in *C. acetobutylicum* P262 has also been connected to butanol toxicity. It has also been proposed that butanol inhibitory concentrations were responsible for initiating the release of cell-free autolysin during the solventogenic phase. Allcock and associates discovered a pleiotropic autolysis-deficient mutant (lyt-J) that was more resistant to the autolysin of both the parent strain and itself and produced less autolysin overall. The lyt-J mutant cellular stability was largely unaffected by the addition of butanol (7 to 16 g/liter), but it did accelerate the breakdown of the parental strain's solvent-producing clostridial stage cells. Furthermore, this mutant strain produced slightly higher concentrations of total solvents and butanol (14.2 g/liter compared with 13.3 g of butanol per liter) and its vegetative cells demonstrated enhanced butanol tolerance.

2.7.2. Butanol Tolerance

Since the amount of solvent that can be produced during AB fermentation seems to be limited by butanol toxicity, it has been hypothesized that increasing the cells' tolerance to butanol will lead to the production of higher concentrations of solvents. There is some evidence to imply that adjusting the growth conditions during the fermentation process may improve tolerance to butanol. It has been demonstrated that increasing the amount of saturated fatty acids in the medium raises the ratio of saturated fatty acids in the membrane. Cell growth and ATPase activity were also increased, and butanol tolerance was up to twofold increased by saturated fatty acid enrichment.

A drop in temperature during the solvent-producing phase may improve butanol tolerance, as evidenced by the finding that membrane fluidity increases with temperature. By lowering the temperature from 30°C during the acidogenic phase of the fermentation to 240°C during the solventogenic phase, this method looks to have served as the foundation for a patented process to increase butanol production. These results also imply that it might be unlikely for thermophilic clostridia strains to produce appreciable amounts of butanol.

To date, there has been little progress in creating mutants that can both tolerate and produce higher butanol concentrations. This may indicate that several mutants will need to be isolated in order to achieve appreciable increases in butanol production, which is not surprising given the complex interactions pertaining to butanol toxicity. Regrettably, strains that produce noticeably higher concentrations of butanol during the nongrowing solvent-producing phase have not been isolated as a result of the selection of mutants that can grow in the presence of inhibitory concentrations of butanol. A butanol-tolerant mutant (SA-1) of *C. acetobutylicum* ATCC 824 was identified by Lin and Blaschek; it shared traits with the lyt-I mutant. At a butanol concentration of 15 g/liter, this strain was able to grow at a rate that was 66% faster than the control, which resulted in negative growth in the parent strain. The total amount of solvent produced decreased as a result of this mutant producing less acetone and more butanol than strain ATCC 824. In addition, this mutant showed higher amylase activity than the ATCC 824, allowing it to use starch more effectively.

2.8. GENETICS AND STRAIN IMPROVEMENT

The performance of the strains used for solvent production needs to be improved for industrial batch fermentation to become economically competitive due to a number of its limitations. Few attempts have been made to enhance the industrial strains through mutation or other forms of genetic manipulation, even though it seems to have been standard procedure to isolate and maintain vigorous solvent-producing strains by the selection of actively sporulating cultures. Current efforts to create systems for the genetic modification of *C. acetobutylicum* and related solvent-producing species have been impeded by the fact that genetic research on obligate anaerobes, such as clostridia, has not progressed as quickly as that on aerobic species.

2.8.1. Mutagenesis

The selection and development of industrially significant strains have traditionally depended heavily on the capacity to generate and isolate mutants. Selecting a potent mutagen is crucial when creating a mutation protocol for a specific microbe. Through mispairing mechanisms involving either template or nucleotide precursors, direct mutagens induce mutations. Indirect mutagens work by causing an error-prone postreplication repair system in *Escherichia coli* that depends on the products of the recA, lexA, and umuC genes. The induction of stable chromosomal mutants in *C. acetobutylicum* was studied by Bowring and Morris. The relative induced mutation frequency of auxotrophic and rifampin-resistant markers was significantly increased by ethyl methanesulfonate and N-methyl-N'-nitro-N-nitrosoguanidine, two direct mutagens. On the other hand, indirect mutagens like mitomycin C and ultraviolet (UV) radiation did not affect *C. acetobutylicum*. Under the test conditions, metronidazole, hydrogen peroxide, and nalidixic acid—other potentially mutagenic agents—were likewise ineffective. Lemmel described the ethyl methanesulfonate mutagenesis conditions for *C. acetobutylicum* ATCC 39236; however, he was unable to induce mutations using UV irradiation, an acridine half-mustard ICR 191, or N-methyl-N'-nitro-N-nitrosoguanidine. Walker hypothesized that Clostridium species might lack the error-prone repair necessary for indirect mutagenesis because UV radiation cannot induce mutations (Bowring et al., 1985).

The preferred mutagen for *C. acetobutylicum* induction and isolation of a range of beneficial mutants that have been applied to physiological research is ethyl methanesulfonate. These comprise sporulation mutants, granulose mutants, capsule mutants, auxotrophic mutants, phage resistance mutants, and solvent production mutants. Bowring and Morris reported that while ethyl methanesulfonate and other directly acting mutagens were successful in inducing mutants, certain types of auxotrophic lesions were isolated significantly more frequently than others. Using effective mutagens in combination with selection processes is a crucial component of any mutation strategy. Conventional methods are widely known for their ability to select and enrich auxotrophic and resistance mutants. However, methods for choosing mutants in the solvent pathway will need to be created.

2.8.2. Genetic Transfer Systems

1. **Bacteriophages**: Serious issues with industrial AB fermentation were brought on by phage infection. McCoy et al. reported the first phage isolation from an anomalous fermentation broth., and it seems that phage infections connected to the industrial fermentation process were rather frequent. Ogata and Hongo have reviewed the properties of several phages that were isolated from clostridia that produce solvent. These phages, which were composed of phages with either long tails that were contractile or noncontractile, or short noncontractile tails, all contained double-stranded DNA. Furthermore, strains of *Clostridium spp.* that produce solvent have also been reported to produce faulty phage like particles and bacteriocins.

Hongo et al. looked into the occurrence of lysogeny and bacteriocinogeny in 106 recently isolated strains of *Clostridium spp.*, the bulk of which were solvent producers. 18 of these strains also produced bacteriocins. Four of these strains contained lysogenic phages that were identical in terms of their host range, serological characteristics, plaque morphology, and UV inactivation kinetics. Clostocins A through E are the five groups into which these bacteriocins were divided. From a solvent-producing strain of C. succinylacetonicum, lysogenic phages and defective phage like particles known as clostocins 0 and M have also been isolated after induction with mitomycin C or UV radiation.

2. **Plasmids and Conjugation**: Relatively large, self-transferable plasmids with a transfer (Tra) region are necessary for bacterial conjugation. Some plasmids that are self-mobilizable have the ability to mobilize other plasmids that lack a Tra region but have compatible mob, nic, and oriT sites. It has been reported that *C. acetobutylicum*, *C. beijerinckii*, *C. butyricum*, and *C. saccharoperbutylacetonicum* contain both large and small plasmids; however, no evidence of their transfer has been shown. Furthermore, none of these plasmids' functions have been linked to their presence in any bacterial strain, suggesting that they are all mysterious. C. perfringens has been shown to harbor plasmids encoding for bacteriocins, caseinase, lecithinase, mercurial and organomercurial compounds, and antibiotic resistance. There are no reports of these plasmids being transferred to or maintained in *C. acetobutylicum* strains, although some of them are self-transferable.

3. **Protoplast Transformation**: Thus far, attempts to transform *C. acetobutylicum* cells using chromosomal or plasmid DNA have not been successful. These methods include the use of treatments that have been shown to facilitate the uptake of DNA by whole cells in other species of bacteria as well as the use of natural competence.

The use of protoplasts for plasmid DNA uptake in gram-positive bacteria has proven to be particularly effective. The creation of processes for the generation and regeneration of protoplasts is necessary for the transformation of protoplasts. Techniques for *C. acetobutylicum* P262, *C. pasteurianum*, and *C. saccharoperbutylacetonicum* have been reported recently for the production and regeneration of stable protoplasts. All of these strains can easily produce protoplasts with the help of lysozyme, although the frequency of regeneration differs. Frequencies exceeding 80% were obtained in *C. acetoblitylicum*, whereas *C. pasteurianum* and C. *saccharoperbutylacetonicum* were reported to have frequencies between 1 to 10%.

4. **Protoplast Fusion**: The fusion of bacterial protoplasts induced by polyethylene glycol has been used in both basic and applied genetics research. Jones along with others used *C. acetobutylicum* P262's high regeneration frequency to generate stable recombinants and segregating biparentals at frequencies of 0.3 to 2.0 percent and 1.4 to 8.3%, respectively. Progeny of both parental types were produced by the segregating biparentals, who carried both parental genomes. These biparentals could be further classified as noncomplementing, partially complementing, and prototrophic complementing biparentals.

The noncomplementing biparentals expressed neither of the parental phenotypes (zero noncomplementing biparentals) or the phenotypes of one of the parental strains (single-parent noncomplementing biparentals). While single-parent noncomplementing biparentals and prototrophic complementary biparentals have been observed in B. subtilis, partially complementing and zero noncomplementing biparentals, which are novel classes of biparentals, have not been reported. Further research aimed at characterizing the organization and structure of the *C. acetobutylicum* genome will be made possible by the demonstration of protoplast fusion and the isolation of recombinants in *C. acetobutylicum* P262. After protoplast fusion, stable chromosomal recombinants were isolated, suggesting that *C. acetobutylicum* can undergo homologous recombination at a frequency comparable to that of aerobic species like *B. subtilis*. In *Clostridium spp.*, the emergence of protoplast fusion may promote interspecific gene and plasmid transfer. This could be used to create hybrid strains that have the right commercial characteristics. It is also possible to shield plasmids from the elevated extracellular nuclease levels generated by *C. acetobutylicum* strains by using plasmid transfer via protoplast fusion. One potential solution to the extracellular nuclease issue is to shield plasmids within liposomes, which can subsequently merge with protoplasts.

2.8.3. Gene Cloning

Nowadays, cloning and gene research in *C. acetobutylicum* are hindered due to the challenges associated with introducing DNA into the organism. Cloning and analyzing the genes of *C. acetobutylicum* in other bacteria is an alternate strategy. This would enable the study of gene functions through site-directed mutagenesis and make it easier to characterize significant structural and regulatory genes. One benefit of cloning genes from genetically poorly characterized strains into well-characterized bacteria like *B. subtilis* and *E. coli* is the ability to analyze gene function and regulation through complementation and the availability of numerous defined mutants.

The expression and stability of *C. acetobutylicum*'s genes in gram-negative aerobes are of interest because the bacterium is a gram-positive obligate anaerobe. It is also possible to use cloned *C. acetobutylicum* DNA as a source of genes to modify and enhance other industrial bacteria.

Many chromosomal genes from Clostridia that produce *Saccharomyces cerevisiae* arolytic solvents have been cloned and expressed in *Escherichia coli*. P-isopropylmalate dehydrogenase and hydrogenase, two chromosomal genes from *C. butyricum*, have been cloned and expressed in *E. coli*. After *E. coli* HB101 leu was transformed, the 3-isopropylmalate dehydrogenase gene was isolated by selecting leu+ transformants. The complementation of an *E. coli hyd* mutant allowed for the cloning of the hydrogenase gene. In comparison to the *C. butyricum* and *E. coli* hyd+ strains, the hydrogenase activity of the *E coli* transformant carrying a *C. butyricum* DNA insert in pBR322 was roughly 3.1–3.5 times higher. Given that pBR322 is a multicopy plasmid, the gene dosage effect was thought to be the cause of the increase in activity (Afschar et al., 1985).

REFERENCES

1. Abou-Zeid, A. A., Fouad, M., & Yassein, M. (1978). Microbiological production of acetone-butanol by *Clostridium acetobutylicum*. *Zentralblatt für Bakteriologie, Parasitenkunde, Infektionskrankheiten, 2 133*, 125–134.
2. Adams, M. W. W., Mortenson, L. E., & Chen, J. S. (1980). Hydrogenases. *Biochimica et Biophysica Acta, 594*, 105–176.
3. Afschar, A. S., Biebl, H., Schaller, K., & Schugerl, K. (1985). Production of acetone and butanol by *Clostridium acetobutylicum* in continuous culture with cell recycle. *Applied Microbiology and Biotechnology, 22*, 394–398.
4. Allcock, E. R., Reid, S. J., Jones, D. T., & Woods, D. R. (1982). *Clostridium acetobutylicum* protoplast formation and regeneration. *Applied and Environmental Microbiology, 43*, 719–721.
5. Bahl, H., Andersch, W., Braun, K., & Gottschalk, G. (1982). Effect of pH and butyrate concentration on the production of acetone and butanol by *Clostridium acetobutylicum* grown in continuous culture. *European Journal of Applied Microbiology and Biotechnology, 14*, 17–20.
6. Barber, J. M., Robb, F. T., Webster, J. R., & Woods, D. R. (1979). Bacteriocin production by *Clostridium acetobutylicum* in an industrial fermentation process. *Applied and Environmental Microbiology, 37*, 433–437.
7. Biggins, D. R., & Dilworth, M. J. (1968). Control of pyruvate phosphoroclastic activity in extracts of *Clostridium pasteurianum* by ADP and acetyl phosphate. *Biochimica et Biophysica Acta, 156*, 285–296.
8. Bowles, L. K., & Ellefson, W. L. (1985). Effects of butanol on *Clostridium acetobutylicum*. *Applied and Environmental Microbiology, 50*, 1165–1170.
9. Bowring, S. N., & Morris, J. G. (1985). Mutagenesis of *Clostridium acetobutylicum*. *Journal of Applied Bacteriology, 58*, 577–584.
10. Compere, A. L., & Griffith, W. L. (1979). Evaluation of substrates for butanol production. *Developments in Industrial Microbiology, 20*, 509–517.
11. Crabbendam, P. M., Neijssel, O. M., & Tempest, D. W. (1985). Metabolic and energetic aspects of the growth of *Clostridium butyricum* on glucose in chemostat culture. *Archives of Microbiology, 142*, 375–382.
12. Daldal, F., & Applebaum, J. (1985). Cloning and expression of *Clostridium pasteurianum* galactokinase gene in *Escherichia coli* K_{12} and nucleotide sequence analysis of a region affecting the amount of the enzyme. *Journal of Molecular Biology, 186*, 533–545.

13. Doerner, P., Lehmann, J., Piehl, H., & Megnet, R. (1982). Process analysis of the acetone-butanol fermentation by quadruple mass spectrometry. *Biotechnology Letters*, *4*, 557–562.

14. Fick, M., Pierrot, P., & Engasser, J. M. (1985). Optimal conditions for long-term stability of acetone-butanol production by continuous cultures of *Clostridium acetobutylicum*. *Biotechnology Letters*, *7*, 503–508.

15. Finn, R. K., & Nowrey, J. E. (1958). A note on the stability of clostridia when held in continuous culture. *Applied Microbiology*, *7*, 29–32.

16. Fond, O., Matta-Ammouri, G., Petitdemange, H., & Engasser, J. M. (1985). The role of acids on the production of acetone and butanol by *Clostridium acetobutylicum*. *Applied Microbiology and Biotechnology*, *22*, 195–200.

17. Forberg, C., & Haggstrom, L. (1985). Control of cell adhesion and activity during continuous production of acetone and butanol with adsorbed cells. *Enzyme and Microbial Technology*, *7*, 230–234.

18. Haggstrom, L., & Molin, N. (1980). Calcium alginate immobilized cells of *Clostridium acetobutylicum* for solvent production. *Biotechnology Letters*, *2*, 241–246.

Chapter 3

Organic Acids

Contents

INTRODUCTION

An organic compound having acidic qualities is called an organic acid. The most prevalent types of organic acids are carboxylic acids, which get their acidity from the carboxyl group (COOH) in them. Sulfonic acids are comparatively stronger acids because they contain the group $-SO_2OH$. With -OH, alcohols can function as acids, although they are typically very weak. The acidity of an acid is determined by the relative stability of its conjugate base. Other groups such as the thiol group (SH), the enol group, and the phenol group can also impart acidity, albeit typically in a weaker way. Organic compounds that contain these groups are commonly referred to as organic acids in biological systems.

3.1. CITRIC ACID

An organic acid called citric acid (CA) is typically present in a range of fruits, including grapefruits, limes, lemons, oranges, and pineapples. It's a natural ingredient that supports healthy kidney and digestive function, energy levels, and detoxification. It's used to cut the sweetness of soft drinks, juices, and other beverages. It tastes slightly tart and refreshing. Because of its antioxidant qualities, citric acid is used in the food and beverage (F&B) industry to preserve food or as an acidifier to improve the flavors and aromas of ice cream, fruit juices, and marmalades. It is utilized in the pharmaceutical sector as an effervescent, blood preservative, pH corrector, antioxidant to preserve vitamins, ointments and cosmetic preparations, and iron citrate tablets as a body's source of iron. It is used as a foaming agent in the chemical industry to soften and treat textiles. Certain metals are used in metallurgy in the form of citrate. CA is used as a phosphate substitute in the detergent industry due to its reduced eutrophic effect. Furthermore, because it naturally lightens and brightens the skin tone, reduces oiliness and breakouts, and regenerates dead skin cells, CA is widely used in face packs and masks. The projection for the global CA market was to reach USD 3.2 billion by 2023, with a Compound Annual Growth Rate (CAGR) of 5.1% over the course of the forecast period. An estimated 736,000 tons of CA are produced worldwide each year, all of which are produced through fermentation. Nearly all of Brazil's CA demand is satisfied by imports. Owing to its wide range of uses, fermentation-based CA production is growing at a high rate of 5% per year, while demand and consumption are also rising steadily. The Joint Food and Agriculture Organization (FAO)/World Health Organization (WHO) Expert Committee on Food Additives approved it, and as a result, it is recognized globally as a generally recognized as safe (GRAS) (Rohr et al., 1996; Carlos et al., 2006). The growing application in numerous industries is the primary driver of the global CA market's expansion. The last century has seen a significant increase in the development of CA production, largely due to the following factors: biochemistry, which offers knowledge of various factors that affect synthesis and blockage of CA production; molecular regulatory mechanisms; and strategies that enhance CA production. Biotechnology also provides proper knowledge of fermentation techniques and product recovery.

3.1.1. Properties

In the metabolism of all aerobic organisms, citric acid (CA) is an intermediate in the CA cycle. It is systematically known as 2-hydroxypropane-1,2,3-tricarboxylic acid by the International Union of Pure and Applied Chemistry (IUPAC). CA is a white, crystalline powder with no color and almost no smell. It can be found in two forms: monohydrate ($C_6H_8O_7H_2O$, molecular weight 210.14) and anhydrous ($C_6H_8O_7$, molecular weight 192.12). While monohydrate CA is soluble in water and sparingly soluble in ether, anhydrous CA is highly soluble in water, freely soluble in ethanol, and sparingly soluble in ether. It melts at 153°C, boils at 310°C, and is solid at room temperature. Above roughly 175°C, it breaks down with the loss of carbon dioxide (CO_2). When dissolved in water, it exhibits a weak acidity but a strong acid taste that influences sweetness and gives a fruity tartness. For this reason, it is frequently used in the F&B industry to balance fruit flavors. Its excellent metal ions chelating properties add to the physicochemical properties that make it ideal for food, cosmetic, nutraceutical, and pharmaceutical applications—whose number attests to its exquisite versatility—while its combination with citrate demonstrates excellent buffering capacity.

3.1.2. Global CA Market

Approximately 0.5 million tons of citrate salts and CA were produced worldwide in 1989. It surpassed 2 million tons in 2015, and until 2020, the global market is predicted to grow at a rate of at least 3.7% annually. China, home to the biggest producers, accounted for 74% of global exports and 59% of global production in 2015. The product was suddenly very abundant, and between 2004 and 2013, production output nearly doubled. This resulted in historically low prices, which peaked in 2015 at USD 700/t. Ready-to-drink beverages and processed foods have become more popular in recent years due to economic expansion, hectic lifestyles, growing urbanization, and rising disposable incomes. As a result, there is now more demand for CA across several industries. Additionally, a move toward products made with natural ingredients has been brought about by growing awareness of the negative effects caused by chemicals used in everyday products. Due to these reasons, the market size for CA is anticipated to reach USD 32.2 billion by 2023 and grow at a compound annual growth rate of 5.1% over the course of the forecast period.

The market has been divided into F&Bs, household cleaners and detergents, pharmaceuticals, and other categories based on the application. Currently, the food and beverage (F&B) industry is the main user of CA for the preparation of various food items, including soups, mayonnaise, juices, creams, and carbonated drinks. The market has been divided into the liquid and anhydrous segments based on the form. At the moment, anhydrous CA holds the lion's share of the market. Geographically speaking, the world's largest market for CA is found in Western Europe. The US, China, the Middle East and Africa, Central and Eastern Europe, Brazil, and India are some other significant markets. China and the United States are the two largest importers and exporters of CA, respectively, according to an analysis of the market's import and export scenario. Research also looks at the market's competitive environment, highlighting Archer Daniels Midland Company and Cargill Inc. as two of the major participants. Cofco Biochemical (Anhui) Co, Ltd., Jungbunzlauer Suisse AG, Tate & Lyle PLC. Xinghua Huangshi Biochemical Co. Ltd., Group Co Ltd. RZBC. Ensign Industry Co., Ltd., Weifang. Gadot Biochemicall Industries Ltd. is the name. S A Citrique Belge N., as well. V. The US market is expected to grow steadily over the next eight years, with a valuation of USD 448.4 million in 2016. Due to the robust demand for functional food additives from the region's thriving processed food and beverage industry, the European region held the largest market share in 2014. With over a quarter of the market share in both volume and value, food is the most popular application. It currently has the fastest rate of growth and is anticipated to do so for the duration of the projection years. The market for CA applications in the personal care and pharmaceutical industries is expanding and expected to grow at a competitive CAGR between 2015 and 2020. By launching CA-based goods like candies, baked goods for diabetics, ice creams, low-calorie jams, low-calorie sugar, dietary drinks and snacks, and low-fat dairy products, the manufacturers are expanding into a number of market niches. It is anticipated that consumer demand will increase due to the growing need for these products.

In order to meet the growing demand for their products, companies in this sector are concentrating on expanding their production capabilities. For example, the large biodegradable ingredient manufacturer Jungbunzlauer announced in December 2017 that it was building a new CA plant in Austria. The WHO estimates that 65–70 million people in North America struggle with digestive system issues. Consequently, this has positively impacted the industry trend by raising the demand for pharmaceuticals and FandBs based on digestion CA in the area. As per the German Nutrition Society (DGE), the proportion of European consumers who are aware of the benefits of CA for preserving digestive health went from 44% in 2015 to 49% in 2016. The group added that more people are consuming yogurt made in California to preserve their digestive health. The growth of the Middle Eastern market is being driven by growing awareness of F&Bs and pharmaceuticals based in California. In an effort to expand, a number of foreign market players are setting up offices and manufacturing facilities in the area. For example, Pfizer opened a new pharmaceutical manufacturing facility in Saudi Arabia in January 2017. Over the course of the forecast period, China's CA market is anticipated to grow at the fastest rate in the Asia Pacific region. The market is primarily driven by factors like the growing geriatric population and the occurrence of lifestyle diseases like mental health disorders, gut health-related problems, and cardiovascular diseases.

3.1.3. Microorganisms Known to Produce CA

Table 3.1 lists the various microorganism types that have been linked to the bio-production of CA, including fungi, yeast, and bacteria. Microorganisms, including *Corynebacterium* sp., *Bacillus licheniformis*, and *Arthrobacter paraffinens*are recognized to generate significant amounts of CA. *Yarrowia*, *Hansenula*, *Pichia*, *Debaromyces*, *Torula*, *Torulopsis*, *Kloekera*, *Saccharomyces cerevisiae aromyces*, *Zygo Saccharomyces cerevisiae aromyces*, and *Candida* are among the yeasts that can produce CA in addition to bacteria. Despite reports of yeast producing CA, one disadvantage of yeast fermentation is the significant production of iso-citric acid (isoCA), an undesirable byproduct.

Table 3.1. *Citric Acid (CA) Producing Microorganism*

Microbes	CA Producing Species	References
Bacteria	*Bacillus licheniformis, Arthrobacter paraffinens, Corynebacterium* sp., *Bacillus subtilis, Brevibacterium flavum, Corynebacterium* sp., and *Penicillium janthinellum*	Kroya Fermentation Industry (1970); Sardinas (1972); Fukuda et al. (1970)
Fungi	*Aspergillus Niger, A. aculeatus, A. awamori, A. carbonarius, A. wentii, A. foetidus, Penicillium janthinelum*	Karow and Waksman (1947); Roukas (1991); El Dein and Emaish (1979); Grewal and Kalra (1995); Chen (1994)

Yeast	*Saccahromicopsis lipolytica, Candida tropicalis, C. Oleophila, C. Guilliermondii, C. Parapsilosis, C. Citroformans, Hansenula anamola, Yarrowia lipolytica, Torulopsis, Hansenula, Debaromyces, Torula, Pichia, Kloekera*, and *Zygosaccharomyces*	Gutierrez et al. (1993); Kapelli et al. (1978); Oh et al. (1973); Omar and Rehm (1980); Rane and Sims (1993); Uchio et al. (1975)

Fungi that are reported to produce CA are Aspergillus. Niger, A. awamori, A. clavatus, A. nidulans, A. fonsecaeus, A. luchensis, A. phoenicus, A. wentii, A. saitoi, A. flavus, Absidia sp., Acremonium sp., Botrytis sp., Eupenicillium sp., Mucor piriformis, Penicillium citrinum, P. janthinellu, P. luteum, P. restrictum, Talaromyces sp., Trichoderma viride, and Ustulina vulgaris (Dhillon et al., 2011).

Source: Behera BC, Mishra R, Mohapatra S. Microbial citric acid: Production, properties, application, and future perspectives. Food Frontiers. 2021; 2: 62–76. https://doi.org/10.1002/fft2.66*Behera BC, Mishra R, Mohapatra S. Microbial citric acid: Production, properties, application, and future perspectives. Food Frontiers. 2021; 2: 62–76.* https://doi.org/10.1002/fft2.66.

3.1.3.1. Improvement of Strains for CA Production

Numerous techniques, including recombinant DNA technology, protoplast fusion, gene cloning, and mutations, are used to improve industrially significant microorganisms, as illustrated in Figure 3.1 (Parekh et al. (2000)). The most straightforward and often applied methods among them are protoplast fusion and random mutagenesis. Physical, chemical, and site-directed mutagenesis are the mutagenic processes used to improve strains. In the commercial fermentation process, the overproduction of industrial products through strain improvement has been taken into consideration. Enhancement of the strain can also be accomplished by Modifying the microorganism's metabolism by causing mutations in them using physical or chemical mutagens. The most often utilized physical mutagens are ultraviolet (UV) and gamma radiations. Chemical mutagens include diethyl sulfonate, ethidium bromide aziridine, N-methyl-N-nitroso-guanidine, N-nitroso-N-methyl urea, and ethyl methanesulfonate. Pontecorvo et al. (1953) were the first to describe strain improvement through the parasexual cycle. In comparison to their parent haploids, diploids showed larger CA yields. The other two approaches, which include the passage method and the single-spore technique, are well-known substitutes for choosing better strains.

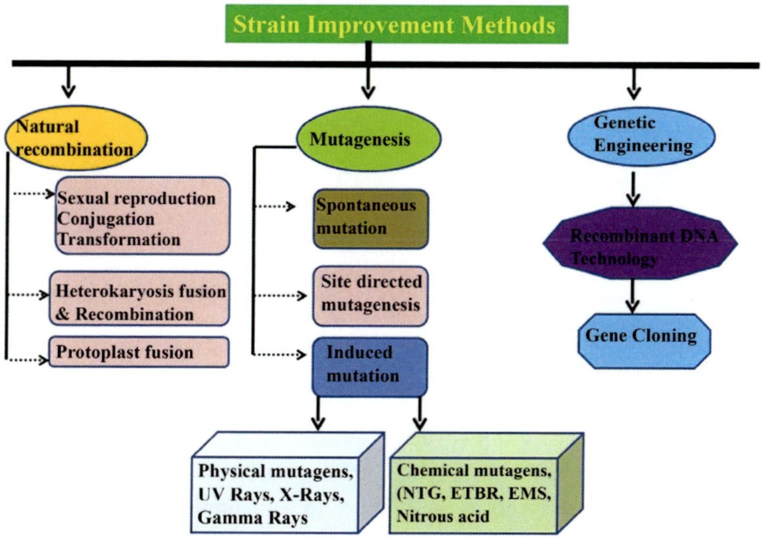

Figure 3.1. *Overview of strain improvement.*

Source: Behera BC, Mishra R, Mohapatra S. (2021). Microbial citric acid: Production, properties, application, and future perspectives. Food Frontiers, 2, 62–76. https://doi.org/10.1002/fft2.66.

3.1.4. Substrates and Pretreatment

When it comes to fermentation conditions, substrate is essential for obtaining the best yield and minimizing costs. Therefore, substrate plays a bigger role in yield and productivity during fermentation. Reductions in fermentation time and yield increases are directly correlated with substrate purity. In order to produce palm oil and crystalline glucose, *Aspergillus Niger* ferments molasses, sucrose, syrups made from beet or cane sugar, and hydrolysate that is produced as a byproduct. In order to remove trace metals from the substrate used in CA fermentation, pretreatment is necessary. It has been discovered that the calcium, magnesium, manganese, iron, and zinc present in cane molasses used in fermentation have the effect of delaying the synthesis of CA. In order to effectively precipitate iron and zinc from substrate prior to fermentation, potassium ferrocyanide is typically used as a chemical pre-treatment. Additionally, starch materials like corn, wheat, and potatoes were pretreated with enzyme hydrolysis to produce CA and lower the concentration of heavy metals below threshold values.

In the production of CA, agricultural waste and by-products are utilized to lower production costs. The agricultural residues that are commonly used are: coffee husk, rice bran, wheat bran, banana peel, carrot waste, cassava bagasse, sugarcane bagasse, tapioca, cheese whey, rice straw, coconut husk, brewery wastes, decomposing fruits, corn cobs, orange peels, pineapple peels, kiwifruit peels, and apple pomaces.

3.1.5. Biochemical Aspect of CA Production

Nicotinamide adenine dinucleotide (NAD) + hydrogen (H) (NADH), pyruvate produced during glycolysis is oxidized and combined with coenzyme A to form CO_2, acetyl coenzyme A (Acetyl-CoA), and CO_2. The oxaloacetate and the resulting acetyl-co-A are then mixed to create citrate (Figure 3.2). Additionally, pyruvate carboxylase (PC) can carboxylate pyruvate generated during glycolysis to form oxaloacetate. Acetyl-co-A and oxaloacetate combine to form CA, which is subsequently converted via a series of reactions that produces two molecules of CO_2 and regenerates the four-carbon oxaloacetate. As a result, one molecule of acetic acid enters the cycle at each turn, two molecules of CO_2 and ATP are created, and one molecule of oxaloacetate is used to form citrate.

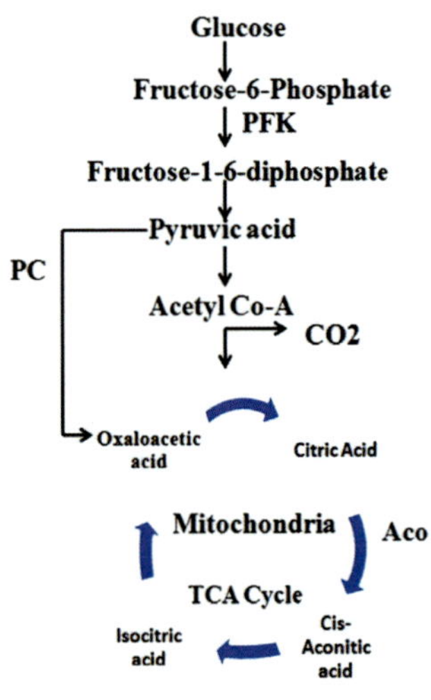

Figure 3.2. *Production of citric acid by Aspergillus Niger.*

Note: PFK = phosphofructokinase, PC = pyruvate carboxylase, ACO = aconitase.

Source: Behera BC, Mishra R, Mohapatra S. (2021). Microbial citric acid: Production, properties, application, and future perspectives. Food Frontiers.; 2: 62–76. https://doi.org/10.1002/fft2.66.

Enzymes are essential to the formation of CA. *A. Niger* uses the pentose phosphate and glycolytic pathways as a pathway to metabolize glucose and build up CA. The enzyme known as citrate synthase is in charge of the reversible catalysis that favors the production of citrate by combining oxaloacetate and acetyl-CoA. Only when aconitase (ACO), isocitrate dehydrogenase, and succinic dehydrogenase were severely inhibited during the tricarboxylic acid cycle (TCA) did CA accumulation occur. However, these enzymes were present during the CA fermentation period, albeit in very small amounts. Therefore, increased biosynthesis could be the cause of CA accumulation rather than inhibited degradation. Invertase, an enzyme attached to a membrane, breaks down sucrose in *A Niger* into glucose and fructose, which are subsequently taken up by glucose transporters and entered the cell. Hexokinase, an enzyme found inside the cell, changed the transported glucose into glucose-6-phosphate and started the glycolysis process.

When an enzyme is inhibited during glycolysis, there is a high flux through the process, which leads to an accumulation of CA by *A. Niger*. For instance, a manganese deficiency or phosphate and nitrogen limitation can prevent the fungus from anabolitically metabolizing, which can lead to protein degradation and an increase in intracellular NH_4^+ concentration. Due to the inhibition of phosphofructokinase (PFK), an essential enzyme that requires magnesium as a cofactor, by this high concentration of intracellular NH_4^+, fructose 6-phosphate is converted to fructose 1,6-bisphosphate during glycolysis (Figure 3.2). When PFK is inhibited, a flux through glycolysis occurs, which causes CA to accumulate. As an illustration, Papagianni et al. (2005) found that the presence of an intracellular ammonium pool that inhibits the PFK enzyme was the cause of *A Niger's* CA accumulation. They discovered during their investigation that ammonium ions entered the cell and mixed with glucose to produce glucosamine, which is then released outside the cell or into the fermentation broth and responsible for PFK inhibition, as opposed to accumulation or deposition inside the cell to form an ammonium pool. Therefore, the precise relationship between ammonium ion concentrations, glucose, TCA cycle enzymes, and CA accumulation is unclear and undoubtedly requires more research.

Isocitrate dehydrogenase is an NADP+ dependent enzyme that is activated by Mg^{2+} and Mn^{2+} and is present in both the mitochondria and the cytoplasm of cells. When α-ketoglutarate and citrate inhibit this enzyme, citric acid builds up more readily. Elevated levels of glucose and ammonium pool suppress the production of α-ketoglutarate dehydrogenase, a crucial enzyme involved in controlling the TCA cycle. Consequently, the CA cycle is impeded, leading to the build-up of CA.

3.1.6. Productions

Depending on the raw material and type of microorganism used, fermentation is the preferred method of producing CA. Two categories of raw materials are used in the production of CA: (i) Those with low ash content, from which the cations could be eliminated using common procedures (e.g., dextrose syrups, crystallized dextrose, cane or beet sugar); and (ii) raw materials containing a high percentage of ash and other non-sugar substances (e.g., cane and beet molasses, hydrolysates of starch that are crude and unfiltered).

Table 3.2. *Comparative Analysis of CA Production by Different Fermentation Technique*

Fermentation	Advantage	Disadvantage	Material	Quantity of CA	Type of Microorganism	References
SSF	Lower energy and cost requirements, higher yield, low-cost natural media, better oxygen circulation, less susceptible to trace elements inhibition, low risk of bacterial contamination, less amount of post-recovery waste	Difficulties in scale-up, difficult control of process parameters higher recovery product costs	Pineapple waste	124[d]	*A. Niger DS-1*	Kumar et al. (2003)
			Apple pomace	51.4[a]	*A. Niger BC-1*	Shojaosadati and Babaeipour (2002)
			Banana peels	180 g[d]	*A. Niger MTCC 282*	Karthikeyan and Sivakumar (2010)
			Coffee husk	150 g[d]	*A. Niger CFTRI 30*	Shankaranand and Lonsane (1994)
			Corn cob	254 g[d]	*A. Niger*	Hang and Woodams (1998)
			Grape pomace	60[a]	*A. Niger NRRL 567*	Hang and Woodams (1985)
			Corn husk	259 ± 10 g[d]	*A. Niger NRRL 2001*	Hang and Woodams (2000)
			Kiwi fruit	60%[a]	*A. Niger NRRL 567*	Hang et al. (1987)

SF	Less effort in operation, installation, and energy cost	Large amount of heat generation, time-consuming and needs large area/ space, sensitive to contamination by *Penicillia*, other *Aspergilli*, yeasts, and lactic acid bacteria	Brewery waste	78.5a	*A. Niger ATCC 9142*	Roukas and Kotzekidou (1986, 1987)
			Turnip whey (supplemented with molasses)	27-46a	*A. Niger NCIM 595*	Chanda et al. (1990)
			Sweet potato starch hydrolysate	45.90±4.2e	*A. Niger IIB-A6*	Anwar et al. (2009)
SMF			Beet molasses	68.7	*Y. lipolytica A-101*	Lesniak et al. (2002)

	Sophisticated control mechanisms, lower labor cost, higher productivity and yield	High media cost, sensitive to trace metal inhibition, high waste water generation	Cane molasses	114 g/L	*A. Niger GCMC 7*	Haq et al. (2004)
			Date syrup	50 ± 1.5%	*A. Niger ATCC 9142*	Roukas and Kotzekidou (1997)
			Corn cobs	603.5[d]	*A. Niger NRRL 2001*	Hang and Woodams (2001)
			Coconut oil	99.6[c]	*C. lipolytica*	Soccol et al. (2006)
			Soybean oil	115[c]	*C. lipolytica N-5704*	Soccol et al. (2006)
			Orange peel	53%a	*A. Niger CECT-2090*	Rivas et al. (2008)

a, e: based on sugar consumed, b, d: based on dry matter, c: based on oils and fatty acids.

Source: Behera BC, Mishra R, Mohapatra S. (2021). Microbial citric acid: Production, properties, application, and future perspectives. Food Frontiers, 2: 62–76. https://doi.org/10.1002/fft2.66.

3.1.6.1. Surface Fermentation

The beginning of the CA fermentation process, or surface fermentation, is a stationary batch fermentation process that takes eight to 12 days to complete. It is done in fermentation chambers with several trays stacked on stainless steel shelves. On the medium's surface in those trays, the fungal mycelium grows. The trays containing the fermentation medium are sterilized, then the medium is inoculated with spore suspension and left to incubate for 24 hours at 28–30°C. The germination of spores began with a drop in medium pH from 6.0–6.5 to 1.5–2.0, which was visible to the unaided eye as continuous mycelium was visible on the surface. Fermentation in this process generates a significant amount of heat, which is managed with appropriate aeration. At concentrations greater than 10%, the fermentation process would prevent the synthesis of CA due to the CO_2 produced. The fermentation chambers are equipped with an efficient air circulation system that passes over the medium surface through a bacteriological filter to regulate temperature and humidity through evaporative cooling, as the chamber needs to be ventilated effectively. In contrast to submerged fermentation, surface fermentation has certain drawbacks, including a lower yield, greater labor costs, and increased maintenance requirements. It is additionally susceptible to modifications in the media's composition.

3.1.6.2. Submerged Fermentation

The liquid broth medium within the nutrient substratum is used in the submerged fermentation process, allowing the organism to proliferate all throughout it. The fermentation process takes five to 12 days to finish in bioreactors. After being injected, the organism grows into pellets that are freely suspended in the medium and have a diameter of about 0.5 cm after one to two days. As a result, the organism can absorb nutrients and oxygen thanks to its large contact surface area. To raise the oxygen levels, agitation equipment mixes and breaks the air bubbles as fast-moving air is introduced into the vessel. The microbes in the medium break down the carbon source either entirely or partially anaerobically. Better control over the fermentation process and optimum utilization of a variety of substrates are two benefits of submerged fermentation. It offers greater yields and is better suited for CA production due to its lower capital, labor, maintenance, and contamination risks.

3.1.6.3. Solid-State Fermentation

Solid-state fermentation involves the growth of microorganisms in an insoluble material with low water content that serves as both a source of nutrients and physical support. The substrate used in solid-state fermentation must be solid and moistened to a moisture content of roughly 70%, have a pH between 4.5 and 6.0, and be between 28 and 30°C in temperature. Four to five days are needed to finish the process. Compared to submerged fermentation, solid-state fermentation offers several benefits, including lower energy requirements, increased yield, reduced risk of contamination, less work in downstream processing, less effluent generation, simpler operation, lower operating costs, and operability under less water. Additionally, since the system is less sensitive to the presence of trace elements than submerged fermentation, inexpensive and widely accessible agro-industrial substrates can be easily used in solid-state fermentation without the need for any pre-treatment. However, there are some drawbacks to this approach, including challenges in scaling up, a process that is not well suited to standardization, challenges in controlling process parameters, and issues with heat build-up. The inadequate heat and oxygen transfer in the substrate prevents it from fully utilizing the nutrients that are available (Behera et al., 2021).

3.1.7. Recovery of CA

At the end of the fermentation process, CA contain a variety of unwanted by-products, including mycelium, other organic acids, mineral salts, proteins, and other contaminants. Therefore, it is necessary to obtain only pure CA. Precipitation, extraction, and purification are the three main processes typically used to recover CA from the fermented broth. It has been noted that during the fermentation of CA, oxalic acid is produced. This can be eliminated by using calcium hydroxide to raise the pH to 3.0 at 72–75°C. The result is the formation of calcium oxalate, which can be further precipitated and removed using a filtration or centrifugation process. Addition of calcium oxide at 90°C and pH close to 7.0 will precipitate out the calcium acid (CA) that was present in the original solution as calcium salt (calcium citrate).

An alternative method for purifying and crystallizing CA from fermentation broth is the solvent extraction method, which uses either water-insoluble amines or a mixture of two or more of these amines, n-octyl alcohol, tridodecylamine, isoalkane, alanine 336 in heptane or xylene, mixture of butylacetate and N, N-disubstituted alkylamide, aliphatic alcohols, ketones, ethers or esters, organophosphorus compounds like tri-n-butylphosphate, alkylsulphoxides, and water-insoluble amines. One benefit of using the solvent extraction method is that it avoids using lime and sulfuric acid (H_2SO_4), which reduces the need to produce gypsum. This procedure involves washing the extract with water to recover the CA from the aqueous solution, which is then crystallized and concentrated. Compressed CO_2 is passed through the concentrated CA solution that has been dissolved in acetone. Remaining contaminants are eliminated by the CO_2's anti-solvent properties, and food-grade CA is produced through easy decolorization and crystallization.

3.1.8. Factors Affecting CA Production

The primary determinants of citric fermentation are the carbon source, the concentration of trace elements, pH, aeration, nitrogen and phosphate limitation, and the shape of the microorganism that produces the acid. When it comes to CA fermentation, the pH of the medium matters most in the beginning and towards the end, prior to the product being recovered. At the beginning of the germination stage, a pH of greater than 5 is necessary for the fungal spores to begin germination in the fermentation medium following inoculation. The spores that are germinating take up ammonia and release protons, which makes the medium more acidic and encourages the formation of CA. However, the substrate that is used determines the medium's initial pH. A pH value from 2.5 to 4.0 was found to be optimum for chemically defined medium, whereas an initial pH value from 6.0 to 7.5 was found to be the best in molasses medium.

However, the pH for the CA recovery stage needs to be less than 2. This low pH facilitates the recovery of CA by preventing the formation of undesirable products like oxalic and gluconic acid and lowering the risk of contamination by other microorganisms.

The TCA cycle enzyme and glycolysis are necessary for the production of CA. The temperature affects these enzymes. These enzyme systems are inactive at higher or lower temperatures, which has an impact on the production of CA. It was discovered that incubation temperatures between 25 and 30°C were better suited for producing CA at high yields and quick rates. Temperatures below 25°C slow the growth of the organism and the rates of fermentation, while temperatures above 30°C cause an increase in the production of oxalic acid and a decrease in the CA yield.

A Niger's extracellular mycelium-bound invertase allows it to quickly hydrolyze sucrose at low pH levels, making sucrose the best carbon source for producing CA compared to glucose, fructose, and lactose. The success of producing CA also depends on the carbon source's concentration. The ideal starting sugar concentration for *A. Niger* strains was between 10 and 14%; however, no CA was produced at less than 2.5 percent, and maximum CA production was noted between 14%–22% of sugar. The reason for this is that a high concentration of sugar has been shown to inhibit α-ketoglutarate dehydrogenase, which maximizes the accumulation of CA. As opposed to this, the mycelium's size and shape both decrease at low sugar concentrations. Additionally, it has been noted that for maximum CA accumulation, immobilized *A. Niger* cells required lower sugar concentrations than free cells.

A concentration of nitrogen higher than 0.25% causes the fermentation medium to become overloaded with oxalic acid, which lowers the yield of CA. Furthermore, a high nitrogen concentration has been shown to decrease the amount of CA produced while increasing the consumption of sugar and fungal growth. In essence, the production of CA involves the use of ammonium salts, such as ammonium nitrate and sulfate, urea, peptone, malt extract, and so forth. Acid ammonium compounds are preferred because they lower pH, which is necessary for the fermentation of citrus fruits. The best kind of salt to use is ammonium sulfate because it lowers the pH of the medium as it is ingested and does not produce the undesirable oxalic acid. Typically rich in nitrogen, molasses is utilized in industrial fermentation without the need for extra ammonium salt supplements. It has been discovered that low concentrations of phosphate are ideal for the synthesis of CA, while high concentrations of phosphate can cause the formation of specific sugar acids, a reduction in CO_2 fixation, and growth stimulation. Potassium dihydrogen phosphate (KH_2PO_4) at a concentration of 0.5–5 g/L has been reported to be the most suitable for optimal CA production. Increased CA accumulation results from the addition of ethanol, which also doubles the activity of citrate synthetase and reduces the activity of ACO. Because alcohols like methanol inhibit metal ions, they have a beneficial effect on the formation of CA. The strain of microorganism being used and the medium's composition determine how much methanol or ethanol is needed. Alcohols increase the production of CA by altering the lipid composition of the cell membrane, which has an impact on growth and sporulation. Lower alcohols added to pure material prevent the formation of CA; however, when these alcohols are added to crude carbohydrates, the production of CA is increased. Maximum CA formation is favored by concentrations of methanol, ethanol, n-propanol, isopropanol, or methylacetate between 1–5%, which counteract the negative effects of metal ions in CA production. Ferrous ions (Fe^{2+}), copper ions (Cu^{2+}), zinc ions (Zn^{2+}), and manganese ions (Mn^{2+}) are the metal ions that are most sensitive to the formation of CA and must therefore be limiting. It has been demonstrated that the yield of CA production is significantly reduced by Mn^{2+} concentrations less than 3 μg/L, while the addition of 10 mg Mn^{2+} per liter reduced the accumulation of CA by 50% in comparison to the control culture. Anabolism of A's cells. A manganese deficiency impairs Niger, and high intracellular ammonium is caused by protein breakdown. The enzyme PFK, which is crucial for converting glucose and fructose to pyruvate, is inhibited by the elevated intracellular ammonium concentration. This results in a flux through the glycolysis pathway and the creation of CA.

Better CA production requires an oxygen concentration above 25%. The critical dissolved oxygen tension has been found to be 9–12% of air saturation during the growth phase and 11–23% of air saturation during the production phase. Even though the dry mass hasn't increased by more than 5%, the dissolved oxygen tension quickly drops to less than 50% of its initial value when the organism begins to develop filaments. As a result, during the CA production process, small, compact pellets are the preferred mycelial form of *A Niger*. The growth limitation that is essential for the production of CA is directly related to low oxygen environments, while strongly aerated cultures (0.3 m^3/kg dry CB/h) increase sporulation, which in turn reduces the accumulation of CA. An essential PC substrate, CO_2 restores the oxaloacetate supply needed for citrate synthase. Pyruvate decarboxylase catalyzes a reaction that produces enough CO_2, but excessive aeration causes some losses. On the other hand, higher CO_2 levels harm the final biomass and citrate concentrations. It has been noted that high partial pressure of CO_2 likely promotes CA accumulation by delaying filamentous fungal spore release. Consequently, the environment with high CO_2 concentrations benefits the synthesis of CA.

It has been noted that the fungal ability to form compact aggregates or pellets during CA fermentation promotes the production of more CA than filamentous form. The main variables that influence the morphology of *A. Niger* are the agitation rate, medium pH, medium composition, and inoculum concentration. At pH values of approximately 2.0 ± 0.2, small aggregates of short filaments form and are linked to an increase in the production of CA. Higher pH (pH > 3.0) resulted in aggregates with longer perimeters and the formation of oxalic acid, while lower pH (pH 1.6) caused the fungus to change from aggregated short filamentous form to bulbous hyphae and produce very little CA. The synthesis of CA is also impacted by lipids like sodium monofluoro acetate and groundnut oil. With no impact on the mycelium's dry weight, lipid can increase the CA yield.

3.1.9. Emerging Application of CA

In the food, sugar, confectionery, and beverage industries, CA is frequently used as a safe acidulant because of its pleasant taste, high water solubility, chelating, and buffering qualities. It adds fruit and berry flavors to carbonated drinks; it acts as a flowing agent in the confectionery industry; it prevents turbidity in wines; it gives dark color and tartness to candies; it synergizes antioxidants in fats, oils, and fat-containing foods; it enhances the flavor of sherbets; and it emulsifies ice cream. Furthermore, nontoxic plasticizers known as triethyl, tributyl, and acetyltributyl are utilized in plastic films to safeguard food items. By stopping frozen fruit's color and flavor from deteriorating, CA's chelating and pH-adjusting qualities are used in the food industry to improve the stability of frozen food products. Additionally, it aids in extending the shelf life of frozen fish and shellfish.

Ascorbic acid (CA) and its salts are widely used in the pharmaceutical industries as oral pharmaceutical liquids, elixirs, and suspensions to buffer and maintain the stability of the active ingredients of the pharmaceutical product because of their sequestering action, which includes stabilizing of ascorbic acid and good buffering capacity. Because it complexes calcium, sodium citrate is frequently used as a blood preservative, preventing clotting. Kidney stones can be avoided by taking acetic acid along with sodium citrate. It can be used as a cross-linker in a variety of applications, including the production of ultrafine protein fibers for biomedical applications, biodegradable films like bio-plastic for environmentally friendly packaging, and bioceramic composites for orthopedic tissue engineering using hydroxyapatite and polyols. CA is helpful in soap, dishwashing detergents, and household detergents because it softens water. Hard water's metal ions, such as Ca^{2+} and Mg^{2+} ions, are chelated to help create foam and function better without the need for water softening.

Because of its superior metal chelating qualities, CA can be used as a cleaning agent. It attaches itself to metals and solubilizes them. Therefore, to get rid of and prevent the accumulation of limescale, cleaning solutions containing CA are applied to boilers in power plants and other similar installations. It was initially demonstrated by Brazilian researchers in 2005 that pectin from apple pomace could be effectively recovered using CA instead of harmful mineral acids. With CA, pectin extraction yield had the highest average value (13.75%). Because of its superior metal chelating capabilities, CA is frequently used for bioremediation of heavy metal-contaminated soils and cleaning radioactively contaminated nuclear sites. It has been noted that using bio-based chemical agents for soil environmental remediation, CA in conjunction with rhamnolipid biosurfactants produces very good outcomes. In addition to being environmentally benign, this combination encourages the ecological restoration of the soil following remediation.

For poliovirus, a mixture of 70% ethanol, 1% urea, and 1% CA was highly effective in inactivating the virus; however, this combination was insufficient for adenovirus and polyomavirus. To effectively deactivate all enveloped viruses, including polyomavirus, norovirus, and adenovirus vaccinia virus, on surfaces, a mixture of 70% ethanol, 2% urea, and 2% CA was used. CA and its esters are used in photography, in oil well treatment and cements, in the textile, paper, and tobacco industries, as well as in the printing plate emulsions in various bleaches, fixers, and stabilizers.

3.1.10. Economic Aspects

The 300 and 150 m^3 surface and submerged fermentation process capacities over nine days of fermentation at 72 and 12 tons per day of productivity were compared. He noted that compared to the surface fermentation process, the building investment costs associated with submerged fermentation are 2.5 times lower. But the costs of equipment at submerged fermentation are much higher, with over 60% of those costs coming from complex components like bioreactors and more advanced instrument control, which are prone to relatively high wear. The total investment costs for the submerged process are approximately 15% less than for surface fermentation for smaller capacities, while the submerged process has investment costs that are approximately 25% lower for larger capacities. The submerged process uses approximately 30% more electricity than surface fermentation, but surface fermentation requires much more labor in highly industrialized nations. Similarly, additional costs for cooling the bioreactors are needed for the submerged process in nations where the temperature of the cooling water is higher than 20°C. Conversely, the submerged fermentation process is susceptible to brief pauses or malfunctions in aeration, leading to yield losses as well as the batch's complete disintegration.

3.1.11. Future Prospects

A key interest is a rise in CA productivity at a reduced cost. Innovations and new technologies are required for online parameter monitoring and control as well as bioreactor designs that can solve the challenges of fermentation process scale-up. In order to cut costs and minimize environmental issues, less expensive substrates for CA production, such as agricultural industry byproducts and residues, can be utilized. A number of high throughput technologies can be used to improve strains in order to find high CA yielding strains. Researchers have been working extremely hard to find solutions to the main problems of recovery and purification of end products. It is best to employ the integrated fermentation and product recovery method, which can be applied to the large-scale production of CA. The creative methods listed above can reduce production and recovery costs while protecting the environment from the hazardous chemicals used in traditional CA production.

The main reason for the broad aspects of CA is its numerous applications across various industries. An efficient industrial production process is desperately needed to meet the growing demand for CA worldwide. This process must be highly sensitive, cost-effective, and dependent on a number of variables, including the strain improvement, quantification, and recovery techniques, raw material selections, fermentation techniques, and the design of suitable bioreactors with precise control over process parameters. By lowering the amount of CA imported from overseas, using agricultural wastes for CA production can save significant foreign exchange in addition to resolving the issue of waste disposal. While major research efforts are required, adoption of advanced technology and metabolic engineering can also aid in overcoming some critical parameters raised during fermentation (Behera et al., 2021).

3.2. ITACONIC ACID

Microorganisms are able to produce itaconic acid, also known as 2-methylidenebutanedioic acid, which is valuable in the commercial sector as a precursor to fuels, chemicals, and polymers. The use of microorganisms to transform biomass into beneficial compounds like itaconic acid is gaining attention. *Aspergillus terreus* is the primary producer of itaconic acid, with a maximum production of 86 g/L.

3.2.1. Definition

Itaconic acid is an unsaturated dicarbonic acid, also referred to as 2-methylidenebutanedioic acid. The US Department of Energy views it as a biomass product with added value since it can be used as a chemical intermediate and polymer precursor. Styrene, 2-methyl-1,4-butanediol, acrylic plastics, acrylate latexes, paint, synthetic fibers, resins, paint, super-absorbent materials, anti-scaling agents, and 3-methytetrahydrofuran are a few of these. Additionally, itaconic acid has demonstrated some antimicrobial properties.

3.2.2. Itaconic Acid Production

Certain microorganisms possess the ability to ferment itaconic acid. These are mostly fungal species, such as *Rhodotorula*, *Ustilago zeae*, and *Ustilago maydis*. With a maximum production of 86 g/L, *Aspergillus terreus* is the primary fungal species. Although this might seem high, it is not as high as the 200 g/L of citric acid that can be produced.

3.2.2.1. Biosynthetic Pathway

Citric acid is converted to cis-aconitate, an intermediate in the citric acid cycle, which yields itaconic acid. Since it is connected to the citric acid cycle, sugars like glucose can be converted to itaconic acid. Glycolysis first breaks down glucose into pyruvate, which is then transformed into acetyl-CoA when CO_2 is released. Following its entry into the citric acid cycle, acetyl-CoA forms cis-aconitate through initial steps. Alternatively, CO_2 can be added to pyruvate to convert it to oxalate, which can then be further transformed into malate, another molecule that is a part of the citric acid cycle. Within *A. terreus*, the enzyme cis-aconitate decarboxylase, or CadA in this species, is encoded by the cadA gene. The enzyme that changes cis-aconitate into itaconic acid is called CadA, and it is a 55 kDa protein with 490 amino acids. Research has indicated that the allylic rearrangement and decarboxylation of cis-aconitate are catalyzed by CadA.

It's interesting to note that research has revealed no mutations in the resulting protein, despite a strain that overproduces itaconic acid having increased transcription of cadA. Additional research using different constitutively expressing promoters and the fungus *Aspergillus Niger*, which lacks CadA naturally, revealed a correlation between the levels of cadA transcription and the amount of itaconic acid produced. For *A terreus* to produce itaconic acid, more than just CadA is needed. There are two additional genes involved: mfsA, which encodes a putative major facilitator superfamily transporter (MFSA), and mttA, which encodes a putative mitochondrial tricarboxylic transporter (MTTA). It is believed that MFSA transports itaconic acid into the extracellular environment, whereas MTTA transports cis-aconitate from the mitochondria to the cytosol. Trans-aconitate, the isomer of cis-aconitate, is used by *U maydis*, another microorganism that makes itaconic acid. Here, aconitate-Δ-isomerase isomerizes cis-aconitate to trans-aconitate, which is then transformed to itaconic acid by trans-aconitate decarboxylase.

3.2.2.2. Producing Itaconic Acid on An Industrial Scale

Itaconic acid must be extracted from the fermentation culture after it is produced, usually by filtration and crystallization, and then it can be dried and decolored. The majority of itaconic acid is recovered during crystallization, and the yield is normally in the range of 80%. Crystallization can be substituted with other processes such as adsorption, liquid-liquid extraction, membrane separation, and precipitation.

The primary microbial species involved in the production of itaconic acid is a terreus. *A terreus* is not the most efficient, though, due to its growth requirements. For instance, it necessitates a constant supply of O_2, which raises NADH levels. Certain essential enzymes involved in the synthesis of itaconic acid are inhibited by high NADH levels. Additionally, *A terreus* produces mycelia, which is readily harmed by vigorous stirring throughout the fermentation process. Itaconic acid can also be produced by genetically modifying *Saccharomyces cerevisiae* and *Escherichia coli*. It was found that the overexpression of CadA from *A terreus* was necessary because the engineered strains exhibited low levels of endogenous cis-aconitate decarboxylase activity. The yield from these modified strains was still low, though.

3.3. FUMARIC ACID

An organic compound with the formula $HO_2CCH{=}CHCO_2H$ is fumaric acid. Fumaric acid is a white solid that is commonly found in nature. It has been used as a food additive and tastes like fruit. It has the E number E297. Fumarates are the name for the salts and esters. Another term for fumarate is the $C_4H_2O^{2-}{}_4$ ion (in solution). The trans isomer of butenedioic acid is fumaric acid, and the cis isomer is maleic acid.

3.3.1. Biosynthesis and Occurrence

It is created by the enzyme succinate dehydrogenase in eukaryotic organisms from succinate in complex 2 of the electron transport chain. Bolete mushrooms (especially *Boletus fomentarius* var.), fumitory (Fumaria officinalis), and other plants contain fumaric acid. pseudo-igniarius), Iceland moss, and lichen. Fumarate is a citric acid cycle intermediate that cells use to convert food into adenosine triphosphate (ATP), which is energy. It is created when the enzyme succinate dehydrogenase oxidizes succinate. The enzyme fumarase then transforms fumarate into malate. When exposed to sunlight, human skin naturally produces fumaric acid. Another byproduct of the urea cycle is fumarate.

3.3.2. Synthesis

The process of catalytic isomerization of maleic acid in low pH aqueous solutions yields fumaric acid. It separates from the solution of the reaction. Maleic anhydride, which is created by catalytically oxidizing benzene or butane, hydrolyzes to produce maleic acid, which is readily available in large quantities.

- **Historic and Laboratory Routes:** First, succinic acid was used to make fumaric acid. Furfural obtained from the processing of maize is oxidized using chlorate in the presence of a vanadium-based catalyst in a conventional synthesis.

3.3.3. Uses

- **Food:** Since 1946, fumaric acid has been used as an acidulant in food. It is permitted for use as a food additive in the USA, Australia, New Zealand, and the European Union. It is designated with the E number E297 and is used as an acidity regulator in food additives. It is typically used in drinks and baking powders where purity standards are met. When making wheat tortillas, fumaric acid is used as a food preservative and as the leavening agent. Similar to malic acid, fumaric acid is typically used to add sourness in place of tartaric acid and occasionally in place of citric acid at a ratio of 1 g to every ~1.5 g of citric acid. In addition to being a part of some fake vinegar flavors, like potato chips with a salt and vinegar flavor, it also functions as a coagulant in pudding mixes that are cooked on the stove. In 2014, the Scientific Committee on Animal Nutrition of the European Commission (DG Health) concluded that fumaric acid is essentially non-toxic, although prolonged use of high doses may result in nephrotoxicity.
- **Medicine:** In the 1950s, fumaric acid was created as a medication in Germany to treat psoriasis, an autoimmune condition. The medication was sold as Fumaderm in Europe by Biogen Idec and contained three esters, mainly dimethyl fumarate. Later on, Biogen would create the primary ester—dimethyl fumarate—as a multiple sclerosis treatment. In a phase 3 trial, the ester dimethyl fumarate

(BG-12, Biogen) significantly decreased relapse and disability progression in patients with relapsing-remitting multiple sclerosis. It initiates the Nrf2 antioxidant response pathway, which is the main line of defense for cells against oxidative stress's cytotoxic effects.

- **Other Uses:** Fumaric acid is utilized as a dye mordant and in the production of polyester resins and polyhydric alcohols. Lambs can reduce their methane production during digestion by up to 70% when fumaric acid is added to their diet.

3.3.4. Reactions

The acid's constituent functional groups can be used to predict some of its chemical characteristics. This weak acid is a good dienophile, forms a diester, and brominates across the double bond.

3.4. L-MALIC ACID

One "building block" chemical for the synthesis of biodegradable polymers is malic acid. The original methods of isolating malic acid involved using eggshells, fruits, and apple juice, but this method was labor-intensive and very costly. Malic acid is used in a variety of ways in commerce. It has been used as an acidulant in foods and drinks. Compared to citric acid, malic acid tastes less bitter when used as an acidulant. Food flavor can also be improved by malic acid. Malic acid is also used commercially in the cleaning of metal, textile finishing, pharmaceuticals, and agriculture. Malic acid is used in agriculture to dissolve aluminum phosphate in the soil. The global market for malic acid is estimated to be between 40,000 and 60,000 metric tons per year, with a 4% annual growth rate. The market is estimated to be worth $130 million. Per reports, the retail cost of one pound of malic acid varies between $1.80 and $2.00. Malic acid is currently synthesized industrially using a petroleum-based process. Maleic anhydride, which is produced when benzene or butane oxidizes, is hydrated under higher pressure and temperature in the commercial production of malic acid. This process creates a racemic mixture of d- and l-malic acid, where the isomers need to be resolved chirally. Using the enzyme fumarate hydratase or *Saccharomyces cerevisiae* cells with the fumarate hydratase overexpressed gene to catalyze the conversion of fumarate to malate is an alternative way to synthesize malic acid. The latest technique for producing malic acid involves the acid hydrolysis of poly(β-l-malic acid) (PMA), which is produced metabolically by microorganisms like *Aureobasidium pullulans*. Malic acid can also be hydrolyzed to produce PMA, or it can be synthesized from malic acid. Because PMA is a water-soluble biopolymer, it may be helpful in the creation of biodegradable polymers for use in medicine and agriculture. Furthermore, it has been reported that low molecular weight PMA inhibits proteases. Research has started on using green chemistry techniques to make malic acid from low-value biomass or by-products of processing. When malic acid is produced using petroleum-based products, the price of the acid could be lowered with the aid of such low-value products. The production of malic acid from biobased sources could more successfully compete with petroleum-based sources for usage in a range of food and non-food products by reducing the cost per pound.

3.4.1. Pathways of Malic Acid Biosynthesis

After investigating potential routes of malic acid biosynthesis, researchers came to the conclusion that there are probably three routes via which microbes produce malic acid (Figure 3.3). According to Figure 3.3, the first pathway is classified as a reductive pathway. Pyruvate is first carboxylated to oxaloacetate, and then oxaloacetate is reduced to malate. This pathway is thought to involve pyruvate carboxylase and malate dehydrogenase. It is believed that carbon dioxide from sparging the medium with CO_2 or adding a carbonate salt to the culture medium is necessary for this pathway. As a result, although this pathway is neutral with regard to ATP, it needs the fixation of CO_2. There is proof that eukaryotes like *Penicillium* spp., *Aspergillus flavus*, *Aspergillus oryzae*, and *Candida glabrata* are involved in this pathway and *Saccharomyces cerevisiae*, as well as in the prokaryotes *Bacillus subtilis* and an *Escherichia coli* strain with altered metabolism. It has been demonstrated that pyruvate carboxylase and malate dehydrogenase activities were higher in a strain of *Saccharomyces cerevisiae* that produced more malic acid. Phosphoenolpyruvate is reduced to malate-by-malate dehydrogenase and transformed to oxaloacetate by phosphoenolpyruvate carboxylase in a strain of the thermophilic soil bacterium *Thermobifida fusca* muC. So far, the highest levels and yields of malic acid have been produced by microorganisms that use the reductive pathway. Two molecules of acetyl-CoA are utilized in the glyoxylate pathway, which is the second pathway of malate biosynthesis by microbes (Figure 3.3). Isocitrate lyase, an enzyme involved in the glyoxylate pathway, transforms isocitrate into succinic acid and glyoxylate. Acetyl-CoA, glyoxylate, and water are the products of a reaction that is catalyzed by the second enzyme in the pathway, malate synthetase, which yields malate and Co-A. The glyoxylate cycle and the synthesis of oxaloacetate depend on the carboxylation of malate. It has been demonstrated that *Saccharomyces cerevisiae aromyces cerevisiae* produces malate by using this pathway. The tricarboxylic acid cycle is involved in the oxidative pathway of malic acid synthesis by microbes, as Figure 3.3 illustrates. After condensation with oxaloacetate, acetyl-CoA enters the tricarboxylic acid cycle as citrate and releases two CO_2 atoms to form malate. It is believed that *Saccharomyces cerevisiae aromyces cerevisiae* can produce excess malate through this pathway, especially if the yeast cells have an overexpression of fumarase. The reductive pathway seems to be the main one that most organisms use out of the three pathways for synthesizing malic acid (West, 2017).

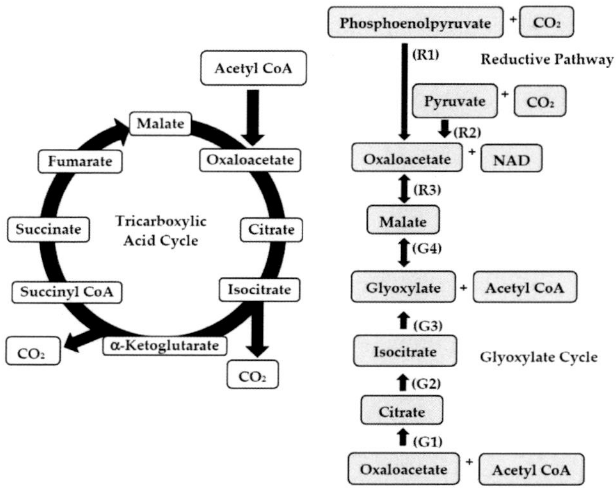

Figure 3.3. *Three metabolic pathways in microorganisms utilized to synthesize malic acid. The tricarboxylic acid cycle which forms malate oxidatively is indicated on the right.*

Note: The reductive pathway (R) of malic acid synthesis involves phosphoenolpyruvate carboxykinase (R1), pyruvate carboxylase (R2) and malate dehydrogenase (R3). The glyoxylate cycle (G) pathway enzymes include citrate synthase (G1), aconitase (G2), isocitrate lyase (G3) and malate synthase (G4). The tricarboxylic acid pathway is mitochondrial (no shading) while the reductive and glyoxylate cycle pathways are cytosolic (shaded in gray).

Source: West, T.P. (2017). Microbial production of malic acid from biofuel-related coproducts and biomass. Fermentation, 3, 14. https://doi.org/10.3390/fermentation3020014.

3.4.2. Microbial Malic Acid Production from Sugars

Glucose is the best fermentable sugar in the presence of a carbonate salt when the medium's nitrogen concentration is limiting, according to research on the microbial synthesis of malic acid from sugars. The carbonate salt has two purposes: it produces CO_2 and acts as a neutralizing agent. Aspergillus species have been demonstrated to produce malic acid in response to fermentable sugars. Prior studies demonstrated malic acid production by *A. flavus* ATCC 13697. This strain grown in a 10 L fermentor (0.5 volume of air/volume of culture/min with a stirring speed of 300 rpm) for 192 h at 25°C from 10% (*w/v*) glucose as the carbon source and 80 g/L calcium carbonate produced 36.4 g/L malic acid. Pyruvate carboxylase was involved in the synthesis of malic acid from pyruvate. With 12% (w/v) glucose as a carbon source and 60 g/L calcium carbonate, ATCC 13697 produced 113 g/L malic acid in a 16-liter fermentor in 190 hours (0.05 volume of air/volume of culture/min at 400 rpm stirring speed producing a 20% dissolved oxygen concentration) at 32°C. The issue surrounding A's extensive production of malic acid. flavus is that it is not considered to be "generally regarded as safe status" due to its ability to produce aflatoxins. It was found that the A is subject to nitrogen-limiting conditions. Peptone, a source of nitrogen, enhanced the production of malic acid from glucose, and oryzae strains induced the production of malic acid when the temperature at which A is incubated. After 168 hours on 114 g/L (w/v) glucose and 90 g/L (w/v) calcium carbonate, oryzae DSM 1863's growth was raised to 35°C, and its production of malic acid in a 2 L fermentor (0.5 volume of air/volume of culture/min with a stirring speed of 300 rpm) increased by a factor of 107. *A oryzae* DSM 1863 was able to produce malic acid using the five-carbon sugar xylose as well, although its molar yield (0.56) was lower than that of glucose (0.82) when used as a carbon source.

Using corn straw hydrolysate (100 g/L glucose and 25 g/L xylose) and the fungus Rhizopus delemar, 60 g/L malic acid was produced in 60 hours at 30°C in a 3 L fermentor (aeration supplied by magnetic stirring). Using 142 g/L (w/v) glucose as a carbon source and 50 g/L (w/v) calcium carbonate, a level of 88.6 or 92 g/L calcium malate was produced for the fungus Penicillium sclerotium K301 after 96 or 72 hours at 28°C (aeration rate of 8 L/min with stirring speed of 300 rpm), respectively. In a medium containing 140 g/L (w/v) glucose, 40 g/L (w/v) calcium carbonate, and 0.5% (v/v) corn steep liquor, *Penicillium viticola* 152, a different species of the fungus, was shown to produce 168 g/L calcium malate in a 10 L fermentor (aeration rate of 8 L/min with stirring speed of 300 rpm) for 96 hours at 28°C. Another method being investigated is the metabolic engineering of different species to create high concentrations of malic acid in a fermentor under carefully regulated circumstances. It was reported that a metabolically engineered *A. oryzae* ATCC 56747 produced approximately four-fold higher levels of malic acid than did the wild type strain. Pyruvate carboxylase (Figure 3.3), malate dehydrogenase (Figure 3.3), and the C4-dicarboxylate transporters were overexpressed in the genetically altered strain. When the dissolved oxygen level was kept above 50%, the metabolically modified ATCC 56747 produced 154 g/L malic acid after 164 hours at 34°C (1 volume of air/volume of culture/

min with an agitation rate set at 500 rpm). Improved malic acid production has also been achieved by genetically modifying strains of *E. coli* (1.42 mol malic acid/mol glucose) and *S. cerevisiae* (0.42 mol malic acid/mol glucose). Despite being cultivated on 188 g/L glucose at 30°C with 500 mM calcium carbonate at pH 6.0, the metabolically altered *Saccharomyces cerevisiae* strain was also genetically altered. The *E, coli* strain yielded more when only 50 g/L of glucose was used at 37°C and 100 mM potassium bicarbonate was present at pH 7.0. Similar to this, C glabrata CCTC M202019's metabolic engineering resulted in a multifold increase in malate production over that of its parent strain by overproducing the activities of pyruvate carboxylase, malate dehydrogenase, and a malate transporter. Malic acid overproduction was noted in a metabolically modified strain of *A oryzae* upon several-fold increase in pyruvate carboxylase and malate dehydrogenase relative to the parent strain.

3.4.3. Malic acid Production from Biofuel-Related Coproducts

It has been demonstrated that coproducts related to biofuels can be utilized to support the production of malic acid by microbes (Table 3.3). However, the amount of malic acid that is produced will vary depending on the type of coproduct and the microbe's tolerance or ability to use it. Thin stillage, a byproduct of dry milling corn to produce ethanol, is one that can be utilized to support microbial production of malic acid (Table 3.3). After the yeast cells are removed by centrifugation, the coproduct known as thin stillage is recovered from the entire stillage. Its composition consists primarily of glycerol and less than 1% nitrogen. To create dried distillers' grains with solubles, it is typically combined with wet distillers' grains and dried. It is not permitted to combine thin stillage with wet distillers' grains and dry them in order to increase the economics of ethanol production. The value of thin stillage as a coproduct has decreased as a result of the marketing of wet distillers' grains alone. There is glycerol in this stillage that is appropriate for microbial fermentation. A previous study looked at how Aspergillus species use thin stillage to support the production of malic acid. The capacity of several strains, including *Aspergillus Niger* ATCC 9029, ATCC 9142, ATCC 10577, and *A. flavus* (ATCC 13697) to synthesize malic acid from thin stillage was tested. *A. Niger* ATCC 9142 and ATCC 10577 produced the highest malic acid yields (0.79 g malic acid/g glucose and glycerol in thin stillage) among the screened strains. It was determined that fungal malic acid production could be facilitated by using the thin stillage, a byproduct of ethanol production.

Table 3.3. *Malic Acid Production by Microorganisms Grown on Biofuel-Processing Coproducts or Hydrolyzed Lignocellulosic Biomass*

Coproduct/ Hydrolyzed Lignocellulosic Biomass	Microorganism	Growth Conditions	Malic Acid (g/L)	Yield (g/g)	References
Corn stover	*T. fusca* muC-16	55°C	21.5	0.43	Khan, Nazir, Wang, Liu, & Chi, (2014)
Corn straw	*R. delemar* HF-119	30°C	60.0	0.48	Li et al., (2014)
	R. delemar HF-121	30°C	121.8	0.97	Li et al., (2014)
Thin stillage	*A. flavus* ATCC 13697	25°C	10.2	0.48	West (2011)
	A. Niger ATCC 9029	25°C	1.0	0.05	West (2011)
	A. Niger ATCC 9142	25°C	16.9	0.79	West (2011)
	A. Niger ATCC 10577	25°C	16.4	0.79	West (2011)
Crude glycerol	*A. Niger* ATCC 9142	25°C	16.5	0.17[1]	West (2011)
	A. Niger ATCC 10577	25°C	20.3	0.20[1]	West (2011)
	A. Niger ATCC 12846	25°C	23.5	0.24[1]	West (2011)
	U. trichophora TZ1	30°C	108.0	0.26	Zambanini et al. (2016)
Syngas (plant biomass)	*C. ljungdahli* DSM 13528/*A. oryzae* DSM 1863	25°C	1.1	0.17	Oswald et al. (2016)
Biooil (plant biomass)	*A. oryzae* DSM 1863	32°C	0.0	0.0	Dorsam et al. (2016)

[1] *The malic acid yields produced by A. Niger ATCC 9142, ATCC 10577 and ATCC 12846 on 10% (v/v) crude glycerol represent unpublished data.*

Source: West, T.P. (2017). Microbial production of malic acid from biofuel-related coproducts and biomass. Fermentation, 3, 14. https://doi.org/10.3390/fermentation3020014.

Another byproduct generated during the synthesis of biodiesel, crude glycerol, may be used for the synthesis of malic acid (Table 3.3). The process of esterifying a vegetable oil with methanol at an alkaline pH with heat yields biodiesel. Crude glycerol is produced as a byproduct of this process. Crude glycerol is made up of methyl esters of fatty acids, glycerol, and fatty acids. The crude glycerol fraction is a byproduct of esterification that makes up about 10% of the total coproduct. With the current global annual production of biodiesel exceeding 30 million tons, significant volumes of crude glycerol are expected to need to be processed as waste. Since crude glycerol is currently valued at $0.05 per pound, it will be inexpensive to use in large quantities as a feedstock for microbial fermentation. It has been reported that bacteria can produce malic acid using crude glycerol as a substrate. It has been demonstrated that the synthesis of fungal malic acid can be supported by crude glycerol obtained from the biodiesel production process. In a recent study, it was discovered that *A. Niger* ATCC 9142, ATCC 10577, and ATCC 12846 could produce malic acid after 192 hours at 25°C on 10% (v/v) crude glycerol. *A. Niger* ATCC 12846 generated the maximum amount of malic acid from crude glycerol among the three strains. Since ATCC 10577 produced more biomass than ATCC 12846 after 192 hours at 25°C, it appears that the capacity to produce malic acid from crude glycerol is not highly correlated with cellular biomass. While *A. oryzae* DSM 1863 was also able to produce 45 g/L of malic acid from 84 g/L (w/v) of glycerol after 353 hours at 35°C; however, it has not been studied if *A. oryzae* DSM 1863 can use crude glycerol as a substrate to produce malic acid. In recent studies, the ability of the fungus *Ustilago trichophora* TZ1 to produce malic acid from glycerol and crude glycerol was examined (Table 3.3). It was observed that the fungal growth rate was inhibited by high concentrations of glycerol or crude glycerol. Calcium carbonate was added to the medium to supply carbon dioxide and keep the pH above 5.4, which is necessary for the fungus to produce malic acid. *U trichophora* produced less malic acid as the amount of crude glycerol in the culture was raised from 100 to 200 g/L (w/v). It was determined that strain TZ1's adaptive evolution would probably be necessary for the fungus to grow on crude glycerol efficiently (West, 2017).

3.4.4. Lignocellulosic Biomass-Based Malic Acid Production

Utilizing plant biomass hydrolysates or biooils to synthesize malic acid is an alternative to using coproducts from the production of biofuel (Table 3.3). The expense of enzymes (cellulases and xylanases) to break down the biomass into fermentable sugars (glucose or xylose) is a barrier to using lignocellulosic biomass rather than a processing coproduct to microbially produce malic acid. Since plant biomass is bio-renewable, the hydrolysis of lignocellulosic biomass to produce glucose from cellulose and xylose from Xylan through physical, chemical, and enzymatic treatments represents a green chemistry approach. The fungus R. *delemar* strain HF-119 has been observed to synthesize malic acid using hydrolyzed corn straw as a substrate (Table 3.3). After being heated to 160°C for 20 seconds, 0.4% sulfuric acid was applied to the corn straw. After treating the solid residue for 72 hours at 52°C at 120 rpm with cellulase and β-glucosidase, glucose and xylose were produced. The corn straw hydrolysate (125 g/L mixed sugars) could yield 120.5 g/L malic acid in 60 hours when the fluoroacetate mutant strain HF-121 was isolated (Table 3.3). Compared to strain HF-119, this strain was found to use glucose and xylose as carbon sources more effectively. Other fungi, like *A. oryzae* DSM 1863, may also use xylose to produce malic acid. Malic acid can be produced from syngas by a sequential process. Firstly, *Clostridium ljungdahli* is grown anaerobically at 37°C in a medium sparged with syngas obtained from the gasification of straw in order to synthesize acetate. In the absence of nitrogen, *A. oryzae* aerobically converts the acetate to 4.34 g/L malic acid (Table 3.3). It was also demonstrated that this two-step method could be used to produce malic acid in a bioreactor. Depending on the growth conditions, *A. oryzae* DSM 1863 has been shown to support the production of malic acid using a pyrolysis oil that is produced from the fast hydrolysis of wheat straw (Table 3.3). The strain was found to be able to withstand 1% to 2% biooil, but fungal malic acid synthesis was not supported by biooil alone. After seven days at 32°C, the strain produced more malic acid when pyrolysis oil (0.5%) was added to a glucose-containing medium. However, when higher biooil concentrations (1–2%) were added, the levels of malic acid in the same medium decreased. It was discovered that the biooil's inhibitors were the cause of the strain's decreased production of malic acid. Research has demonstrated that a particular strain of the filamentous soil bacterium *T. fusca* muC at 55°C could generate malic acid from cellulose. The bacterium has the ability to break down cellulose across a broad pH range because it produces a cellulase. Cell growth and malate production were stimulated at 250 rpm and 55°C when yeast extract was added to the medium. The pyruvate carboxylase-containing metabolically modified strain *T. fusca* muC-16 was found to have a histidine phosphocarrier protein as a repressor of malic acid production (West, 2017).

*3.4.5. Malic Acid Derived from Poly(*β-l-Malic Acid) (PMA) Production

The most efficient and cost-effective way to create malic acid could be through the fungus that produces PMA, which can then be converted to malic acid. It has been demonstrated that the acid hydrolysis of PMA produced by *A. pullulans* can also yield malic acid. Using 2 M sulfuric acid, PMA is hydrolyzed at 85°C to pure malic acid. In contrast to using sodium carbonate as a neutralizing agent during the fungal production of PMA, the molecular weight of the PMA increased when calcium carbonate was used. It was shown that four strains of *A. pullulans* supported PMA when treated with alkaline peroxide in corn or wheat straw. PMA production by the strains was increased within 7 days at 25°C if 3% (w/v) calcium carbonate and hydrolytic enzymes (such as cellulase, xylanase, and glucosidase) were added to the medium. Under ideal circumstances, a *A. pullulans* NRRL 50383 produced more than 20 g/L PMA within 7 days at 25°C. The ZX-10 mutant strain that was separated from *A. pullulans* NRRL Y-2311-1, after 140 hours at 25°C in a stirred-tank 5 L bioreactor, generated PMA (76.2 g/L) and malic acid (87.6 g/L) from 180 g/L glucose. 0.49 g of malic acid/g of glucose was the yield of malic acid from glucose. After 72 hours at 25°C in a 5 L fermentor, another strain of *A. pullulans*, called YJ6-11, produced 28.6 g/L PMA or 32.4 g/L malic acid from a corncob hydrolysate (made by treating with 1% sulfuric acid and treating with cellulase and xylanase). Glucose, xylose, and arabinose were among the 90 g/L mixed sugars present in the corncob hydrolysate. Utilizing strain YJ6-11 has the benefit of allowing it to produce PMA at a similar rate using either glucose or xylose. This strain was further adapted to corncob hydrolysate using an aerobic fibrous bed bioreactor. The hydrolysate (96 g/L glucose, 54% xylose, 22.5% furfurals, and 12.8% acetic acid) was fermented for 864 hours with a new isolate (CCTCC M2012223) of the original strain being isolated. Through adaptation to increasing concentrations of corncob hydrolysate containing the growth inhibitors furfural, 5-hydroxymethyfurfural, formic acid, and acetic acid, the strain CCTCC M2012223 was isolated. The resulting evolved strain exhibited greater resistance to known inhibitors and their ability to inhibit growth. Another study discovered that nitrogen availability affected the amount of PMA produced by strain CCTCC M2012223. When the isolated strain was grown on the hydrolysate (110 g/L sugar mix) and ammonium sulfate (2 g/L) in the fibrous bed reactor at 25°C with aeration for 120 hours, PMA production peaked at 36 g/L. Free cells (stirred tank bioreactor) or immobilized cells (in an aerobic fibrous bed bioreactor) of *A. pullulans* CCTCC M2012223 could produce PMA in a medium containing raw sweet potato hydrolysate (120 g/L carbohydrate), yeast extract and citrate. Using free cells of the strain, 29.6 g/L PMA was produced after 120 h at 25°C.

After 120 hours at 25°C, 29.6 g/L PMA was produced using strain-free cells. After 156 hours at 25°C in the immobilized cell system, the strain produced 57.5 g/L PMA. PMA can also be produced by *A. pullulans* ZX-10 on sugarcane molasses (44% sucrose, 6% glucose and 5% fructose). During batch fermentation, the strain produced 52.6 g/L PMA after 187.5 hours at

25°C. After 340 hours at 25°C, the strain in a 5 L fermentor produced 116.3 g/L PMA. *A. pullulans* ZX-10 was also able to produce 31.3 g/L malic acid from 27.2 g/L PMA under fed-batch conditions by using a soybean hull hydrolysate (26.8 g/L carbohydrate) supplemented with corn steep liquor (10 g/L, w/v) at 25°C for 168 hours. In the same study, *A. pullulans* ZX-10 was able to grow for approximately 260 hours under fed-batch conditions on soybean molasses (26.8 g/L carbohydrate) in order to convert 62.06 g/L PMA into 71.9 g/L malic acid. To create this industrially valuable organic acid, microbial production of PMA from agricultural biomass appears to be just as efficient as direct microbial fermentation of malic acid.

In conclusion, there is a chance to create the platform chemical malic acid using low-value biomass or processing coproducts like thin stillage, crude glycerol, or straw. From the standpoint of "green chemistry," using biobased methods instead of petroleum-based chemicals to synthesize malic acid could help lessen dependency on such chemicals. The development of strains and processes are the two most urgent problems that must be solved for the biobased production of malic acid. For the price of biobased malic acid to be competitive with that of petroleum-based malic acid, both elements must be met. Only microbes with elevated malic acid production in the presence of growth inhibitors in plant biomass hydrolysates or processing coproducts can lead to new environmentally friendly microbial-based approaches. Process development will be necessary to efficiently bioconvert biomass hydrolysates or coproducts into malic acid using these microbes in order to effectively compete with the existing petroleum-based production methods economically (West, 2017).

3.5. KOJIC ACID

The chemical compound kojic acid has the formula $HOCH_2C_5H_2O_2OH$. It is a naturally occurring 4-pyrone derivative that is produced by a variety of fungal species, most notably Aspergillus oryzae, also known as koji in Japanese. It serves as a chelation agent. In the process of fermenting rice to make sake, the Japanese rice wine, kojic acid is produced as a byproduct. It is used in food and cosmetics to preserve or alter the color of materials. It is a mild inhibitor of the formation of pigment in plant and animal tissues. It combines with ferric ions to form a vivid red complex.

3.5.1. Biosynthesis

Studies using ^{13}C-labeling have identified at least two routes leading to kojic acid. Dehydratase enzymes normally convert glucose to kojic acid. Pentoses are also potential precursors; in this scenario, dihydroxyacetone is used as a bridge.

3.5.2. Applications

Kojic acid can be used in cosmetics to lighten skin, on cut fruits to prevent oxidative browning, and in seafood to preserve pink and red colors. For instance, the latter is applied in the treatment of skin conditions such as melasma. Additionally, kojic acid possesses antifungal and antibacterial qualities.

3.5.3. Chemical Reactions

Kojate is created when the ring-OH group deprotonates from kojic acid. Iron(III) and kojiate chelate to form the red complex $Fe(HOCH_2C_5OH_2O_2)_3$. The biological solubilization of ferric iron by kojic acid may be based on this type of reaction. Kojic acid is a multifunctional molecule with a wide range of organic chemistry. When thionyl chloride is applied, the hydroxymethyl group yields the chloromethyl derivative.

3.5.4. Safety

Certain animal research suggests that kojic acid might have a mild carcinogenic effect. It is proven safe at the level used in cosmetics and is not thought to reach carcinogenic thresholds in human skin.

3.6. LACTIC ACID

An organic acid is lactic acid. Its chemical formula is $C_3H_6O_3$. When it is solid, it has a white color and is miscible with water. It transforms into a colorless solution when dissolved. Production uses both natural sources and synthetic materials. Because lactic acid has a hydroxyl group next to the carboxyl group, it is an alpha-hydroxy acid (AHA). It is employed as a synthetic intermediate in numerous biochemical and organic synthesis industries. Lactate, also known as the lactate anion, is the conjugate base of lactic acid. Lactoyl is the term for the derived acyl group. It can ionize in solution by losing a proton, resulting in the lactate ion $CH_3CH(OH)CO_2^-$. Lactic acid has a pKa that is one unit lower than that of acetic acid, which indicates that it is 10 times more acidic. The increased acidity is a result of the α-hydroxyl and carboxylate group forming an intramolecular hydrogen bond.

Lactic acid has two enantiomers and is chiral. D-lactic acid, also referred to as (R)-lactic acid or (−)-lactic acid, is the mirror image of L-lactic acid, also known as (S)-lactic acid or (+)-lactic acid. Racemic lactic acid, sometimes known as DL-lactic acid, is a combination of the two in equal proportions. Hygroscopic acid is lactic acid. At temperatures between 16 and 18°C (61 and 64 °F), which is its melting point, DL-lactic acid is miscible with both water and ethanol. The melting points of D- and L-lactic acids are higher. Milk fermentation results in lactic acid, which is typically racemic, though some bacterial species only produce D-lactic acid. On the other hand, lactic acid that is created in animal muscles through anaerobic respiration has the (L) enantiomer and is sometimes referred to as sarcolactic acid, which comes from the Greek word *sarx*, which means flesh.

During regular metabolism and exercise, the enzyme lactate dehydrogenase (LDH) continuously converts pyruvate into L-lactate through a fermentation process. The concentration of lactate does not rise until the rate of production surpasses the rate of removal, which is determined by several factors such as the presence of monocarboxylate transporters, the concentration and isoform of LDH, and the oxidative capacity of tissues. Blood lactate concentrations range from 1–2 mM Tooltip millimolar at rest to over 20 mM during vigorous exercise and up to 25 mM afterward. L-lactic acid plays various biological functions, one of which is as the main endogenous agonist of the Gi/o-coupled G protein-coupled receptor (GPCR) known as hydroxycarboxylic acid receptor 1 (HCA1).

Lactic acid bacteria are the agents of lactic acid fermentation in industry, whereby simple carbohydrates like glucose, sucrose, or galactose are converted to lactic acid. These bacteria can also proliferate in the mouth, and the acid they generate is what causes cavities in teeth. One of the key ingredients in lactated Ringer's solution and Hartmann's solution in medicine is lactate. These IV fluids are made up of distilled water, sodium and potassium cations, lactate, and chloride anions, usually in concentrations that are isotonic with human blood. Fluid resuscitation following blood loss from trauma, surgery, or burns is the most frequent use for it. Human tissues produce lactic acid when the supply of oxygen cannot keep up with the demand. When there is a restriction in the blood flow, tissue ischemia takes place. It could also happen more frequently during high-demand situations, like during strenuous exercise. An oxygen debt caused by lactic acidosis can be paid back when tissue oxygenation increases.

3.6.1. Production

Industrial lactic acid production involves chemical synthesis from acetaldehyde or bacterial fermentation of carbohydrates. As of 2009, fermentation accounted for 70–90% of the lactic acid produced. Microbial fermentation can produce racemic lactic acid, which is a 1:1 mixture of D and L stereoisomers or mixtures containing up to 99.9% L-lactic acid. While fermentation can produce D-lactic acid on an industrial scale, it is far more difficult.

3.6.1.1. Fermentative Production

Fermented milk products are obtained industrially by fermentation of milk or whey by *Lactobacillus* bacteria: *Lactobacillus acidophilus*, *Lacticaseibacillus casei* (*Lactobacillus casei*), *Lactobacillus delbrueckii* subsp. *bulgaricus* (*Lactobacillus bulgaricus*), *Lactobacillus helveticus*, *Lactococcus lactis*, *Bacillus amyloliquefaciens*, and *Streptococcus salivarius* subsp. *thermophilus* (*Streptococcus thermophilus*).

Almost any carbohydrate source that contains C5 (penttose sugar) and C6 (hexose sugar) can be used as a starting material for the industrial production of lactic acid. Raw sugar, beet juice, pure sucrose, and glucose derived from starch are commonly utilized. There are two types of bacteria that produce lactic acid: heterofermentative species that produce one mole of lactate from one mole of glucose along with carbon dioxide and acetic acid/ethanol, and homofermentative bacteria such as *Lactobacillus casei* and *Lactococcus lactis* that produce two moles of lactate from one mole of glucose.

3.6.1.2. Chemical Production

Industrial acetaldehyde-hydrogen cyanide reactions yield lactonitrile, which is then hydrolyzed to produce racemic lactic acid. Ammonium chloride is a byproduct of hydrolysis with hydrochloric acid; Musashino, a Japanese company, is one of the last major producers of lactic acid using this process. It is also feasible to synthesize enantiopure and racemic lactic acids from other starting materials, such as glycerol, vinyl acetate, etc., through the use of catalytic techniques.

3.6.2. Uses

3.6.2.1. Polymer Precursor

It is possible to dehydrate two lactic acid molecules to produce lactone lactide. Catalysts cause lactide to polymerize into biodegradable polyesters called atactic or syndiotactic polylactide (PLA). One plastic that comes from sources other than petrochemicals is PLA.

3.6.2.2. Pharmaceutical and Cosmetic Applications

In pharmaceutical technology, lactic acid is also used to convert otherwise insoluble active ingredients into water-soluble lactates. Its disinfectant and keratolytic qualities, as well as its ability to modify acidity, make it useful in topical preparations and cosmetics. Due to their ability to descal calcium compounds, bacteria that produce lactic acid have demonstrated potential in lowering oxaluria.

3.6.2.3. Foods

Wholesome food production and preservation depend heavily on lactic acid bacteria. Lactic acid fermentations are generally low-cost and fuel-efficient because they require little to no heat during preparation. Foods that have undergone lactic acid fermentation are crucial to feeding people on every continent.

3.6.2.3.1. Fermented Food

Sour milk products like kumis, Laban, yogurt, kefir, and some types of cottage cheese are the main sources of lactic acid. Lactic acid causes the casein in fermented milk to coagulate, or curdle. Another factor contributing to the sour taste of sourdough bread is lactic acid. Lactic acid may be listed under the term "carbohydrate" (or "carbohydrate by difference) in nutritional information lists because this term is often used to refer to anything besides water, protein, fat, ash, and ethanol.

Certain sour beers, like Belgian lambics, are intentionally made with lactic acid. This is most frequently created naturally by different bacterial strains. Unlike yeast, which ferments sugar into ethanol, these bacteria convert sugars into acids. Yeast and bacteria are allowed to fall into the open fermenters once the wort has cooled. Common beer style brewers would make sure that the fermenter is kept free of these kinds of bacteria. Flanders red, American wild ale, and Berliner Weisse are some other sour.

3.6.2.3.2. Separately Added

Approved for use as a food additive in the EU, US, Australia, and New Zealand, it is identified by its INS number 270, also listed as E number E270. Lactic acid finds application in food preservation, curing, and flavoring. It is a decontaminant used in the processing of meat and an ingredient in processed foods. Commercial lactic acid production involves chemical synthesis or fermentation of carbohydrates like lactose, glucose, or sucrose. Sources of carbohydrates include cane sugar, corn, and beets.

3.6.2.4. Forgery

In the past, lactic acid has been used to help erase official documents' ink so that it can be altered during forgeries.

3.6.2.5. Cleaning Products

Some liquid cleaners use lactic acid as a descaling agent to get rid of deposits like calcium carbonate that are left behind by hard water.

3.7. GLUCONIC ACID

With the molecular formula $C_6H_{12}O_7$ and the condensed structural formula $HOCH_2(CHOH)_4CO_2H$, glutamic acid is an organic compound. In a neutral aqueous solution, it is a white solid that forms the gluconate anion. Gluconates are the salts of gluconic acid. Due to the oxidation of glucose, gluconate species such as gluconate esters, gluconate salts, and glucononic acid are widely distributed in nature. Gluconates are injectable forms of certain medications.

3.7.1. Chemical Structure

A six-carbon chain with five hydroxyl groups arranged in the same manner as in the open-chained form of glucose makes up the chemical structure of gluconic acid. The chain ends in a carboxylic acid group. It is one of the 2,3,4,5,6-pentahydroxyhexanoic acid's 16 stereoisomers.

3.7.2. Production

The aerobic oxidation of glucose in the presence of the enzyme glucose oxidase usually results in the production of gluconic acid. Hydrogen peroxide and gluconolactone are the products of the conversion. In water, the lactone hydrolyzes spontaneously to produce gluconic acid.

$$C_6H_{12}O_6 + O_2 \rightarrow C_6H_{10}O_6 + H_2O_2$$

$$C_6H_{10}O_6 + H_2O \rightarrow C_6H_{12}O_7$$

variations on the fermentation-based oxidation of glucose (or another substrate containing carbohydrates). or catalysis by noble metals. Hlasiwetz and Habermann first synthesized gluconic acid in 1870 by chemically oxidizing glucose. Boutroux used the fermentation of glucose to prepare and isolate gluconic acid in 1880.

3.7.2.1. Microbial Production

One well-established biotechnological method for producing gluconic acid on an industrial scale is microbial production. Gluconic acid is a multipurpose substance that finds use in food processing, pharmaceuticals, and industrial operations.

Microbial Production Process:

1. **Microorganism Selection:**
   - **Bacteria:** Certain strains of bacteria, such as *Aspergillus Niger* (a fungus), *Bacillus* species, and *Pseudomonas* species, are commonly used. *Aspergillus Niger* is particularly noted for its high gluconic acid yield.
   - **Yeasts and Molds:** Although certain molds and yeasts can also produce gluconic acid, bacteria are frequently chosen for large-scale manufacturing.
2. **Substrate Preparation:**
   - **Carbon Source:** Generally, a carbon source like glucose, sucrose, or lignocellulosic biomass is used to initiate the production process. The most popular substrate is glucose since it can be transformed into gluconic acid with ease.
   - **Medium:** The nutrients required for microbial development and the synthesis of gluconic acid are included in a suitable growth medium that is prepared.
3. **Fermentation Process:**
   - **Inoculation:** The selected microorganism is inoculated into the fermentation medium.
   - **Conditions:** The aeration, temperature, and pH levels during fermentation are all carefully calibrated to optimize the production of gluconic acid. This usually calls for temperatures between 25 and 30°C and a slightly acidic pH (about 3–5) for *Aspergillus Niger*.
   - **Duration:** A few days to a week can pass during the fermentation process, depending on the microorganism and the volume of product being produced.
4. **Gluconic Acid Recovery:**
   - **Separation:** After fermentation, the broth containing gluconic acid is harvested. Cells and other particulates are usually removed by filtration or centrifugation.
   - **Purification:** The gluconic acid is then purified, often through processes like ion-exchange chromatography or precipitation.
5. **Concentration and Crystallization:**
   - **Concentration:** The purified gluconic acid solution is concentrated, if necessary, by evaporation.

- **Crystallization:** Gluconic acid can be crystallized out of the solution, depending on the desired end product.

3.7.3. Occurrence and Uses

Fruit, honey, and wine all naturally contain gluconic acid. As a food additive (E574), it is now known as an acidity regulator. The gluconate anion chelates Ca^{2+}, Fe^{2+}, K^{+}, Al^{3+}, and other metals, including lanthanides and actinides. Additionally, it dissolves mineral deposits in cleaning products, particularly in alkaline solutions. Injections of zinc gluconate are used to neuter male dogs. To slow down the cement hydration reactions and extend the cement setting time, glutaconate is also used in building and construction as a concrete admixture, or retarder. To prevent excessive heat and the accompanying cracking, it permits a longer time for the concrete to be laid or for the cement hydration heat to be distributed over a longer duration. When the weather is warm, retarders are added to concrete to help cast thick, large slabs of concrete in successive layers that are adequately mixed. An aqueous solution of glutaric acid is used as a medium for organic synthesis.

3.7.3.1. Medicine

In order to treat electrolyte imbalance, gluconate is most frequently used in medicine as a biologically neutral carrier of Zn^{2+}, Ca^{2+}, Cu^{2+}, Fe^{2+}, and K^{+} Injections of calcium gluconate may be used in more severe cases to prevent necrosis of deep tissues and to treat hypocalcemia in hospitalized patients. Calcium gluconate gel is used to treat burns caused by hydrofluoric acid. Another electrolyte found in some solutions, like plasmalyte A, which is used for intravenous fluid resuscitation, is glutamate. Quinine gluconate, an intramuscular injection used to treat malaria, is a salt of quinine and gluconic acid. Injections of ferrous gluconate have been suggested as a treatment for anemia in the past.

3.7.4. Market

In the global fermentation market, organic acids are the third largest category, behind amino acids and antibiotics. 2009 will see an increase in the market value of organic acid to $3 million. Because it is used in so many different fields, citric acid is the market leader for organic acids. Comparatively speaking, the gluconic acid market is smaller. Nonetheless, 60 000 tonnes are produced globally each year, and it is sold as 50% technical grade aqueous solution (by mass) on the market. Because of its characteristics and uses, sodium gluconate is the principal product of gluconic acid derivatives. Pfizer Inc. is the US manufacturer of gluconic acid and its salt. Manhattan, Bristol-Meyers Co. Premier Malt Products Inc., New York. Wisconsin. Among the producers of gluconate in Europe are Pfizer in Ireland, Roquette Frères in France, and Benckiser in Germany. Japan's two gluconate producers are Fujisawa and Kyowa Hakko. Among the derivatives of gluconic acid, calcium gluconate is another significant product that can be purchased as a dietary supplement in the forms of tablets, powder, and liquid.

3.8. ERYTHORBIC ACID

Erythorbic acid is a stereoisomer of ascorbic acid, also known as isoascorbic acid or D-araboascorbic acid (vitamin C). It is produced by reacting sodium methoxide with methyl 2-keto-D-gluconate. It can also be produced using strains of Penicillium that have been chosen specifically for this property, or from sucrose. E315 is the E number for it, and processed foods frequently use it as an antioxidant. Studies in humans have been carried out to look into the nutritional benefits of erythorbic acid.

3.8.1. Microbial Production of Oxogluconic Acids

Oxogluconic acids are significant keto acids with a variety of industrial and scientific uses, such as 2-keto-D-gluconic acid (2-KGA) and 5-keto-D-gluconic acid (5-KGA). They are produced microbially through particular fermentation processes and are intermediates in the gluconic acid metabolic pathway.

1. **Microorganism Selection:**
   - **Bacteria:** Oxogluconic acids are known to be produced by a variety of bacteria, including strains of *Pseudomonas* species and *Aspergillus Niger*. These acids can be produced at higher levels through metabolic and genetic engineering.
   - **Fungi:** Although bacterial systems are frequently chosen due to their higher yields and more controlled production, some fungi can also produce oxogluconic acids.
2. **Substrate Preparation:**
   - **Carbon Source:** Carbon sources such as glucose or other sugars are frequently utilized, much like in the synthesis of gluconic acid. The yield of oxogluconic acids can be affected by the substrate selection.
   - **Medium:** A nutrient-rich medium is prepared to support microbial growth and production. This often includes sources of nitrogen, vitamins, and minerals.

3. **Fermentation Process:**
   - **Inoculation:** The selected microorganism is inoculated into the prepared fermentation medium.
   - **Conditions:** The pH, temperature, and aeration levels during fermentation are all optimized. For instance, *Aspergillus Niger* frequently needs temperatures between 25 and 30°C and slightly acidic conditions (pH 3–5).
   - **Duration:** Fermentation can take several days to a week, depending on the microorganism and the desired concentration of oxogluconic acids.
4. **Conversion Process:**
   - **Direct Production:** Some microorganisms produce oxogluconic acids directly from glucose or gluconic acid.
   - **Indirect Production:** Oxogluconic acids can occasionally be created as byproducts of the conversion of glucose to other molecules; in these situations, extracting and purifying the acids will need extra steps.
5. **Product Recovery:**
   - **Separation:** After fermentation, the broth is harvested, and cells are removed by filtration or centrifugation.
   - **Purification:** The oxogluconic acids are purified, often using techniques such as ion-exchange chromatography or other separation methods.
   - **Concentration:** The purified solution may be concentrated, depending on the desired form of the final product.

3.8.2. Applications of Oxogluconic Acids

- **Biochemical Research:** Used as intermediates in various biochemical pathways and as research tools.
- **Industrial Processes:** Employed in the production of specific chemicals and as chelating agents.
- **Pharmaceuticals:** Potential applications in drug development and as excipients in pharmaceutical formulations.

3.8.3. Advantages of Microbial Production

- **Eco-Friendly:** Microbial processes are generally more environmentally friendly compared to chemical synthesis.

- **High Specificity:** Microbial fermentation can offer high specificity for the production of oxogluconic acids.
- **Cost-Effective:** Utilization of renewable resources as substrates can make the process cost-effective.

3.8.4. Challenges and Optimization

- **Yield and Productivity:** Enhancing productivity and yields requires genetic engineering of microorganisms and optimization of fermentation conditions.
- **Purification:** Oxogluconic acid purification from fermentation broths can be difficult and necessitate the use of sophisticated methods.

The goals of this field's research and development are still to increase the microbial production of oxogluconic acids' efficiency and broaden their industrial applications.

3.9. TARTARIC ACID

An organic acid that occurs naturally, tartaric acid finds extensive use in food, medicine, and other industrial processes. Because of its potential cost savings over chemical synthesis and sustainability, the production of it through microbial fermentation is of economic interest.

3.9.1. Microbial Production Process

- **Microorganisms:** The use of particular strains of bacteria or fungi that can metabolize substrates to produce tartaric acid is a common practice in the microbial production of tartaric acid. For example, specific lactic acid bacteria and Aspergillus species are employed in the production process.
- **Utilization of Substrates:** Usually, a carbon source, such as glucose or other sugars, is used to begin the production process. The process's ability to be economically profitable can be greatly impacted by the use of cheap and plentiful substrates.
- **Fermentation Conditions:** Improving the pH, temperature, and aeration of the fermentation process is essential to increasing yield and lowering production costs.

3.9.2. Economic Considerations

- **Cost of Raw Materials:** The cost and selection of the substrate have a significant impact on the overall economics of producing tartaric acid. Costs can be decreased by using waste materials or agricultural byproducts as substrates.
- **Energy and Operating Expenses:** In order to maintain ideal conditions, fermentation processes need energy. The price of labor, energy, and equipment maintenance for fermentation adds to the total cost of production.
- **Production Scale:** Large-scale fermentation can result in economies of scale, which lowers the price of tartaric acid per unit. Large-scale fermentation facility investments may ultimately prove to be financially beneficial.
- **By-Product Utilization:** The production of tartaric acid can become more economically viable through the effective use of by-products or co-products. For instance, some production costs may be offset by other valuable compounds produced during fermentation.

3.9.3. Market Demand and Applications

- **Food Industry:** Tartaric acid is utilized in a variety of food products as an acidulant and as a stabilizing agent in cream of tartar. In the food industry, demand determines economic viability and production volumes.
- **Pharmaceuticals:** As a chelating agent and in pharmaceutical formulations, tartaric acid is utilized. One potential source of consistent demand is the pharmaceutical industry.
- **Industrial Uses:** Tartaric acid is used as a component in some industrial processes and in the production of different chemicals.

3.9.4. Advantages of Microbial Production

- **Sustainability:** Compared to chemical synthesis, microbial production of tartaric acid is more environmentally friendly, following sustainability trends and possibly lowering regulatory expenses.
- **Renewable Resources:** Microbial fermentation can be made more economical and sustainable by utilizing renewable resources.

3.9.5. Challenges and Future Directions

- **Optimization:** In order to increase yield and lower expenses, current research focuses on improving microbial strains and fermentation conditions.
- **Process Improvement:** Technological developments in process engineering and fermentation can improve economic viability.
- **Market Competition:** The viability of microbial production economically also rests on its ability to compete with other production techniques, such as chemical synthesis.

The optimization of the production process to balance the costs of raw materials, energy requirements, and scale efficiencies is a key component of tartaric acid's economic microbiology. Enhancing the economic viability of tartaric acid production requires advances in fermentation technology, microbial strains, and by-product utilization. As the need for environmentally friendly and sustainable production techniques increases, microbial production presents a viable substitute for conventional chemical synthesis (Zhao et al., 2018).

REFERENCES

1. Anwar, S., Ali, S., & Sardar, A. A. (2009). Citric acid fermentation of hydrolysed raw starch by *Aspergillus Niger* IIB-A6 in stationary culture. *Sindh University Research Journal (Science Series)*, *4*(1), 1–8. https://doi.org/10.18201/surj.4.1.1.
2. Behera, B. C., Mishra, R., & Mohapatra, S. (2021). Microbial citric acid: Production, properties, application, and future perspectives. *Food Frontiers*, *2*, 62–76. https://doi.org/10.1002/fft2.66.
3. Bentley, R. (2006). From miso, sake and shoyu to cosmetics: A century of science for kojic acid. *Natural Product Reports*, *23*(6), 1046–1062. https://doi.org/10.1039/b603758p.
4. Carlos, R. S., Vandenberghe, L. P. S., Rodrigues, C., & Pandey, A. (2006). New perspectives for citric acid production and application. *Food Technology and Biotechnology*, *44*(2), 141–149. https://hrcak.srce.hr/109832 (accessed on 8 November 2024).
5. Chanda, S., Chakrabarty, S., & Matai, S. (1990). Citric acid production in liquid waste from a leaf protein production plant: Effects of sugar and potassium ferrocyanide. *Biological Wastes*, *34*, 77–81. https://doi.org/10.1016/0269-7483(90)90145-I.
6. Dhillon, G. S., Brar, S. K., Verma, M., & Tyagi, R. D. (2011). Recent advances in citric acid bio-production and recovery. *Food and Bioprocess Technology*, *4*, 505–529. https://doi.org/10.1007/s11947-010-0399-0.
7. Dorsam, S., Kirchhoff, J., Bigalke, M., Dahmen, N., Syldatk, C., & Ochsenreither, K. (2016). Evaluation of pyrolysis oil as carbon source for fungal fermentation. *Frontiers in Microbiology*. https://doi.org/10.3389/fmicb.2016.01039.
8. El Dein, S. M. N., & Emaish, G. M. I. (1979). Effect of various conditions on production of citric acid from molasses in presence of potassium ferrocyanide by *A. aculeatus* and *A. carbonarius*. *Indian Journal of Experimental Biology*, *17*, 105–106.
9. Fukuda, H., Susuki, T., Sumino, Y., & Akiyama, S. (1970). Microbial preparation of citric acid. German Patent No. 2,003,221.
10. Grewal, H. S., & Kalra, K. L. (1995). Fungal production of citric acid. *Biotechnology Advances*, *13*, 209–234. https://doi.org/10.1016/0734-9750(95)00002-8.
11. Gutierrez, N. A., McKay, I. A., French, C. E., Brooks, J., & Maddox, I. S. (1993). Repression of galactose utilization by glucose in the citrate-producing yeast *Candida guilliermondii*. *Journal of Industrial*

Microbiology, *11*, 143–146. https://doi.org/10.1007/BF01583714.

12. Hang, Y. D., & Woodams, E. E. (1985). Grape pomace: A novel substrate for microbial production of citric acid. *Biotechnological Letters*, *7*, 253–254. https://doi.org/10.1007/BF01042372.
13. Hang, Y. D., & Woodams, E. E. (1998). Production of citric acid from corncobs by *Aspergillus Niger*. *Bioresource Technology*, *65*(3), 251–253. https://doi.org/10.1016/S0960-8524(98)00015-7.
14. Hang, Y. D., & Woodams, E. E. (2000). Corn husks: A potential substrate for production of citric acid by *Aspergillus Niger*. *Lebensmittel-Wissenschaft und-Technologie*, *33*, 520–521. https://doi.org/10.1006/fstl.2000.0711.
15. Hang, Y. D., & Woodams, E. E. (2001). Enzymatic enhancement of citric acid production by *Aspergillus Niger* from corn cobs. *Lebensmittel-Wissenschaft und-Technologie*, *34*, 484–486. https://doi.org/10.1006/fstl.2001.0784.
16. Hang, Y. D., Luh, B. S., & Woodams, E. E. (1987). Microbial production of citric acid by solid-state fermentation of kiwi fruit peel. *Journal of Food Science*, *52*, 226–227. https://doi.org/10.1111/j.1365-2621.1987.tb14014.x.
17. Haq, I., Sikander, A., Qadeer, M. A., & Iqbal, J. (2004). Citric acid production by selected mutants of *Aspergillus Niger* from cane molasses. *Bioresource Technology*, *93*, 125–130. https://doi.org/10.1016/j.biortech.2003.10.018.
18. Kapelli, O., Muller, M., & Fiechter, A. (1978). Chemical and structural alterations at cell surface of *Candida tropicalis*, induced by hydrocarbon substrate. *Journal of Bacteriology*, *133*, 952–958. https://doi.org/10.1128/jb.133.2.952-958.1978.
19. Karow, E. O., & Waksman, S. A. (1947). Production of citric acid in submerged culture. *Industrial and Engineering Chemistry*, *39*, 821–825.
20. Karthikeyan, A., & Sivakumar, N. (2010). Citric acid production by koji fermentation using banana peel as a novel substrate. *Bioresource Technology*, *101*(14), 5552–5556. https://doi.org/10.1016/j.biortech.2010.02.063.
21. Khan, I., Nazir, K., Wang, Z. P., Liu, G. L., & Chi, Z. M. (2014). Calcium malate overproduction by *Penicillium viticola* 152 using the medium containing corn steep liquor. *Applied Microbiology and Biotechnology*, *98*, 1539–1546. https://doi.org/10.1007/s00253-014-5906-0.
22. Kroya Ferment. Ind. (1970). Citric acid prepared by fermentation. British Patent No. 1187610. Retrieved from: http://www.scielo.br/pdf/babt/v42n3/v42n3a01.pdf (accessed on 8 November 2024).

23. Kumar, D., Jain, V. K., Shanker, G., & Srivastava, A. (2003). Citric acid production by solid state fermentation using sugarcane bagasse. *Process Biochemistry*, *38*, 1731–1738. https://doi.org/10.1016/S0032-9592(02)00252-2.

24. Lesniak, W., Pietkiewicz, J., & Podgorski, W. (2002). Citric acid fermentation from starch and dextrose syrups by a trace metal resistant mutant of *Aspergillus niger*. *Biotechnological Letters*, *24*, 1065–1067. https://doi.org/10.1023/A:1016030513270.

25. Li, X., Liu, Y., Yang, Y., Zhang, H., Wang, H., Wu, Y., Zhang, M., Sun, T., Cheng, J., Wu, X., Pan, L., Jiang, S., & Wu, H. (2014). High levels of malic acid production by the conversion of corn straw hydrolyte using an isolated *Rhizopus delemar* strain. *Biotechnology and Bioprocess Engineering*, *19*, 478–492. https://doi.org/10.1007/s12257-013-0867-3.

26. Omar, S. H., & Rehm, H. J. (1980). Physiology and metabolism of two alkane oxidizing and citric acid producing strains of *Candida parapsilosis*. *European Journal of Applied Microbiology and Biotechnology*, *11*, 42–49.

27. Oswald, F., Dorsam, S., Veith, N., Zwick, M., Neumann, A., Ochsenreither, K., Syldatk, C. (2016). Sequential mixed cultures: From syngas to malic acid. *Frontiers in Microbiology*. https://doi.org/10.3389/fmicb.2016.01489.

28. Papagianni, M., Wayman, F. M., & Mattey, M. (2005). Fate and role of ammonium ions during fermentation of citric acid by *Aspergillus niger*. *Applied and Environmental Microbiology*, *71*, 7178–7186. https://doi.org/10.1128/aem.71.11.7178-7186.2005.

29. Parekh, S., Vinci, V. A., & Strobel, R. J. (2000). Improvement of microbial strains and fermentation processes. *Applied Microbiology and Biotechnology*, *54*, 287–301. https://doi.org/10.1007/s002530000403.

30. Pontecorvo, G., Roper, J. A., & Forbes, E. (1953). Genetic recombination without sexual reproduction in *Aspergillus niger*. *Journal of General Microbiology*, *8*, 198–210. https://doi.org/10.1099/00221287-8-1-198.

31. Rane, K. D., & Sims, K. A. (1993). Production of citric acid by *Candida lipolytica* Y-1095: Effect of glucose concentration on yield and productivity. *Enzyme and Microbial Technology*, *15*, 646–651. https://doi.org/10.1016/0141-0229(93)90063-8.

32. Rivas, B., Torrado, A., Torre, P., Converti, A., & Dominguez, J. M. (2008). Submerged citric acid fermentation on orange peel autohydrolysate. *Journal of Agricultural and Food Chemistry*, *56*, 2380–2387. https://doi.org/10.1021/jf073388r.

33. Rohr, M., Kubicek, C. P., & Kominek, J. (1983). Citric acid. In G. Reed & H. J. Rehm (Eds.), *Biotechnology* (Vol. 3, pp. 419–454). Verlag-Chemie.
34. Rohr, M., Kubicek, C. P., & Kominek, J. (1996). Citric acid. In H. J. Rehm & G. Reed (Eds.), *Biotechnology* (Vol. 6, 2nd ed., pp. 307–345). Wiley-VCH. https://doi.org/10.3109/07388558509150788.
35. Roukas, K., & Kotzekidou, P. (1986). Production of citric acid from brewery waste by surface fermentation using *Aspergillus niger*. *Journal of Food Science*, *51*, 225–228. https://doi.org/10.1111/j.1365-2621.1986.tb10876.x.
36. Roukas, K., & Kotzekidou, P. (1987). Influence of some trace metals and stimulants on citric acid production from brewery waste by *Aspergillus niger*. *Enzyme and Microbial Technology*, *9*, 291–294. https://doi.org/10.1016/0141-0229(87)90006-8.
37. Roukas, K., & Kotzekidou, P. (1997). Pretreatment of date syrup to increase citric acid production. *Enzyme and Microbial Technology*, *21*, 273–276. https://doi.org/10.1016/S0141-0229(97)00041-0.
38. Roukas, T. (1991). Production of citric acid from beet molasses by immobilized cells of *Aspergillus niger*. *Journal of Food Science*, *56*, 878–880. https://doi.org/10.1111/j.1365-2621.1991.tb05409.x.
39. Sardinas, J. L. (1972). Fermentative production of citric acid. French Patent No. 2113668. Retrieved from: https://www.scielo.br/pdf/babt/v42n3/v42n3a01.pdf (accessed on 8 November 2024).
40. Shankaranand, V. S., & Lonsane, B. K. (1994). Coffee husk: An inexpensive substrate for production of citric acid by *Aspergillus niger* in a solid-state fermentation. *World Journal of Microbiology & Biotechnology*, *10*, 165–168. https://doi.org/10.1007/BF00360879.
41. Shojaosadati, S. A., & Babaeipour, V. (2002). Citric acid production from apple pomace in multi-layer packed bed solid-state bioreactor. *Process Biochemistry*, *37*, 909–914. https://doi.org/10.1016/S0032-9592(01)00294-1.
42. Soccol, C. R., Vandenberghe, L. P. S., Rodrigues, C., & Pandey, A. (2006). A new perspective for citric acid production and application. *Food Technology and Biotechnology*, *44*, 141–149. https://hrcak.srce.hr/109832 (accessed on 8 November 2024).
43. Steiger, M. G., et al. (2013). Biochemistry of microbial itaconic acid production. *Frontiers in Microbiology*. https://doi.org/10.3389/fmicb.2013.00023.

44. Uchio, R., Maeyashiki, I., Kikuchi, K., & Hirose, Y. (1975). Citric acid production by yeast. Japanese Kokai, *76*, 63,059. Retrieved from: https://www.scielo.br/pdf/babt/v42n3/v42n3a01.pdf (accessed on 8 November 2024).
45. van der Straat, L., et al. (2014). Expression of the *Aspergillus terreus* itaconic acid biosynthesis cluster in *Aspergillus Niger*. *Microbial Cell Factories*, *13*, 11. https://doi.org/10.1186/1475-2859-13-11.
46. West, T. P. (2011). Malic acid production from thin stillage by *Aspergillus* species. *Biotechnology Letters*, *33*, 2463–2467.
47. West, T. P. (2015). Fungal biotransformation of crude glycerol into malic acid. *Z. Naturforsch. C*, *70*, 165–167.
48. West, T. P. (2017). Microbial production of malic acid from biofuel-related coproducts and biomass. *Fermentation*, *3*(2), 14. https://doi.org/10.3390/fermentation3020014.
49. Zambanini, T., Kleineberg, W., Sarikaya, E., Buescher, J. M., Meurer, G., Wierckx, N., & Blank, L. B. (2016). Enhanced malic acid production from glycerol with high-density *Ustilago trichophora* TZ1 cultivations. *Biotechnology for Biofuels*, *9*, 135. https://doi.org/10.1186/s13068-016-0549-z.
50. Zhao, M., et al. (2018). Itaconic acid production in microorganisms. *Biotechnology Letters*, *40*, 1–12. https://doi.org/10.1007/s10529-017-2500-5.

Chapter **4**

Acetic Acid: Vinegar

Contents

INTRODUCTION

Acetic acid is also referred to as vinegar acid, methane carboxylic acid, ethanoic acid, and ethylic acid. The distinctive smell of vinegar is caused by acetic acid, a fermentation byproduct. About 4–6 percent of vinegar is acetic acid in water. Glacial acetic acid is pure acetic acid with very little water present; more concentrated solutions are used in laboratories. Skin contact with diluted solutions, such as vinegar, is safe, but skin burns when exposed to concentrated solutions. Glacial acetic acid corrodes metal, burns skin, and can result in long-term eye damage. The following products are made with acetic acid: rubber, plastics, dyes, insecticides, cellulose acetate, vinyl acetate monomer, acetic esters, and photographic chemicals. The production of vitamins, hormones, antibiotics, organic compounds, and food additives (acidulants) are among its other commercial applications. It is also employed in a variety of textile printing techniques.

During the metabolism of food, acetates, or salts of acetic acid, are frequently found in the tissues of both plants and animals. Most tissues can quickly break down acetate, which can result in the intermediate ketone production. The body uses acetate as a building block in many different human and animal tissue preparations to create neutral lipids, saturated and unsaturated fatty acids, steroids, and sterols. Most people are not affected by the low concentrations found in vinegar and other foods. Acetic acid is a potent irritant of the eyes, skin, and mucous membranes at higher concentrations that may be found in a laboratory or factory. Long-term skin contact with concentrated acetic acid can cause tissue damage. Exposure to high concentrations of acetic acid vapors through inhalation can result in irritation of the nose, throat, and eyes. Individuals who have a high level of occupational exposure may experience pharyngitis, bronchitis, conjunctivitis, and exposed tooth erosion (incisors and canines).

4.1. ACETIC ACID: CHEMICAL COMPOUND

Acetic acid, also known as ethanoic acid in formal terminology, is a colorless, acidic liquid organic compound with the chemical formula CH_3COOH (sometimes written as CH_3CO_2H, $C_2H_4O_2$, or $HC_2H_3O_2$). A minimum of 4% of vinegar's volume is made up of acetic acid, which is vinegar's primary ingredient other than water. It has been a part of vinegar for at least the third century BC, according to historical accounts. The second most basic carboxylic acid, after formic acid, is acetic acid. Used mostly in the creation of synthetic fibers and fabrics, polyvinyl acetate for wood glue, and cellulose acetate for photographic film, it is a significant chemical reagent and industrial chemical in a variety of fields. Diluted acetic acid is a common descaling agent used in homes. The food additive code E260 governs the use of acetic acid in the food industry as a condiment and acidity regulator. The acetyl group in biochemistry, which is derived from acetic acid, is essential to all living forms. When attached to coenzyme A, it plays a crucial role in the metabolism of fats and carbohydrates. As of 2023, there will be around 17.88 million metric tons (t/a) of acetic acid required globally. The carbonylation of methanol yields the majority of the acetic acid produced worldwide. Workers may suffer long-term respiratory injuries from inhaling it, as well as accidental skin damage, as a result of its production and subsequent industrial use.

4.1.1. Nomenclature

The IUPAC name that is most frequently used and preferred is the trivial name acetic acid. The substitutive nomenclature is used to construct the systematic name ethanoic acid, which is a recognized IUPAC name. The word "acetic acid" comes from the Latin word "acetum," which means vinegar and is related to the word "acid."

Anhydrous (water-free) acetic acid is known as glacial acetic acid. The name derives from the solid ice-like crystals that form with agitation, slightly below room temperature at 16.6°C (61.9 °F). This is similar to the German Eisessig (ice vinegar) name. Since acetic acid can never be completely free of water in an atmosphere, glacial acetic acid's melting point is lowered by 0°C when 0.1 percent water is present.

Acetic acid is commonly represented by the symbol AcOH (or HOAc), where Ac is the pseudoelement symbol for the acetyl group $CH_3{-}C({=}O){-}$. AcO− is the representation of the conjugate base, acetate ($CH_3COO{-}$). The ion left over after acetic acid loses its H+ content is called acetate. Acetate can also refer to an ester of acetic acid or a salt that contains this anion. To avoid confusion among organic chemists, the symbol Ac for the acetyl functional group should not be confused with the symbol Ac for the element actinium. Acetic acid is written as $CH_3{-}C(O)OH$, $CH_3{-}C({=}O){-}OH$, CH_3COOH, and CH_3CO_2H to better represent its structure. The acronym HAc is occasionally used to refer to acid-base reactions, where Ac stands for acetate rather than acetyl. The chemical formula of the carboxymethyl functional group, which is obtained by subtracting one hydrogen from the methyl group of acetic acid, is $-CH_2{-}C({=}O){-}OH$.

4.1.2. History

Acetic acid-producing bacteria are found all over the world, which is why vinegar was discovered early in civilization as the natural byproduct of beer and wine exposed to air. Theophrastus, a Greek philosopher, wrote about the use of vinegar on metals to create paints that were useful for art, such as white lead (lead carbonate) and verdigris, a green mixture of copper salts that included copper(II) acetate. This indicates the use of acetic acid in alchemy dating back to the third century BC. By boiling soured wine, the ancient Romans created Sapa, an extremely sweet syrup. The Roman nobility suffered from lead poisoning in part because Sapa, which was made in lead pots, was high in lead acetate, a sweet material also known as sugar of lead or sugar of Saturn. The process of ketonic decarboxylation, or dry distillation of lead acetate, was disclosed by the German alchemist Andreas Libavius in the 16th century. This process produced acetone. Since vinegar contains water, which significantly alters the properties of acetic acid, chemists have long held the view that glacial acetic acid and vinegar acid are two distinct substances. The identicality was demonstrated by French chemist Pierre Adet. Acetic acid was first produced in 1845 by German chemist Hermann Kolbe using inorganic substances. The steps involved in this reaction sequence were aqueous chlorination of trichloroacetic acid, pyrolysis of carbon disulfide to tetrachloroethylene, electrolytic reduction of tetrachloroethylene to acetic acid, and so on.

By 1910, the pyroligneous liquor—a byproduct of wood distillation—was the primary source of glacial acetic acid. After treating the calcium acetate produced by the treatment of milk of lime with sulfuric acid, the acetic acid was extracted. Approximately 30% of the 10,000 tons of glacial acetic acid that Germany was producing at the time was utilized to make indigo dye. For a long time, methanol carbonylation seemed like an appealing precursor to acetic acid because both methanol and carbon monoxide are commodity raw materials. As early as 1925, Henri Dreyfus of British Celanese created a pilot plant for the carbonylation of methanol. Nevertheless, commercialization of these routes was discouraged due to a lack of feasible materials capable of containing the corrosive reaction mixture at the required high pressures (200 atm or higher). German chemical manufacturer BASF invented the first commercial methanol carbonylation process in 1963. This process made use of a cobalt catalyst. A catalyst based on rhodium (cis−$[Rh(CO)_2I_2]^-$) was found in 1968 that could function effectively at lower pressure and produce almost no by-products. The first rhodium-catalyzed methanol carbonylation plant was constructed by the American chemical company Monsanto Company in 1970, and this process went on to become the most common way to produce acetic acid. BP Chemicals introduced the Cativa catalyst ($[Ir(CO)_2I_2]^-$) into the market in the late 1990s. Its enhanced efficiency is attributed to iridium. The iridium-catalyzed glacial acetic acid production process, dubbed the Cativa process, is more environmentally friendly and has largely replaced the Monsanto process, frequently in the same production facilities.

4.1.2.1. Interstellar Medium

A group under the direction of David Mehringer used the former Millimeter Array at the Owens Valley Radio Observatory and the Berkeley-Illinois-Maryland Association array at the Hat Creek Radio Observatory to discover interstellar acetic acid in 1996. It was initially found in the Sgr B2 Large Molecule Heimat source, also known as the Sagittarius B2 North molecular cloud. Since single dish radio telescopes were at least partially responsible for all previous ISM molecular discoveries made in the millimeter and centimeter wavelength regimes, acetic acid holds the distinction of being the first molecule discovered in the interstellar medium using only radio interferometers.

4.1.3. Properties

4.1.3.1. Acidity

The hydrogen center in the carboxyl group (−COOH) in carboxylic acids such as acetic acid can separate from the molecule by ionization:

$CH_3COOH \rightleftharpoons CH_3CO{-}2 + H^+$

This proton (H^+) release is what gives acetic acid its acidic properties. A weak monoprotic acid is acetic acid. Its pK_a value in an aqueous solution is 4.76. Acetate is its conjugate base (CH_3COO^-). Only 0.4 percent of the acetic acid molecules are dissociated in a 1.0 M solution, which is roughly the concentration of household vinegar. Only in extremely thin (. $< 10^{-6}$ M) solution, acetic acid is > 90% separated.

4.1.3.2. Structure

Hydrogen bonds hold the individual molecules in chains that make up solid acetic acid. Dimers are detectable at 120°C (248°F) in the vapour phase. Dimers are also present in the liquid phase of diluted solutions containing non-hydrogen-bonding solvents and, to a lesser extent, in pure acetic acid; however, hydrogen-bonding solvents cause them to break down. The dimer's dissociation entropy is estimated to be 154–157 J mol^{-1} K^{-1}, while its dissociation enthalpy is 65–66 points kJ/mol. Similar interactions involving hydrogen bonds between molecules occur with other carboxylic acids.

4.1.3.3. Solvent Properties

Liquid acetic acid, like ethanol and water, is a hydrophilic (polar) protic solvent. It dissolves polar and non-polar substances alike, including oils, with a relative static permittivity (dielectric constant) of 6.2. It also dissolves polar solutes, such as inorganic salts and sugars. Water, chloroform, and hexane are among the polar and non-polar solvents with which it is miscible. Over longer n-alkanes, acetic acid becomes less soluble in alkanes and is not miscible at all with higher alkanes (beginning with octane). Because of its solvent and miscibility qualities, acetic acid is a valuable industrial chemical that can be used, for instance, as a solvent in the synthesis of dimethyl terephthalate.

4.1.3.4. Biochemistry

Acetic acid is typically completely ionized to acetate at physiological pHs. Formally derived from acetic acid, the acetyl group is essential to all living forms. Usually, acetyl-CoA synthetase enzymes bind it to coenzyme A, where it plays a crucial role in the metabolism of fats and carbohydrates. In contrast to the fatty acids, which are longer-chain carboxylic acids, acetic acid is absent from naturally occurring triglycerides. The majority of the acetate produced by cells for acetyl-CoA is produced straight from pyruvate or ethanol. On the other hand, glycerine triacetate, also known as artificial triglyceride triacetin, is a common food additive that can be found in topical medications and cosmetics. The body breaks down this additive into acetic acid and glycerol. Particularly the genus Acetobacter and *Clostridium acetobutylicum*, acetic acid bacteria are responsible for the production and excretion of acetic acid. Foodstuffs, water, and soil all contain these bacteria, and as fruits and other foods deteriorate, acetic acid is naturally produced. In humans and other primates, acetic acid is also a part of the lubricant used in the vagina, where it seems to have some antibacterial properties.

4.1.4. Production

In the industrial setting, acetic acid is made synthetically and through bacterial fermentation. Carbonylation of methanol, as described below, produces about 75% of the acetic acid used in the chemical industry. Although the biological route only makes up about 10% of global production, it is still significant for vinegar production because many food purity regulations demand that vinegar used in food products come from biological sources. The isomerization of methyl formate, the transformation of syngas into acetic acid, and the gas phase oxidation of ethylene and ethanol are additional processes. Using an ice bath and fractional freezing, acetic acid can be purified. The acetic acid will precipitate out while the water and other contaminants stay liquid. An estimated 5 Mt/a (million tons per year) of virgin acetic acid were produced worldwide as of 2003–2005, with the United States producing about half of that total. Japanese production was 0.7 Mt/a, while European production was about 1 Mt/a and declining. An additional 1.5 million tons were recycled annually, increasing the global market to 6.5 million tons per year. From 2010 to 2023, the global production increased from 10.7 million metric tons to 17.88 million metric tons. BP Chemicals and Celanese are the two companies that produce the most virgin acetic acid. Svensk Etanolkemi [sv], Samsung, Eastman, Millennium Chemicals, and Sterling Chemicals are some of the other significant prducers

4.1.4.1. Methanol Carbonylation

Most acetic acid is produced by methanol carbonylation. In this process, methanol and carbon monoxide react to produce acetic acid according to the equation:

The process involves iodomethane as an intermediate, and occurs in three steps. A metal carbonyl catalyst is needed for the carbonylation (step 2).

1. $CH_3OH + HI \rightarrow CH_3I + H_2O$
2. $CH_3I + CO \rightarrow CH_3COI$
3. $CH_3COI + H_2O \rightarrow CH_3COOH + HI$

There are two related methods for carbonylating methanol: the Cativa process, which is catalyzed by iridium, and the Monsanto process, which is catalyzed by rhodium. The former method has mostly been replaced by the later, more efficient, and environmentally friendly method. Both processes require catalytic amounts of water, but the Cativa process uses less, suppressing the water-gas shift reaction and producing fewer byproducts in the process. Rhodium catalysis can also be used to produce acetic anhydride in plants by changing the process parameters.

4.1.4.2. Acetaldehyde Oxidation

Acetaldehyde was oxidized to produce most acetic acid before the Monsanto process was commercialized. This process is still the second most significant one in the manufacturing process, even though it typically cannot compete with methanol's carbonylation. Acetylene can be hydrated to produce acetaldehyde. Early in the 20th century, this technology was in vogue. The chemical equation, which uses butane as an example, states that light naphtha components are easily oxidized by oxygen or even air to produce peroxides, which then break down to produce acetic acid.

$2\ C_4H_{10} + 5\ O_2 \rightarrow 4\ CH_3CO_2H + 2\ H_2O$

Metal catalysts, such as the manganese, cobalt, and chromium naphthenate salts, are needed for these kinds of oxidations. The usual reaction takes place at pressures and temperatures that are intended to be as high as possible without causing the butane to lose its liquid state. Reaction temperatures of 150°C (302°F) and 55 atm are typical. However, butanone, ethyl acetate, formic acid, and propionic acid are examples of side-products that could arise. These byproducts have commercial value as well, and where necessary, the reaction conditions can be changed to yield more of these byproducts. But the process gets more expensive when acetic acid has to be separated from these byproducts. Acetaldehyde can be oxidized by the oxygen in the air, just as butane can be oxidized under similar circumstances and with the same catalysts.

$$2\ CH_3CHO + O_2 \rightarrow 2\ CH_3CO_2H$$

This reaction can yield more than 95% acetic acid when using contemporary catalysts. The main byproducts are formaldehyde, formic acid, and ethyl acetate; these three substances can all be easily separated by distillation and have lower boiling points than acetic acid.

4.1.4.3. Ethylene Oxidation

Using the Wacker process, acetaldehyde can be made from ethylene and oxidized as previously described. More recently, a less expensive single-stage ethylene to acetic acid conversion was commercialized by the chemical company Showa Denko, which established an ethylene oxidation plant in Ōita, Japan, in 1997. A palladium metal catalyst supported on a heteropoly acid, such as silicotungstic acid, catalyzes the reaction. The same metal catalyst is used in a related procedure on silica and silicotungstic acid.

$$C_2H_4 + O_2 \rightarrow CH_3CO_2H$$

It is thought to be competitive with methanol carbonylation for smaller plants (100–250 kt/a), depending on the local price of ethylene.

4.1.4.4. Oxidative Fermentation

Vinegar has been manufactured for the majority of human history by acetic acid bacteria belonging to the genus Acetobacter. These bacteria can make vinegar from a range of alcoholic foods if they are given enough oxygen. Apple cider, wine, fermented grain, malt, rice, or potato mashes are examples of feeds that are frequently used. The overall chemical process that these bacteria support is:

$$C_2H_5OH + O_2 \rightarrow CH_3COOH + H_2O$$

vinegar can be made from a diluted alcohol solution inoculated with Acetobacter and stored in a warm, airy location. By giving the bacteria a better supply of oxygen, industrial vinegar-making techniques quicken this process. It's likely that mistakes made during the winemaking process led to the initial batches of vinegar produced by fermentation. Acetobacter will outnumber the native yeast on the grapes if the must is fermented at an excessively high temperature. Vineyard owners soon discovered that they could make vinegar in the sweltering summer months before the grapes were ripe and prepared for making wine by using other organic materials, as the demand for vinegar for hygienic, medicinal, and culinary uses grew. However, since the vintners were not familiar with the procedure, this approach was slow and not always effective.

The German or fast method was one of the earliest contemporary commercial processes, having been implemented in Germany in 1823. In this method, fermentation occurs inside a tower filled with charcoal or wood shavings. The feed that contains alcohol is slowly pushed to the top of the tower, where forced or natural convection brings in fresh air from below. The time it took to prepare vinegar was reduced from months to weeks by this process's improved air supply. Submerged tank culture, initially reported by Otto Hromatka and Heinrich Ebner in 1949, is currently used to produce the majority of vinegar. This process involves bubbling air through the solution while alcohol ferments to vinegar in a tank that is constantly stirred. With current implementations of this technique, 15% acetic acid vinegar can be made in a 24-hour batch process, and 20% in a 60-hour fed-batch process.

4.1.4.5. Anaerobic Fermentation

Species of anaerobic bacteria, including members of the genus *Clostridium* or *Acetobacterium*, can convert sugars to acetic acid directly without creating ethanol as an intermediate. The overall chemical reaction conducted by these bacteria may be represented as:

$$C_6H_{12}O_6 \rightarrow 3CH_3COOH$$

These acetogenic bacteria produce acetic acid from one-carbon compounds, including methanol, carbon monoxide, or a mixture of carbon dioxide and hydrogen:

$$2CO_2 + 4\ H_2 \rightarrow CH_3COOH + 2H_2O$$

Clostridium may be able to produce acetic acid more quickly than ethanol-oxidizers like Acetobacter because of its capacity to metabolize sugars directly or to produce acetic acid from less expensive inputs. But compared to Acetobacter, Clostridium bacteria are less acid-tolerant. Compared to Acetobacter strains, which can produce vinegar at concentrations of up to 20%, even the most acid-tolerant Clostridium strains can only produce vinegar at a few percent concentration. Right now, producing vinegar with Acetobacter is still more economical than using Clostridium and then condensing it. Because of this, even though acetogenic bacteria have been understood since 1940, their industrial applications are limited to a small number of specialized uses.

4.1.5. Uses

Chemical reagents such as acetic acid are used in the synthesis of chemical compounds. The creation of vinyl acetate monomer is the largest single application of acetic acid, and it is closely followed by the synthesis of ester and acetic anhydride. Vinegar contains a relatively small amount of acetic acid.

4.1.5.1. Vinyl Acetate Monomer

Acetic acid's main purpose is for the synthesis of vinyl acetate monomer (VAM). In 2008, it was calculated that this application accounted for approximately one-third of global acetic acid production. The process involves ethylene, acetic acid, and oxygen, catalyzed by palladium, which takes place in a gaseous state.

$$2H_3C{-}COOH + 2C_2H_4 + O_2 \rightarrow 2H_3C{-}CO{-}O{-}CH{=}CH_2 + 2H_2O$$

Vinyl acetate can be polymerized to polyvinyl acetate or other polymers, which are components in paints and adhesives.

4.1.5.2. Ester Production

The major esters of acetic acid are commonly used as solvents for inks, paints and coatings. The esters include ethyl acetate, *n*-butyl acetate, isobutyl acetate, and propyl acetate. They are typically produced by catalyzed reaction from acetic acid and the corresponding alcohol:

$$CH_3COO{-}H + HO{-}R \rightarrow CH_3COO{-}R + H_2O,\ R = \text{general alkyl group}$$

For example, acetic acid and ethanol give ethyl acetate and water.

$$CH_3COO{-}H + HO{-}CH_2CH_3 \rightarrow CH_3COO{-}CH_2CH_3 + H_2O$$

However, the majority of acetate esters are made via the Tishchenko reaction from acetaldehyde. Furthermore, ether acetates are employed as solvents for wood stains, acrylic lacquers, nitrocellulose, and varnish removers. First, ethylene oxide or propylene oxide is combined with alcohol to create glycol monoethers, which are then esterified with acetic acid. The three main products are propylene glycol monomethyl ether acetate (PMA, also known as PGMEA in semiconductor manufacturing processes, where it is used as a resist solvent), ethylene glycol monoethyl ether acetate (EEA), and ethylene glycol monobutyl ether acetate (EBA). Approximately 15% to 20% of the acetic acid produced worldwide is used in this application. It has been demonstrated that ether acetates, such as EEA, are detrimental to human reproduction.

4.1.5.3. Acetic Anhydride

Acetic anhydride is the result of two molecules of acetic acid condensing. About 25 to 30% of the acetic acid produced worldwide is used in the production of acetic anhydride, which is a significant application. At 700–750°C, the primary process involves dehydrating acetic acid to produce ketene. The anhydride is then produced by reacting ketene with acetic acid.

$$CH_3CO_2H \rightarrow CH_2{=}C{=}O + H_2O$$

$$CH_3CO_2H + CH_2{=}C{=}O \rightarrow (CH_3CO)_2O$$

One acetylation agent is acetic anhydride. As a result, cellulose acetate, a synthetic fabric also used for photographic film, is its main use. In addition, acetic anhydride is used as a reagent in the synthesis of heroin and other substances.

4.1.5.4. Use as Solvent

In order to purify organic compounds through recrystallization, acetic acid is widely utilized as a polar protic solvent. When terephthalic acid (TPA), the precursor to polyethylene terephthalate (PET), is being produced, acetic acid is utilized as a solvent. TPA was produced in 2006 using about 20% of acetic acid. Carbocation-related reactions, like Friedel-Crafts alkylation, frequently use acetic acid as a solvent. In the commercial production of synthetic camphor, for instance, acetic acid serves as a nucleophile and a solvent during the Wagner-Meerwein rearrangement of camphene to isobornyl acetate. To estimate weakly alkaline substances like organic amides, analytical chemistry uses glacial acetic acid. In this medium, the amide functions as a strong base because glacial acetic acid is a much weaker base than water. A solution of an extremely strong acid, like perchloric acid, in glacial acetic acid can then be used to titrate it.

4.1.6. Medical Use

Since the 1800s, cancer has been treated with intratumoral acetic acid injections. In many developing nations, acetic acid is part of the cervical cancer screening process. The test is positive if, after about a minute, an area of white appears on the cervix from the application of the acid. When applied as a 1% solution, acetic acid works well as an antiseptic against a variety of pathogens, including enterococci, staphylococci, pseudomonas, and streptococci. It can be applied to skin infections brought on by strains of pseudomonas that are resistant to common antibiotics. Iontophoresis, which uses diluted acetic acid, is a treatment for rotator cuff disease, however it is not well supported by evidence. It is included in the World Health Organization's List of Essential Medicines for the treatment of otitis externa.

4.1.6.1. Foods

There are 349 kcal (1,460 kJ) in 100 g of acetic acid. Acetic acid content in vinegar is usually at least 4% by mass. Jurisdiction-specific legal limits on the amount of acetic acid differ. Vinegar can be used to pickle vegetables and other foods or used directly as a condiment. Table vinegar usually has a lower acetic acid content (4–8%), whereas concentrated solutions are used in commercial food pickling. Although vinegar is the oldest and most well-known use of acetic acid, its global usage is not as great as that of industrial applications.

4.1.7. Risks, Side Effects and Interactions

The pH of acetic acid, which indicates how basic or acidic a substance is, is approximately 2.04. Despite being categorized as a weak acid, it is highly caustic and, if applied topically, can burn or irritate the skin. It is usually advised to dilute vinegar with water before consumption or topical application due to potential hazards associated with acetic acid. For example, it is recommended to dilute apple cider vinegar with one cup of water for every 1–2 tablespoons before use. Large-scale, undiluted consumption may have adverse effects, such as skin irritation, tooth enamel erosion, and a burning sensation in the throat. If you take any medications or have underlying medical conditions, consult your doctor before using any type of vinegar medicinally to avoid possible interactions. To determine your tolerance, start with a low dose and gradually increase it. If you experience negative side effects, such as low blood sugar or digestive issues, consider reducing your dosage or discontinuing use (Krapac et al. 2024; Tan 2005).

4.2. CHEMISTRY AND VARIETIES OF VINEGAR

Vinegar is a sour beverage produced by fermenting a variety of diluted alcoholic beverages to produce an acetic acid-containing liquid. A range of ingredients can be used to make vinegar, including industrial alcohol (distilled white vinegar), malted barley or oats (malt vinegar), and apples or grapes (wine or cider vinegar). Additionally, vinegars made from rice, sugars, beer, and other ingredients are available. However, vinegar was most likely first produced from wine as a commercial product (French vin, meaning “wine”; aigre, meaning “sour”). Any liquid that can be transformed into alcohol in two steps can be used to make vinegar. Yeast enzymes transform the sugar present in fruit juice or other liquids into carbon dioxide gas and alcohol. Through the action of the bacteria Acetobacter, the alcohol thus produced combines with atmospheric oxygen to form acetic acid and water. Despite its long history, vinegar production technology developed slowly, with the majority of advancements coming from improved aeration techniques. The oldest method still in use, the Orleans process, used a barrel with a capacity of approximately 50 gallons (200 L). A small amount of vinegar containing a mass of vinegar bacteria, known as the mother of vinegar, was added to the barrel along with a mash of wine or another alcoholic beverage to initiate the reaction. Aeration was made possible by drilling one or two tiny air holes above the liquid's surface. There was a wooden spigot near the bottom where the finished vinegar was extracted. To prevent disrupting the bacterial surface layer, great care was used when adding fresh raw ingredient charge to the barrel. Hermann Boerhaave, a Dutch technologist, discovered early in the 18th century that the rate at which acid was produced during the vinegar-making process was directly related to the surface area exposed to air. As a result, later techniques tried to add more air to the casks. Continuous aeration, or pumping air bubbles through the mash, was developed in the 20th century.

The main applications of vinegar are in food flavoring and in pickling—the process of preserving—fish, meat, fruit, and vegetables. Vinegar is frequently flavored with tarragon, garlic, onions, or other herbs and spices when used as a condiment. When combined with seasonings and oil, it becomes the traditional cold sauce known as vinaigrette, which is then served with cold cooked meats, fish, and vegetables as well as used as a dressing on vegetable salads. Cucumbers and other vegetables are pickled in vinegar, which is also a popular ingredient in marinades.

4.2.1. Vinegars and Other Fermented Condiments

A liquid that is safe for ingestion and has a predetermined concentration of water and acetic acid is called vinegar. It is made from raw materials that come from various agricultural sources and contain starch and sugars that go through an acetic and alcoholic double fermentation process. One of the earliest fermented foods, vinegar has a history that stretches back to approximately 2000 BC. Despite this, it was long regarded as the inferior fermented food product. Its high acetic acid content helps to preserve food, so even though it lacks nutritional value, its value in the context of healthy eating habits seems exceptional.

Alcoholic fermentation combined with acetification is a two-step process that yields vinegar and involves the bioconversion of various sources of carbohydrates. Various agricultural raw materials with an edible source of carbohydrates (fermentable sugars) can be used as a starting material. The process by which acetic acid bacteria convert ethanol into acetic acid is known as "acetification," as opposed to the more widely used term "acetous fermentation," because it is oxygen-dependent. Yeasts are the first microorganisms to convert sugar to alcohol. The most popular raw ingredients used to make vinegar are listed in Table 4.1. They come from a wide range of sources and are mostly made of plant materials, with a few exceptions like whey and honey. Thus, balsamic, cider, malt, rice, fruit, white distilled alcohol, and white and red wine are common retail vinegar varieties. Vinegar or white distilled wine is used to make flavored vinegars and herbal specialty vinegars. Garlic, basil, and tarragon are common flavorings. In order to create a sweet-and-sour flavor, fruit or fruit juice can be added to specialty fruit vinegars, which are also made from wine or white distilled. Among other flavors, vinegars with flavors like apple and raspberry are still quite popular. The range of raw materials utilized in the manufacturing of vinegar is also exceptional; it includes everything from agricultural surpluses and byproducts to premium substrates for the most rare and valuable vinegars, like aceto balsamico tradizionalle (Italy) and sherry vinegar (Spain). Nonetheless, "white" vinegar, which is vinegar made straight from diluted alcohol and used by the food industry.

Table 4.1. Aromatic Profiles of Vinegars

Compounds (mg/L)	**Treatments (Mean ± SD)**				**O.S.** †	**Sig.**
	Control	**Elderflower**	**Rosemary**	**Thyme**		
Alcohols						
1-Nonanol	0.99 ± 0.10	1.10 ± 0.16	0.95 ± 0.13	0.72 ± 0.24	2	n.s.
2,3-Butanediol	1.29 ± 1.12	1.84 ± 0.16	1.51 ± 0.64	0.98 ± 0.87	3	n.s.
2-Phenylethanol	0.76 ± 0.63	1.05 ± 0.01	0.95 ± 0.05	0.97 ± 0.05	2	n.s.
Amyl vinyl carbinol	0.03 ± 0.00 [b]	0.06 ± 0.00 [b]	1.95 ± 0.17 [a]	1.68 ± 0.25 [a]	7	***
Isoamyl alcohol	0.97 ± 0.11	1.05 ± 0.06	0.92 ± 0.04	0.91 ± 0.06	7	n.s.
alcohols	4.04	5.10	6.28	5.26		n.s.
Acids						
Isovaleric acid	0.74 ± 0.53	1.04 ± 0.08	0.88 ± 0.13	0.63 ± 0.31	7	n.s.
2-Methylbutanoic acid	0.37 ± 0.48	1.16 ± 0.18	0.91 ± 0.138	0.93 ± 0.33	7	n.s.
acids	1.11	2.20	1.79	1.56		n.s.
Esters						

2-Phenylethyl acetate	1.13 ± 0.13 [a]	0.89 ± 0.01 [b]	0.91 ± 0.04 [b]	1.08 ± 0.03 [a]	2	***
3-(Methylthio) propyl acetate	1.14 ± 0.19	0.96 ± 0.10	0.99 ± 0.09	0.86 ± 0.11	1	n.s.
Butyl acetate	1.17 ± 0.24	0.96 ± 0.08	0.96 ± 0.04	0.96 ± 0.05	3	n.s.
Diethyl succinate	0.98 ± 0.09	1.14 ± 0.03	0.89 ± 0.06	0.74 ± 0.46	3	n.s.
Ethyl heptanoate	1.09 ± 0.17	1.17 ± 0.15	0.93 ± 0.10	0.94 ± 0.07	1, 3	n.s.
Ethyl hexanoate	2.14 ± 0.32 [a]	0.75 ± 0.05 [bc]	0.96 ± 0.12 [b]	0.44 ± 0.05 [c]	3	***
Isoamyl acetate	1.22 ± 0.17	0.99 ± 0.00	0.96 ± 0.04	1.07 ± 0.13	3	n.s.
esters	8.87 [a]	6.86 [b]	6.60 [b]	6.09 [b]		**
Terpenes						
γ-Terpinen	0.037 ± 0.10 [c]	0.09 ± 0.01 [c]	2.11 ± 0.17 [a]	1.71 ± 0.11 [b]	3	***
2-Carene	0.04 ± 0.01 [c]	0.10 ± 0.01 [c]	1.95 ± 0.10 [a]	1.52 ± 0.11 [b]	3	***
Camphor	n.d.	n.d.	2.57 ± 0.11 [a]	0.06 ± 0.00 [b]	6	***
D-Carvone	n.d.	0.01 ± 0.00 [c]	1.53 ± 0.08 [a]	0.56 ± 0.03 [b]	1,5	***
Eucalyptol	0.02 ± 0.00 [b]	0.03 ± 0.00 [b]	2.10 ± 0.09 [a]	0.06 ± 0.00 [b]	1,6	***
Fenchol	0.58 ± 0.14 [b]	0.46 ± 0.02 [b]	2.29 ± 0.16 [a]	0.42 ± 0.03 [b]	3,6	***
Isoterpinolene	0.13 ± 0.04 [d]	0.50 ± 0.03 [c]	2.24 ± 0.21 [a]	0.80 ± 0.03 [b]	1	***
Linalool	0.10 ± 0.01 [c]	1.34 ± 0.06 [a]	1.27 ± 0.14 [ab]	1.05 ± 0.18 [b]	2,3,4	***
p-Cymene	0.03 ± 0.00 [c]	0.03 ± 0.00 [c]	1.06 ± 0.06 [b]	2.74 ± 0.07 [a]	1,3	***
Terpinen-4-ol	0.01 ± 0.00 [c]	0.02 ± 0.00 [c]	2.08 ± 0.18 [a]	1.35 ± 0.16 [b]	1,2,5	***

α-Terpineol	0.17 ± 0.02 ^{b}	0.29 ± 0.01 ^{b}	2.57 ± 0.19 ^{a}	0.32 ± 0.04 ^{b}	2,4	***
endo-Borneol	0.01 ± 0.00 ^{c}	0.02 ± 0.00 ^{c}	2.51 ± 0.15 ^{a}	0.68 ± 0.04 ^{b}	6	***
Linalool oxide	0.38 ± 0.09 ^{b}	2.58 ± 0.23 ^{a}	0.63 ± 0.14 ^{b}	0.55 ± 0.13 ^{b}	2	***
p-Thymol	n.d.	0.23 ± 0.01 ^{b}	0.20 ± 0.03 ^{b}	5.82 ± 0.16 ^{a}	5	***
β-Ocimene	0.13 ± 0.02 ^{c}	1.46 ± 0.04 ^{a}	1.14 ± 0.17 ^{b}	1.08 ± 0.08 ^{c}	1,4	***
Hotrienol	0.05 ± 0.00 ^{b}	4.05 ± 0.26 ^{a}	0.08 ± 0.02 ^{b}	0.10 ± 0.02 ^{b}	1,2,3	***
terpenes	1.68 ^{d}	11.21 ^{c}	26.34 ^{a}	18.82 ^{b}		***
Other						
Acetoin Acetate	0.85 ± 0.50	1.03 ± 0.01	0.90 ± 0.10	0.94 ± 0.13	3	n.s.
Anisole	0.01 ± 0.01 ^{b}	4.34 ± 0.10 ^{a}	0.02 ± 0.00 ^{b}	0.01 ± 0.01 ^{b}	5	***
Benzaldehyde	1.11 ± 0.16 ^{a}	1.17 ± 0.01 ^{a}	0.93 ± 0.07 ^{b}	0.93 ± 0.03 ^{b}	3	**
Benzeneacetal-dehyde	0.94 ± 0.27 ^{b}	1.72 ± 0.05 ^{a}	0.72 ± 0.02 ^{b}	0.71 ± 0.05 ^{b}	1	***
others	2.92 ^{b}	8.26 ^{a}	2.57 ^{b}	2.58 ^{b}		***

*† Odorant series: 1—herbaceous; 2—floral; 3—fruity; 4—sweet; 5—spicy; 6—camphor-like; 7—unpleasant. Means ± SD with different superscript letters (a, b, c, d) for each compound in the same row differ significantly ($p \leq 0.05$); n.s. non-significant; **. $p \leq 0.01$; ***. $p \leq 0.001$; n.d. not detected.*

Source: Krapac M, Major N, Plavša T, Jeromel A, Tomaz I, Poljuha D. (2024). Enrichment of white wine vinegar with aromatic plants: the impact on aromatic, polyphenolic, and sensory profiles. Applied Sciences, 14(16):6909. https://doi.org/10.3390/app14166909.

Acetic acid bacteria are the primary producers of vinegar, despite the fact that oenologists fear them for tainting wine. Acetic acid bacteria can be seen under a microscope alone, in pairs, in aggregates and chains, or as ellipsoidal or cylindrical bacteria that are either Gram-negative or Gram-variable (Figure 4.1). Although oxygen is typically used as the final electron acceptor in the aerobic respiratory metabolism of acetic acid bacteria, other compounds can also occasionally serve this function, enabling the bacteria to survive in conditions that are almost anaerobic, like those found during wine fermentation. With over 40 species, Acetobacter, Gluconacetobacter, and Komagataeibacter are the three main genera involved in acetification; however, fewer than half of these species have been reported in vinegars. The "mother of vinegar," an ill-defined concoction of various species involved in the production of vinegar, is typically used to inoculate wines with acetic acid bacteria. Many attempts have been made to obtain a single, clearly defined species of acetic acid bacteria for the production of vinegar; however, it has been determined that the most suitable inoculum to be used as an inoculum for the production of vinegar is a mixture of at least two species, especially those above 5 percent (w/v) acetic acid (one as a "starter" and the other as a "finisher," with varying acetic acid sensitivities). Apart from generating vinegar, acetic acid bacteria play a crucial role in the process of fermenting cocoa and coffee.

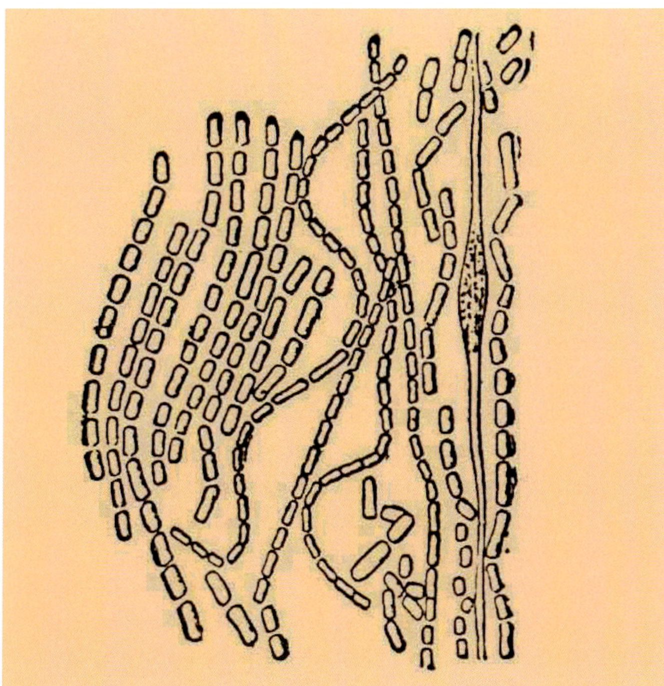

Figure 4.1. *Acetic acid bacteria.*

Source: By Ex-zee – Own work, Public Domain, https://commons.wikimedia.org/w/index.php?curid=5182784.

To make vinegar, there are two primary methods of elaboration: (i) traditional methods, which involve converting ethanol to acetic acid through a static surface culture of acetic acid bacteria placed at the interface between the liquid and air; and (ii) submerged culture methods, which use bioreactors with continuous air flow to convert alcohol to acetic acid at a much faster rate in the air-liquid interfaces of the air bubbles. A top concern for the vinegar industry is optimizing the parameters controlling the submerged process. Because they produce vinegars with exceptional organoleptic complexity, traditional methods typically yield higher-quality vinegars.

Like wine, vinegar has been used for seasoning and preservation since the majority of ancient civilizations were aware of it. Vinegar is not the same as simple wine spoilage, even though it is a naturally occurring process that occurs in wines and musts that come into contact with air. Since it is a byproduct of alcoholic fermentation, it has become increasingly popular as a salad dressing and an ingredient in sauces and other culinary products, including marinades. The Romans used it extensively as a beverage (called oxicrat, made of vinegar, water, and eggs), flavoring (the vinegar-based Apicio recipes were legendary), medicine, and cosmetics. The Greeks used it as a medicine. The processing of grapevines dates back to ancient Egypt, circa 8000 BC. Historical records list several therapeutic uses for vinegar. One such use is that it was suggested for cholera diffusion disinfection. These days, vinegar is only used in food technology and gastronomy, as well as being used as a sauce ingredient and for pickling and dressing foods. The demand for natural and organic foods that are high in nutrients has led to a rise in the production of novel and inventive products with marketable health benefits and a variety of applications that are emerging globally.

To create a dense, dark product, slow acetification of rice and other cereals (such as sorghum, wheat, and others) combined with an exceptional aging process and subsequent concentration is how black vinegars from China or "kurosu" from Japan are made. Yeasts and bacteria grow during a solid-state fermentation process without the presence of water to produce this vinegar. Higher concentrations of amino acids and organic acids are present in kurosu because it is made from unpolished rice. In Japan, small amounts of these vinegars are even accepted as health drinks. But drinking vinegar straight isn't very common. Acetic acid, which gives vinegar its strong scent and sour taste, is its primary ingredient. The acetic acid concentration of various vinegars available worldwide typically falls between four and seven acetic degrees (w/v; 4–7 g acetic acid/100 ml of vinegar). Vinegar also contains other organic acids (such as citric, malic, and lactic acids), alcohols, esters, vitamins, mineral salts, amino acids, and substances with antioxidant qualities (primarily polyphenols) (Tan, 2005; Krapac et al., 2024).

4.2.2. Varieties

The source materials for making vinegar are varied – different fruits, grains, alcoholic beverages, and other fermentable materials are used.

4.2.2.1. Fruit

Fruit wines are used to make fruit vinegars; typically, no additional flavoring is added. Apple, blackcurrant, raspberry, quince, and tomato are among the common flavors of fruit vinegar. The final product usually retains the flavors of the original fruits. In contrast to nonfruit vinegars that are flavored with fruits or have fruit infusions, fruit vinegars are primarily produced in Europe, where there is a market for expensive vinegars made exclusively from particular fruits. Asia produces a number of varieties. In South Korea, gam sikcho, or persimmon vinegar, is widely used. Wolfberry vinegar and jujube vinegar, also known as zaocu or hongzaocu, are made in China. Made from cider or apple must, apple cider vinegar has a golden-brown hue. The mother of vinegar may be present when it is sold unfiltered and unpasteurized. For consumption, it can be made sweeter (usually with honey) or diluted with fruit juice or water. A significant quantity of waste in the form of kiwifruit pomace and malformed or rejected fruit, which can make up to 30% of the crop, is a byproduct of commercial kiwifruit growing. Making kiwifruit vinegar, which has been sold commercially in China in 2008 and New Zealand since at least the early 1990s, is one use for pomace. Middle Eastern cuisines use raisins to make their vinegar. It has a mild flavor, a medium brown color, and is somewhat hazy. Date-based vinegar is a common Middle Eastern product, especially in Eastern Arabia.

4.2.2.2. Palm

Made from fermented coconut sap or coconut water, coconut vinegar is widely used in Southeast Asian cooking, particularly in the Philippines (where it's referred to as sukang tuba), as well as in some Indian and Sri Lankan dishes, particularly Goan cuisine. It is a white, hazy liquid with an especially harsh, acidic flavor that is tinged with yeast. Other varieties of vinegar are made from palm sap in the Philippines. They are byproducts of the production of tubâ (palm wine), just like coconut vinegar. Nipa palm vinegar (sukang nipa or sukang sasa) and kaong palm vinegar (sukang kaong or sukang irok) are two of the most commonly produced varieties. These are the four primary traditional vinegar varieties in the Philippines, along with coconut and cane vinegar, and they play a significant role in Filipino cooking. The sap from the nipa palm's leaf stalks is used to make nipa palm vinegar. Its flavor has citrus overtones and gives off a very musky scent. The sap from the kaong palm's flower stalks is used to make kaong palm vinegar. It is frequently used as salad dressing because it is sweeter than any other type of vinegar from the Philippines. Although it is not as common as coconut, nipa, and kaong vinegars, buri palm sap is also used to make vinegar. Although kaong palm vinegar is made in Indonesia and Malaysia as well, it is not as popular as it is in the Philippines due to the relative lack of growth in the palm wine sector in these nations with a majority of Muslims.

4.2.2.3. Balsamic

The Italian provinces of Modena and Reggio Emilia are home to the production of balsamic vinegar, an aromatic, aged vinegar. Traditional balsamic vinegar, the original product, is made from the must, or concentrated juice, of white Trebbiano grapes. The finest grades are aged in successive casks made of oak, mulberry, chestnut, cherry, juniper, and ash wood. It is dark brown, rich, sweet, and complex. Traditional balsamic vinegar, which was once an expensive product exclusively accessible to the upper classes of Italy, is aged for 12 to 25 years and is labeled tradizionale or DOC to indicate its protected designation of origin status. In the late 20th century, a less expensive commercial version known as aceto balsamico di Modena, or balsamic vinegar of Modena, gained international recognition and accessibility. It is generally produced by combining concentrated grape juice with a potent vinegar, coloring it, and then adding sugar and caramel to make it slightly sweeter. A product from grapes is used to make balsamic vinegar.

4.2.2.4. Cane

In the Philippines, sugarcane juice vinegar is a traditional and widely consumed beverage, especially in the northern Ilocos Region (referred to as sukang Iloko or sukang basi). It ranges in color from dark yellow to golden brown and tastes mild, almost like rice vinegar, but with a slightly fresher flavor. It is no sweeter than any other vinegar because it doesn't contain any leftover sugar. It is frequently referred to as sukang maasim (Tagalog for sour vinegar) in the Philippines. Ilocos cane vinegars are made in two different methods.

Other nations, such as the United States and France, also produce cane vinegar. In recent years, a white variety has gained popularity in Brazil, where it is the most affordable kind of vinegar available. To cut costs, it is now typical for different varieties of vinegar—made from rice, wine, and apple cider—to be marketed combined with cane vinegar. In some regions of northern India, sugarcane juice is used to make sugarcane sirka. People use iron nails to secure cane juice in ceramic pots during the summer. Wild yeast is responsible for the fermentation process. The cane juice is transformed into a dark brown vinegar by the addition of acetate and ferrous oxide. The sirka is used to spice curries and preserve pickles.

4.2.2.5. Grains

Malting barley, which converts the starch in the grain to maltose, produces malt vinegar from ale, also known as alegar. After that, the maltose is used to brew an ale, which is then allowed to mature into vinegar. Usually, it has a light-brown hue. Malt vinegar is a common seasoning for French fries in the UK and Canada, and it is a traditional accompaniment to fish and chips when combined with salt. Some places that sell fish and chips swap it out for non-brewed condiments. A common and traditional flavoring for crisps is a combination of salt and vinegar; in certain varieties, this requires the vinegar to be converted to sodium acetate or sodium diacetate to prevent the product from becoming too moist during manufacturing.

A mature product made from rice, wheat, millet, sorghum, or a mix of these, is called Chinese black vinegar. It tastes complex and malty with an inky black color. Certain Chinese black vinegars might have extra sugar, spices, or caramel color because the recipe isn't set in stone. The most well-known type, Zhenjiang vinegar, is from the eastern Chinese province of Jiangsu, specifically the city of Zhenjiang. Another well-liked variety of Chinese vinegar made solely from sorghum and other grains is called Shanxi mature vinegar. In Shanxi province, some old-fashioned vinegar workshops continue to make handmade vinegar that has been aged for at least five years and has a high acidity. According to the most recent national standard, only vinegars aged for at least three years and produced in Taiyuan and select Jinzhong counties are regarded as genuine Shanxi mature vinegar. Kurozu is the name given to a slightly lighter variety of black vinegar made from rice that is produced in Japan.

Southeast and East Asian cuisines are the ones where rice vinegar is most widely used. It comes in three different colors: black, red, and white (light yellow). When making salad dressings and sushi rice, the Japanese like to use a mild rice vinegar. Red yeast rice is typically used to color red rice vinegar. The most well-known country for black rice vinegar (made from black glutinous rice) is China, though it is also extensively utilized in other East Asian nations. The flavor of white rice vinegar is relatively flat and simple, with a mild acidity. Certain types of rice vinegar have additional flavorings or are sweetened in some way (Tan, 2005; Krapac et al., 2024).

4.2.2.6. Spirits

Spirit vinegar is the name given to the stronger type (5 to 24% acetic acid) that is made from sugar cane or acetic acid that is produced artificially. For a product to be referred to as spirit vinegar, it needs to be produced through double fermentation and come from an agricultural source. Alcohol ferments first to acetic acid and then to sugar in the first fermentation. In the UK, products derived from artificially produced acetic acid are permitted to refer to themselves as non-brewed condiments rather than vinegars. The making of Jerez's sherry wines is associated with sherry vinegar. It is created solely by fermenting wine with acetic acid, giving it a dark mahogany color. It's rich in aromas, with a hint of woodiness, and concentrated, making it perfect for flavoring different dishes and vinaigrettes. The most widely used vinegar in Cyprus, Israel, and Southern and Central Europe is wine vinegar, which is produced from red or white wine. The variation in quality is significant, just like with wine. Superior wine vinegars have a rich, nuanced flavor and can age in wood for up to two years. Generally speaking, wine vinegar is less acidic than white or cider vinegar. Champagne, sherry, or pinot gris are some examples of the more expensive wine varieties used to make wine vinegars.

The term "distilled vinegar," which is used in the US but is actually produced by fermenting distilled alcohol, is a bit misleading (it is known as spirit vinegar in the UK and white vinegar in Canada). A colorless solution of 5 to 8% acetic acid in water, with a pH of roughly 2.6, is obtained by diluting the fermentate. Known by different names such as white vinegar, virgin vinegar, or distilled spirit, this is used not only for cleaning, medical, and laboratory applications but also for baking, pickling, and cooking. Barley malt, or corn in the US, is the most widely used starting ingredient in some areas due to its low cost. It can occasionally be made from petroleum. Though it's also used as a substitute for brown or light malt vinegar in the UK, distilled vinegar is mostly used in cooking. Cleaning can also be done with white distilled vinegar; in fact, some varieties are sold expressly for this use.

4.2.3. Culinary Uses

Vinegar is a common ingredient in food preparation, especially in salad dressings, vinaigrettes, and pickling liquids. It is a component of sauces like mayonnaise, ketchup, mustard, and hot sauce. Chutneys are sometimes made with vinegar. It is frequently used as a condiment either by itself or in combination with other condiments. Marinades frequently include vinegar. As with hot and sour soup, vinegar is occasionally added to soups. Because vinegar is acidic, it keeps for a very long time without needing to be refrigerated because it is practically spoiled at this point.

4.2.3.1. Beverages

Vinegar is used to make a variety of drinks, such as posca from ancient Rome. Sekanjabin, a traditional Persian beverage, is akin to oxymel, an ancient Greek beverage made with vinegar and honey. Other preparations, referred to as shrubs in colloquial terms, include combining small amounts of fruity vinegar with sugar water or honey water, or preparing syrup by soaking fruit or mint in vinegar for several days, then straining off the solid parts and adding a significant amount of sugar. Some people would rather boil the shrub as a last resort. The popularity of these recipes has significantly decreased as carbonated drinks, like soft drinks, have become more popular.

4.2.3.2. Diet and Metabolism

Preliminary research indicates that consuming 2–4 tablespoons of vinegar may cause small reductions in post-meal levels of blood glucose and insulin in people with diabetes.

4.2.3.3. Nutrition

Red wine or distilled vinegar contains 95% water and no fat or protein. Distilled vinegar provides 75 kJ (18 kcal) of food energy in a 100 mL (3+1⁄2 US fl oz) reference amount, with no micronutrients in significant content. Red wine vinegar and apple cider vinegar have the same composition (and lack of nutritional content), while balsamic vinegar is made up of 77% water and 17% carbohydrates. It has a calorie count of 370 kJ (88 kcal) per 100 mL and no protein, fat, or micronutrients.

4.2.4. Non-Culinary Uses

4.2.4.1. Folk Medicine

Vinegar has been used in folk medicine since ancient times, but there isn't any solid clinical research to back up claims that it can treat diabetes, reduce weight, prevent cancer, or act as a probiotic. It may assist type 2 diabetics in lowering insulin and glucose after meals, according to a meta-analysis and systematic review that was conducted later. Common jellyfish stings can be treated with vinegar to deactivate the nematocysts, though the results are not as strong as with hot water. This also holds true for the Portuguese man-o-war, which is actually a siphonophore rather than a jellyfish despite popular belief. Certain vinegar treatments carry health risks. There have been reports of esophageal injury caused by apple cider vinegar, and since vinegar products sold for medicinal purposes are neither standardized nor regulated, their acidity and content can vary greatly.

4.2.4.2. Cleaning

White vinegar is a common cleaning solution for homes. For the majority of uses, diluting with water is advised to ensure safety and prevent surface damage while cleaning. It can remove mineral deposits from glass, coffee makers, and other smooth surfaces due to its acidic nature. It's well known that vinegar works well for cleaning glass and stainless steel. In the United Kingdom, a traditional and enduring method for cleaning grease-smeared windows and mirrors is to sprinkle malt vinegar onto crumpled newspaper. Silver, bronze, brass, and copper can all be polished with vinegar. It works wonders as a solvent to remove adhesive from price tags and epoxy resin. It's said to be a powerful drain cleaner. When vinegar is used in dishwashers and washing machines, it can harm the rubber seals and hoses, which can result in leaks. Consumer Reports' testing indicates that vinegar is useless in a dishwasher for either removing hard-water film or as a rinse aid. The chief spokesman for the American Cleaning Institute, Brian Sansoni, claims that vinegar isn't very helpful for stains, such as food and bloodstains, that have already penetrated clothing. Additional surfaces and objects in the home that vinegar can harm include water tanks, flooring, stone countertops, knives, screens on electronics, clothes irons, and rubber parts of different small appliances. Commonly used in small appliances, lower-grade stainless steel, copper, and aluminum are among the common metals that can be harmed by vinegar.

4.2.4.3. Herbicide

20% acetic acid vinegar can be used as an herbicide, but acetic acid is not absorbed into root systems so the vinegar will only kill the top growth and perennial plants may reshoot.

4.2.5. Reactions, Byproducts and Contaminants

The majority of commercial vinegar solutions that are sold to consumers for use in their homes don't contain more than 5%. Solutions stronger than 10% should be handled carefully since they are caustic and harmful to skin. Mother of vinegar can appear when a vinegar bottle is opened. It can be eliminated by filtering and is thought to be harmless. Certain types of vinegar may contain vinegar eels (*Turbatrix aceti*), a type of nematode, unless the vinegar is kept covered. These can be found in naturally fermenting vinegar and feed on the mother of vinegar. The carbon cycle is completed when vinegar and baking soda are mixed because the bicarbonate ion in the baking soda reacts with the acetic acid to form carbonic acid, which then breaks down into carbon dioxide and water. With the vinegar's water content, sodium acetate stays in solution.

4.2.6. Health Benefits

- **Kills Bacteria:** Because vinegar contains acetic acid, it has been utilized for a very long time as a natural disinfectant. Strong antibacterial qualities allow acetic acid to effectively eradicate a number of particular bacterial strains. In fact, an in vitro study conducted in 2014 discovered that acetic acid could stop the growth of Mycobacteria, the genus of bacteria that causes leprosy and tuberculosis. Because vinegar contains acetic acid, it may also inhibit the growth of bacteria, according to other research.
- **Reduces Blood Pressure:** High blood pressure is a major risk factor for heart disease in addition to putting additional strain on the heart muscle and causing it to gradually weaken over time. Apart from dietary and exercise modifications, encouraging studies have indicated that acetic acid may also aid in blood pressure regulation.
- **Decreases Inflammation:** The immune system relies heavily on acute inflammation to protect the body from disease and infection. Long-term maintenance of high levels of inflammation, however, can be harmful to health because research indicates that inflammation may play a role in the emergence of chronic illnesses like cancer and heart disease. It is believed that acetic acid lowers inflammation, which aids in illness prevention. After giving mice acetic acid for 10 weeks, numerous inflammatory markers were significantly reduced. Another 2019 study discovered that by altering the activity of a particular immunity-related protein, acetic acid helped reduce the inflammatory response in mice.
- **Supports Weight Loss:** Some research suggests that acetic acid could help support weight control by aiding in weight loss. For example, a 2017 report out of Malaysia noted that acetic acid vinegar reduced food intake and body weight in mice.
- **Promotes Blood Sugar Control: The** benefits of apple cider vinegar for supporting blood sugar regulation have been extensively researched. One of the main ingredients in apple cider vinegar, acetic acid, has been linked in studies to some of the vinegar's potent blood sugar-lowering effects. Because vinegar with acetic acid slows down the stomach's emptying, it has been shown in one

study to lower blood sugar and insulin levels when consumed with high-carb meals. Similar results were obtained from another in vitro study, which found that acetic acid inhibited the activity of multiple enzymes involved in the metabolism of carbohydrates, potentially reducing the absorption of carbohydrates and sugar in the small intestine (Tan, 2005; Krapac et al., 2024).

4.3. TECHNOLOGY OF VINEGAR PRODUCTION

A liquid condiment that has undergone fermentation, vinegar has a distinct flavor and scent. The amount of acetic acid in it varies from 4 to 18%, depending on how vinegar is ultimately used. The main flavoring and antimicrobial ingredient in vinegar is acetic acid. The acetic acid content of table vinegar, which is consumed unaltered, is only 4 percent, while pickling vinegar can have up to 18% acetic acid. Vinegar is mostly used in pickles, sauces, salad dressings, tomato products, and other processed foods for flavoring and preservation. The predominant flavor of vinegar is typically characterized as tart, sour, or mildly acidic. There are two varieties of vinegar. e. either brewed and synthetic or natural. The process of making natural or brewed vinegar involves fermenting an alcohol-containing solution (just acetic acid fermentation) or any appropriate medium that contains sugar or starch, such as fruits, malt, molasses, jaggary, etc. It shouldn't include any acetic acid at all, as fermentation does not produce it. It may also contain acids that are produced later on during fermentation and aging, such as citric and tartaric acids, in addition to acetic acid. Spices and caramel can be added or left out to add flavor and color, respectively.

It should comply with the following standards as laid down by FSSR-2011:

- Acidity (m/v) as acetic acid should not be less than 3.75%.
- Total solids should not be less than 1.5%.
- Total ash content should not be less than 0.18%.
- Sulphuric acid or other mineral acid, foreign substances and colorants other than caramel should not be present.
- It should be packed in rigid containers, not containing less than 90% of the water capacity of the container at 20°C.

Conversely, fermentation is not a step in the production of synthetic vinegar. To achieve the required grain strength, acetic acid is diluted (1 percent acetic acid equals 10 grain strength, which is a measure of percent acetic acid in the product). The same standards as natural vinegar must also be met, but there are no restrictions on the amount of total solids or total ash content. It needs to be clearly marked "Synthetic-prepared from Acetic acid."

4.3.1. Source

Any material containing at least 10% fermentable sugar and starch is suitable for the production of vinegar. Potential sources for brewed or natural vinegar production fall under the following categories:

- **Fruits:** Fruit juice is a simple way to obtain the fermentable sugar found in fruits. It is also possible to raise the sugar or total solids content by adding more sugar to the fortification. Apples (apple cider vinegar), grapes (wine or grape vinegar), oranges, jamuns, peaches, pears, pineapples, apricots, berries, coconuts, dates, and bananas are among the fruits that are frequently used to make vinegar. It can also make use of byproducts from the fruit processing industry and imperfect fruits that are rejected during grading or processing.
- **Vegetables:** Starch content is higher in vegetables than in sugars. Diastase must be used to hydrolyze or *Saccharomyces cerevisiae* arify the starch for it to readily ferment into alcohol and then acetic acid. Purple sweet potato vinegar is one example of the long-standing use of tuber-like potatoes in the production of potato vinegar.
- **Cereals:** Malt from cereals aids in the production of vinegar. One important source is malted barley, which can be used either whole or in combination with other cereals (malted or unmalted). Malt supplies the hydrolyzed or *Saccharomyces cerevisiae* arified starch found in cereals, which microorganisms can easily use for fermentation. Rice vinegar made from both red and white rice is one type of vinegar derived from non-malted cereal. A product made from cereals such as wheat, rice, millet, sorghum, or an appropriate combination of all these grains is known as East Asian black vinegar, or Chinese black vinegar. It has a deep, malty flavor and is inky black in color.
- **Grasses:** It includes sugarcane which is used for the preparation of cane vinegar.
- **Other Sources:**
  - **Molasses:** It is a thick, dark-colored byproduct of the process used to make sugar from beetroot, grapes, or sugarcane. It can also be utilized as a source of leftover sugars to make vinegar from molasses. Before it can be used for fermentation, it must undergo pretreatment, which entails neutralization with food-grade acid (typically citric acid) and dilution (16% TSS).
  - H**oney:** Low grade honey is usually used for honey vinegar.

- **Beer:** Alcohol present in beer can also be used for the formation of vinegar. It is light golden in color and has malty and very sharp taste.
- **Wine:** White, red and sherry wine can also be used as a raw material for vinegar production.
- **Raisins:** Most raisin vinegar is made and consumed in the Middle East. Its features include a slight flavor, brown color, and cloudiness. Another significant type of vinegar is flavored and spirit vinegar.

4.3.2. Spirit Vinegar

Grain vinegar, distilled malt vinegar, distilled vinegar, white vinegar, and alcohol vinegar are some other names for it. Before the fermentation of acetic acid, there is a further step of distilling the alcoholic ferment. Spirit vinegar is created by fermenting acetic acid with the distillate, or distilled alcohol, that has been obtained thus far. The majority of Manor Vinegar's production, which is made in state-of-the-art acetators, is spirit vinegar. Contact us right now if you'd like more information about the truly amazing process. Spirit vinegar is mostly used as an ingredient in food because it is made from pure ethanol. Spirit vinegar is also used as a less expensive substitute for malt vinegar by adding malt extract to color the liquid. This kind of vinegar is used in the food preparation processes of most manufacturers of sauces and condiments. Although it is typically supplied in large bulk tankers, spirit vinegar can also be found in smaller IBC containers upon request. As part of our sustainability program, we prefer to offer a stronger vinegar that travels fewer miles on the road and has very competitive prices, even though the strength of the vinegar ranges from 20 to 5%.

4.3.3. Flavored Vinegar

Flavored vinegar accounts for over 90% of the market share for vinegar. It is created by completely allowing the flavor of the flavoring substance to emerge after it is steeped in regular vinegar for a certain amount of time.

- **Balsamic Vinegar:** It is an Italian specialty that is renowned for its distinct flavors. It is made from wine grapes and gets its flavor from aromatic woods such as mulberry, durmast, chestnut, and cherry. Longevity aids in the development of its distinct and nuanced flavor.
- Spices and Condiments like garlic, whole chilies and pepper are used to impart flavor to vinegar.
- **Herbs:** Mostly Mediterranean herbs like thyme, tarragon, oregano

and other herbs like aloe vera and basil are used as flavoring substance.

- **Sinamik:** It is the spiced version of cane or coconut vinegar and flavors are obtained from blend of Thai chilli, onions and garlic.
- **Fruits:** Blood orange and pear are two common fruit flavors. Vinegar is also infused with whole raspberries, blueberries or figs.
- **Sweetened Vinegar:** Rice wine vinegar can be combined with sugar and herbs like ginger, cloves and other spices.

4.3.4. Organic Vinegar

It consists of an assortment of vinegars made from local fruits or organically produced fruit juices obtained from farmers who have earned their organic certification. These fruits impart to vinegar a distinct flavor characteristic to that fruit. But because the fruit varies every year, there isn't consistency in the flavor of organic vinegar.

4.3.5. Mechanization of Vinegar Production

A starting material that contains starch and has the ability to ferment both acetic acid and ethanol. Vinegar is most likely used in many different product development applications. However, the initial step of production involves inoculating fresh material with yeast to convert sugar to alcohol. The remaining alcohol undergoes a reaction with acetic acid when acetic acid bacteria (AAB) are added, leading to the completion of fermentation. Despite the fact that there are many methods for producing vinegar, only two are most likely used in the industry (Figure 4.2).

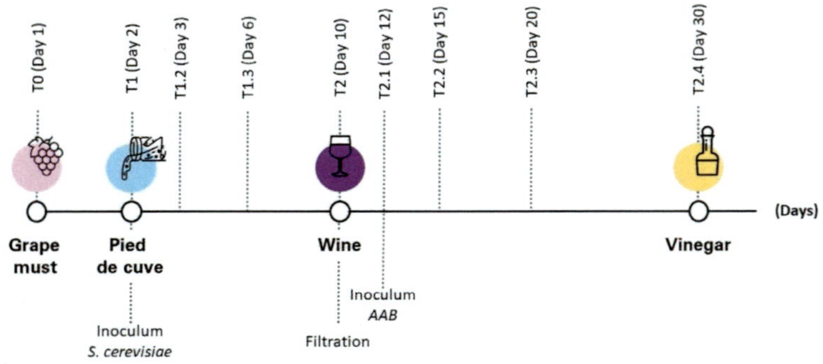

Figure 4.2. *Process of vinegar production.*

Source: Genzardi D, Greco G, Núñez-Carmona E, Sberveglieri V. (2022). Real time monitoring of wine vinegar supply chain through MOX sensors. Sensors, 22(16):6247. https://doi.org/10.3390/s22166247.

Tesfaye et al. (2002) suggested that a large number of contemporary pieces of equipment be brought to the market to improve both production and the time interval at the commercial level. As acetic acid bacteria (AAB) are added to this system, the rate at which ethanol is converted to acetic acid actually accelerates. One of the first two is the "Surface method," which uses wood shavings in the fermentation process. The culture of AAB survives and causes the rate of fermentation to accelerate (supplies oxygen by the surface). The second approach is known as "submerged fermentation," and its unique feature is a steady oxygen supply that controls the rate of large-scale commercial production.

4.3.6. Manufacturing

General methods of vinegar production involve the following steps:

- **Raw Material:** A suitable source has alcohol, starch, or sugar (at least 10%). For substrates containing starch to become fermentable sugar, diastase must hydrolyze them. On the other hand, acetic acid fermentation can be applied directly to substrates that contain alcohol. Fruit serves as a substrate for anaerobic fermentation by yeast, providing nutrients and mineral salts in addition to sugars.
- Anaerobic fermentation by Yeast:

$$C_6H_{12}O_6 \xrightarrow{\text{yeast}} 2C_2H_5OH + 2CO_2$$

The three yeasts that are most frequently used for this purpose are *Saccharomyces cerevisiae, Saccharomyces* aromuces *malei*, and *Saccharomyces aromyces ellipsoideus*. After pasteurization, the procedure consists of inoculating a pure strain of desirable yeast and eliminating any remaining undesirable microbes. Keep the temperature between 25 and 27°C to promote yeast growth. In an oxygen-free environment, yeast performs anaerobic fermentation to create alcohol in about three weeks from available sugars. After 1–2 weeks of sedimentation, the alcohol ferment is strained into a clean container to begin the acetification process.

- Aerobic fermentation by Acetobacter species:

$$C_2H_5OH + O_2 \xrightarrow{\text{Acetic acid bacteria}} CH_3COOH + H_2O$$

The most crucial stage in the production of vinegar is this one, which entails turning the alcohol in the substrate into acetic acid. Fermentation is carried out in a dark, 7–8 percent alcoholic solution at 21–27°C because acetic acid bacteria are less tolerant of light and concentrated alcoholic solutions. To start and encourage fermentation, mother vinegar—a slimy, non-toxic, unfiltered, and unpasteurized vinegar component primarily made of soluble cellulose and acetic acid bacteria—is added in a 1:10 ratio to alcoholic ferments.

- **Yield:** This fermentation process can yield acetic acid as the primary product, up to 40%. Sugar metabolites are among the other products that are produced; these are either volatilized away or change into other compounds as we age.
- **Separation of Vinegar (Acetic acid Solution):** After maximum amount of acetic acid is reached, it is separated leaving mother vinegar at the bottom.
- **Aging:** The resulting vinegar is then kept for a while longer to settle suspended particles and create flavor compounds by combining its chemical components. To obtain clear vinegar with a distinct and nuanced flavor, aging is desired. Typically, the time frame is four to eight months.
- **Filtration and Clarification:** Clear vinegar is separated or clarified through filtration.
- **Distillation:** Carried out to obtain higher amounts of acetic acid (%) in the solution.
- **Color:** Caramel color is added in certain types of vinegar to give it desired color.
- **Pasteurization:** Prepared vinegar is then pasteurized in sterilized bottles at 71–77°C for 15–20 min to stop fermentation and to maintain the strength of vinegar.
- Packaged and Storage.

Three factors influence vinegar quality: manufacturing process, type of raw material, and ageing period.

The aforementioned general process has been modified to include the following steps:

Vinegar is produced using the traditional method, also known as the Orleans method, which involves slow fermentation that typically takes three to six months. By adjusting the rate of exposure to oxygen, the rate of fermentation can be changed. This serves as the cornerstone for changing how vinegar is produced. Hence, a quicker method of making vinegar involves improving oxygen exposure through submerged acetification and trickling generator. The idea is that an increased oxygen supply will promote the growth of acetobacter, which will accelerate the production of acetic acid and, as a result, shorten the time it takes for vinegar to form.

4.3.6.1. Slow Process: Orleans Process

This is a traditional French method where our acetic acid fermentation is carried out in wooden barrels. Acetification takes place for one to three months at 21 to 29°C. Following fermentation, 25–35% of the vinegar is removed, and the same volume of alcoholic liquid is then added.

4.3.6.2. Fast or Quick Process

- **Generator Process:** Created in 1823 by Schutzenbatch. A tank filled with wood shavings is allowed to contain alcoholic ferment while air is circulated in the other direction. This process is repeated until the appropriate volume of acetic acid is generated. That takes three to seven days. Upon completion, 75% of the vinegar is extracted, and an equivalent volume of alcohol is introduced into the system to initiate a fresh batch. CFTRI has created a quick "vinegar generator" process for making coconut vinegar. A vinegar generator is an easy-to-assemble, low-maintenance device with a good yield and no power supply requirements. Other raw materials that can be utilized with it include fruit peels, sugarcane juice, jaggary, and pineapple waste. Fresh coconut water that has matured and been fortified with 10% sugar is used to make coconut vinegar. After that, *Saccharomyces cerevisiae* is added to it to initiate a 4- to 5-day alcoholic fermentation process. After separation, the clear liquid is run through a vinegar generator and inoculated with mother vinegar, which contains acetobacter bacteria. The components of a vinegar generator include a feed vat that receives raw materials for alcohol fermentation, an acidifier that contains "corn cobs," a porous material that is used to oxidize or acetify incoming alcohol ferment into vinegar that contains acetic acid, and a receiving vat that collects partially acetified vinegar and recycles it through the generator until the acetic acid content reaches 4%. The finished product is aged prior to being packaged and consumed.
- **Submerged Acetification:** reported by Hromatka. It is the most widely used technique for making vinegar commercially and yields a larger amount of acetic acid per batch. Using a venturi or turbine system, air is forced through the system. The constant aeration used in this technique speeds up the rate at which oxygen is exposed, which speeds up fermentation.

4.3.7. Uses

In addition to its culinary applications, vinegar has numerous industrial, cleaning, medical, and hygienic uses. Certain medical conditions such as scurvy, cholera, diphtheria, high fever, dysentery, urinary infections, tonsillitis, hoarseness, external inflammations, joint injuries, apathy, obesity, hay fever, asthma, rashes, food poisoning, heartburn, poor vision, brittle nails, and bad breath have all demonstrated clear benefits from curcumin. Backaches, bursitis, headaches, leg cramps, muscle sprains, coughs, respiratory congestion, sinus infections, sore throats, and sunburns have all been shown to benefit from using apple cider vinegar (Figure 4.3).

Figure 4.3. *Use of vinegar.*

Source: Krapac M, Major N, Plavša T, Jeromel A, Tomaz I, Poljuha D. (2024). Enrichment of white wine vinegar with aromatic plants: the impact on aromatic, polyphenolic, and sensory profiles. Applied Sciences, 14(16):6909. https://doi.org/10.3390/app14166909.

4.3.8. Aromatic Characters of Vinegar and Quality Evaluation

The wide range of values that make up vinegar's aromatic characteristics come from both physicochemical and sensory factors. Both sensory analysis and differential analytical attributes are estimated for the purpose of evaluating the quality of vinegar. Aromatic compounds contribute to the high quality of vinegar. Numerous evaluation techniques are included in the process of determining the vinegar's quality, i.e., vinegar, which is made by submerged fermentation in which wooden carks are added to the fermentation medium, is subjected to principal component analysis, cluster analysis (CA), linier discriminant analysis (LDA), and other analyses. To assess methanol, 1-propanol, ethyl propionate, 3-methyl-1-butanol, 2-methyl-1-butanol, acetoin, praline, and total acidity-oxidation gas chromatography (GS) and high-performance liquid chromatography (HPLC) are employed. There are five different types of organic acids in wine vinegar. White vinegar contains roughly 14 volatile compounds, whether they are related to aging or not. The most likely compounds found in both wine and distilled vinegar are acetic acid and ethyl acetate, respectively.

4.3.9. Security of Vinegar

People have been using vinegar as an ingredient or component of food for thousands of years because of these and other reasons, which make vinegar use generally safe. There have only been a few documented cases of vinegar's required negative effects on human health, primarily the gastrointestinal system. In one instance, it was discovered that a 39-year-old woman who consumed 1 tsp of rice vinegar experienced class B damage to her esophagus and heart as well as oropharyngeal inflammation. Although vinegar consumption rarely results in injuries, it does serve as a warning to other consumers because inflammation in the esophageal region lowers the risk of developing cancer. However, after looking through previous studies, researchers also discovered that vinegar can be used in a different way to treat esophageal cancer. Inadvertent aspiration or consumption of vinegar can result in laryngospasm and subsequent vasovagal syncope, which can be spontaneously diagnosed. For the past six years, a 28-year-old woman has been regularly consuming 250 mL of apple cider vinegar, which has led to hypokalemia. These days, a lot of research has shown that vinegar may be able to treat a wide range of illnesses. However, before ingesting vinegar, make sure you follow the recommended dosage and research its historical use (Tan, 2005; Krapac et al., 2024).

REFERENCES

1. Adebayo-Oyetoro, A. O., Adenubi, E., Ogundipe, O. O., Bankole, B. O., & Adeyeye, S. A. O. (2017). Production and quality evaluation of vinegar from mango. *Cogent Food and Agriculture*, *3*(1), 1278193. https://doi.org/10.1080/23311932.2017.1278193.
2. Bjornsdottir, K., Breidit, J. F., & McFeeters, R. F. (2006). Protective effect of organic acids on survival of *Escherichia coli* O157 in acidic environments. *Applied and Environmental Microbiology*, *72*, 660–664. https://doi.org/10.1128/AEM.72.1.660-664.2006.
3. Budak, H. B., & Guzel-Seydim, Z. B. (2010). Antioxidant activity and phenolic content of wine vinegars produced by two different techniques. *Journal of the Science of Food and Agriculture*, *90*, 2021–2026. https://doi.org/10.1002/jsfa.4030.
4. Chou, C. H., Liu, C. W., Yang, D. J., Wu, Y. H. S., & Chen, Y. C. (2015). Amino acid, mineral, and polyphenolic profiles of black vinegar, and its lipid lowering and antioxidant effects *in vivo*. *Food Chemistry*, *168*, 63–69. https://doi.org/10.1016/j.foodchem.2014.07.039.
5. Conzuelo-Quifada, A. E., Rodriguez-Cuevas, S. A., & Labastida-Almendaro, S. (2003). Treatment of large lower genital tract condylomata acuminate with local excision plus topical acetic acid: A preliminary study. *Journal of Reproductive Medicine*, *48*, 506–508.
6. Derakhshandeh-Rishehri, S. M., Heidari-Beni, M., Feizi, A., Askari, G. R., & Entezari, M. H. (2014). Effect of honey vinegar syrup on blood sugar and lipid profile in healthy subjects. *International Journal of Preventive Medicine*, *5*(12), 1608.
7. Fernández-Pérez, R., Torres, C., Sanz, S., & Ruiz-Larrea, F. (2010). Strain typing of acetic acid bacteria responsible for vinegar production by the submerged elaboration method. *Food Microbiology*, *27*, 973–978. https://doi.org/10.1016/j.fm.2010.05.003.
8. Gullo, M., De Vero, L., & Giudici, P. (2009). Succession of selected strains of *Acetobacter pasteurianus* and other acetic acid bacteria in traditional balsamic vinegar. *Applied and Environmental Microbiology*, *75*(8), 2585–2589. https://doi.org/10.1128/AEM.02794-08.
9. Haruta, S., Ueno, S., Egawa, I., Hashiguchi, K., Fujii, A., Nagano, M., Ishii, M., & Igarashi, Y. (2006). Succession of bacterial and fungal communities during a traditional pot fermentation of rice vinegar assessed by PCR-mediated denaturing gradient gel electrophoresis. *International Journal of Food Microbiology*, *109*, 79–89. https://doi.org/10.1016/j.ijfoodmicro.2005.12.016.

10. Krapac, M., Major, N., Plavša, T., Jeromel, A., Tomaz, I., & Poljuha, D. (2024). Enrichment of white wine vinegar with aromatic plants: The impact on aromatic, polyphenolic, and sensory profiles. *Applied Sciences*, *14*(16), 6909. https://doi.org/10.3390/app14166909.
11. Srivastava, R. P., & Kumar, S. (2006). Vinegar. In *Fruit and Vegetable Preservation: Principles and Practices* (pp. 207–213). International Book Distributing Company.
12. Tan, S. C. (2005). *Vinegar Fermentation* (M.S. thesis, Louisiana State University, Agricultural and Mechanical College at Lafayette).
13. Tesfaye, W., Garcia-Parrilla, M. C., & Troncoso, A. M. (2002). Sensory evaluation of Sherry wine vinegar. *Journal of Sensory Studies*, *17*, 133–144. https://doi.org/10.1111/j.1745-459X.2002.tb00362.x.

Chapter 5

Production of Nucleotides by Micro-Organisms

Contents

INTRODUCTION

Organic compounds called nucleotides are made up of a phosphate, a pentose sugar, and a nitrogenous base. They function as the monomeric building blocks of the nucleic acid polymers, ribonucleic acid (RNA) and deoxyribonucleic acid (DNA), which are both vital biomolecules found in all Earthly life forms. In addition to being produced by the liver from common nutrients, nucleotides can also be acquired through diet. Three subunit molecules make up nucleotides: a phosphate group made up of one to three phosphates, a five-carbon sugar (ribose or deoxyribose), and a nucleobase. In DNA, uracil is used in place of thymine, while in RNA, the four nucleobases are guanine, adenine, cytosine, and thymine.

Additionally, nucleotides are essential for basic cellular metabolism. They supply the nucleoside triphosphates, adenosine triphosphate (ATP), guanosine triphosphate (GTP), cytidine triphosphate (CTP), and uridine triphosphate (UTP) necessary for the various energy-demanding cellular processes such as the synthesis of amino acids, proteins, and cell membranes, intracellular and intercellular movement of the cell, cell division, etc. Moreover, nucleotides are integrated into significant cofactors of enzymatic reactions (e.g., cyclic guanosine monophosphate, or cGMP, and cyclic adenosine monophosphate, or cAMP) and participate in cell signaling. g. NAD, FMN, coenzyme A, FAD, and NADP+). Radionucleotides can be produced in experimental biochemistry by radiolabeling nucleotides with radionuclides. Moreover, 5-nucleotides are added to food as flavor enhancers to improve the umami taste; these are frequently yeast extracts.

5.1. BIOSYNTHESIS OF PURINE 5'-RIBONUCLEOTIDES

While nucleotides can be used to prepare synthetic oligo- and polynucleotides and exhibit a variety of pharmacological effects of potential medical and nutritional interest, their ability to improve food flavor is the main driver of interest in their industrial production (Kuninaka et al., 1974; Margalith, 1964). Three purine ribonucleoside 5'-monophosphates have the ability to enhance flavor: guanylic acid (GMP), inosinic acid (IMP), and xanthylic acid (XMP). These compounds are listed in decreasing order of potency. Adenylic acid (AMP), the other purine 5'-ribonucleotide found naturally, does not show flavor potentiation. Nucleosides, pyrimidine nucleotides, free purine bases, and 2' and 3' isomers are also inactive. Therefore, the only active forms are 6-hydroxy purine ribonucleoside 5'-monophosphates and their corresponding deoxyribonucleotides. There is an asynergistic flavor enlarging effect when combined with monosodium glutamate (MSG). Nucleotides are several orders of magnitude more potent than MSG on a weight basis.

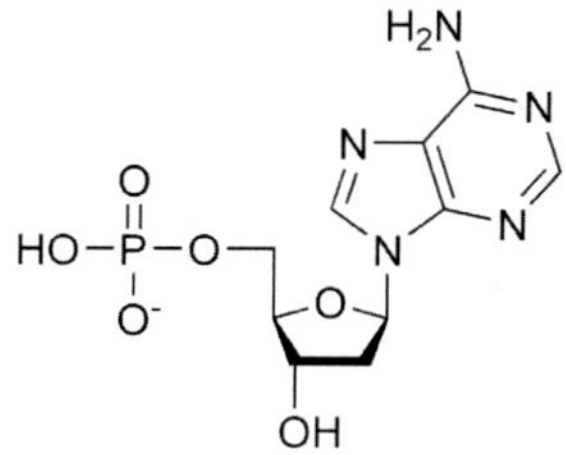

Source: Public Domain, https://commons.wikimedia.org/w/index.php?curid=106155.

A review of the fermentation process for producing purine 5'-ribonucleotides was conducted a few years ago, but a lot of new information has since been published. Similar to the previous advancements, research in this field is still primarily the work of the Japanese. The synthesis of pyrimidine ribonucleotides is facilitated by the significant economic significance of purine 5'-ribonucleotides. However, it should be noted that knowledge of pyrimidine 5'-ribonucleotide biosynthesis and regulation is extensive. Moreover, descriptions of dysregulated mutants and biosynthesis pathways have been made. Bioconversions that result in the production of sugar nucleotides are likewise outside the purview of this review.

5.1.1. Purine Biosynthesis

Purine biosynthesis occurs in the cytosol of all cells. The purine ring is built up in a series of 11 enzyme catalyzed steps. Each enzyme is oligomeric, which means it contains several monomers. Intermediate products that are produced during the reaction are not released.

5.1.1.1. Contextualizing Purines

A review of the fermentation process for producing purine 5'-ribonucleotides was conducted a few years ago, but a lot of new information has since been published. Similar to the previous advancements, research in this field is still primarily the work of the Japanese. The synthesis of pyrimidine ribonucleotides is facilitated by the significant economic significance of purine 5'-ribonucleotides. However, it should be noted that knowledge of pyrimidine 5'-ribonucleotide biosynthesis and regulation is extensive. Moreover, descriptions of dysregulated mutants and biosynthesis pathways have been made. Bioconversions that result in the production of sugar nucleotides are likewise outside the purview of this review.

These perform a wide range of biochemical functions that include:

1. Driving thermodynamically unfavorable reactions
2. Forming the central cofactors of metabolism (such as NAD+ and FAD+)
3. Forming the building blocks of our genetic blueprint, DNA (Mendz et al., 1994; Mendz et al., 1996; Mendz et al., 1997).

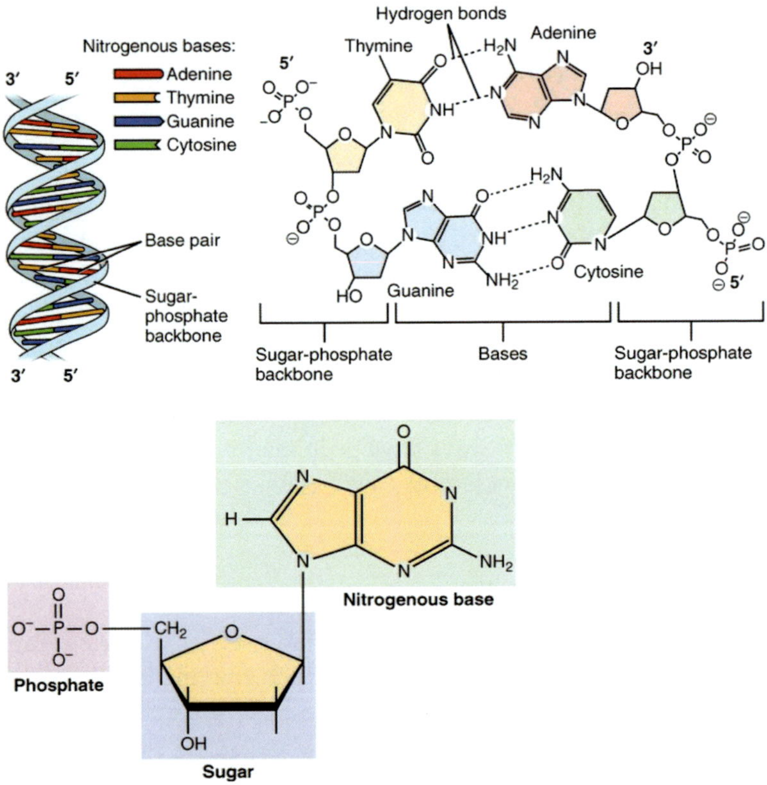

Figure 5.1. *The structure of nucleotides depicting how the base and pentose sugar (nucleoside, in yellow blue and green) may be attached to either one, two or three phosphate groups. A nucleoside attached to one phosphate (red) is a nucleoside monophosphate.*

Source: By OpenStax – https://cnx.org/contents/FPtK1zmh@8.25:fEI3C8Ot@10/Preface, CC BY 4.0, https://commons.wikimedia.org/w/index.php?curid=30131206.

The addition of e second phosphate group (red) forms a nucleoside diphosphate, and finally, the addition of a third phosphate forms a nucleoside triphosphate. Where the phosphate group proximal to the nucleoside the site of the phosphate-ester bond.

5.1.1.2. Structure of Purines

Purine biosynthesis is complex. The purine skeleton is a 6-membered pyrimidine ring fused to a 5-membered imidazole ring (*See* Figure 5.1). Each ring contains two nitrogen (N) atoms, with the remaining 5 positions in each ring occupied by carbon (C), which is attached to a Hydrogen (H).

- Hydrogen can be replaced by different atoms or groups to form distinct purines. The 4 Ns originate from different amino acids and the remaining 5 Cs derive from one-carbon containing groups.

This was discovered in 1948 by John Buchanan who fed pigeons isotopically labelled compounds to determine the positions of the labelled atoms in the uric acid they secreted. The name of the compound that provides each of the C and N atoms are labelled in Figure 5.1.

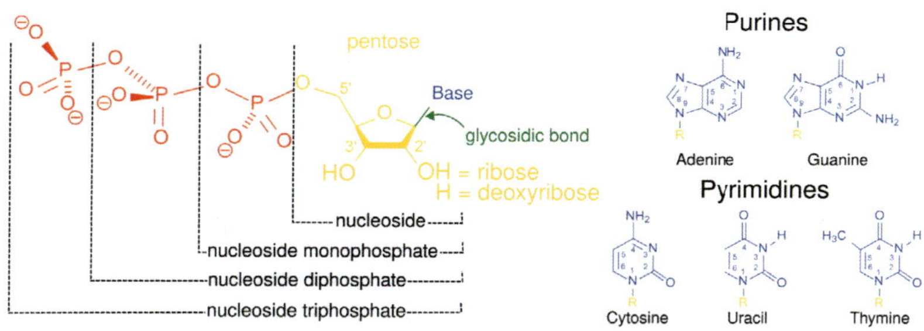

Figure 5.2. *The results of John Buchanan's studies demonstrated that N1 of purines arises from the amino group of aspartate; C2 and C8 originate from a C1 containing compound called formate; N3 and N9 are contributed by the amide (NH_2) group of glutamine; C4, C5, and N7 descend from Glycine and C6 comes from HCO_3^-.*

Source: By Boris (PNG), SVG by Sjef – en: Image: Nucleotides.png, Public Domain, https://commons.wikimedia.org/w/index.php?curid=2933261.

5.1.2. Synthesis of Purine Ribonucleotides

Every cell's cytosol is where purine biosynthesis takes places. 11 stages with the assistance of an enzyme are required to form the purine ring. Since each enzyme is oligomeric, it consists of multiple monomers. Reaction intermediate products are produced but are not released. Rather, they are transferred to the enzyme that comes after them in the pathway. 5-phosphoribosyl-alpha-pyrophosphate (PRPP), a significant compound, is produced in the first step of this pathway. Additionally, pyrimidine nucleotide biosynthesis uses this compound as a precursor. It supplies these ribonucleotides' phospho-ribose units. Ribose-5-phosphate (R5P), a byproduct of the pentose phosphate pathway, is the source of PRPP. Purines are therefore composed of a sequence of addition reactions to a sugar (Mendz et al., 1994, 1996, 1997).

5.1.2.1. Purine Synthesis Yields Inosine Monophosphate

In the first step of purine biosynthesis, ribose phosphate pyrophosphokinase activates the ribose by reacting it with ATP to form 5-phosphoribosyl-alpha-pyrophosphate (PRPP).

Step 2 is the committed step of purine biosynthesis. In this reaction, amidophosphoribosyl transferase catalyzes the displacement of PRPP's pyrophosphate group by glutamine's amide nitrogen. This reaction is the pathway's flux-controlling step i.e., the rate at which the biosynthetic pathway outputs product. It is shown in Figure 5.3.

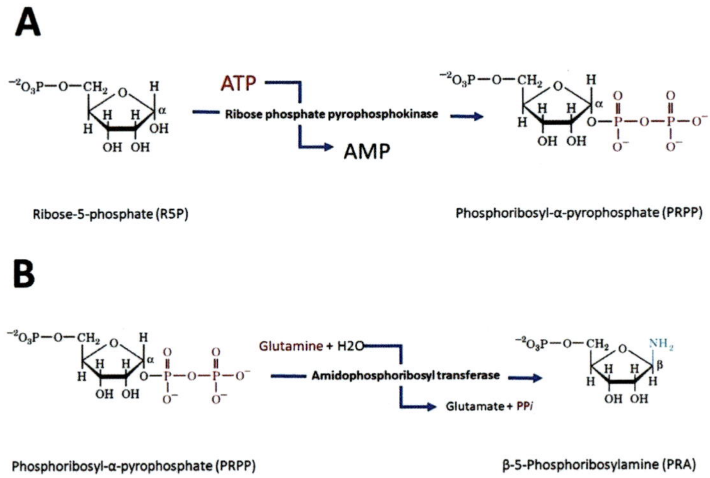

Figure 5.3. *(A) Ribose-5-phosphate activation is the first step. Ribose-5-phosphate, a byproduct of the pentose phosphate pathway, serves as the precursor for purine biosynthesis. Ribose phosphate pyrophosphokinase initiates the first step of purine biosynthesis by reacting ribose with ATP, which powers the reaction, to create 5-phosphoribosyl-alpha-pyrophosphate (PRPP); and (B) The second step, controls flux. Beta-5-phosphoribosylamine is formed when the pyrophosphate group of PRPP is replaced by the amide nitrogen of glutamine through the action of amidophosphoribosyl transferase. ATP is also responsible for this step.*

Source: By BorisTM at English Wikipedia – Transferred from en.wikipedia to Commons by Leptictidium using CommonsHelper., Public Domain, https://commons.wikimedia.org/w/index.php?curid=5827893.

Following the remaining 9 steps, the first purine derivative that is synthesized is inosine monophosphate (IMP). This can be seen in Figure 5.4.

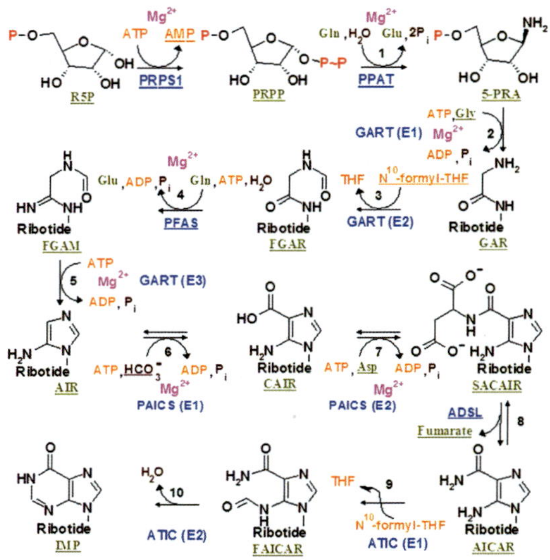

Figure 5.4. *The metabolic pathway for the de novo biosynthesis of IMP. Here the purine residue is built up on a ribose ring in 11 enzyme catalyzed reactions.*

Source: By BorisTM at English Wikipedia – Transferred from en.wikipedia to Commons by Dcirovic., Public Domain, https://commons.wikimedia.org/w/index.php?curid=16522209.

IMP is the precursor for the purine nucleotide, adenosine, and guanosine monophosphate (AMP and GMP). Each is synthesized in a two-reaction pathway with bifurcates at the level of IMP:

- Further phosphate additions to generate diphosphate and triphosphate nucleosides may follow completion of monophosphate synthesis. These reactions are carried out by kinases.

Kinases are so-called due to their property of transferring phosphate groups from a high-energy phosphate molecule to specific substrates. The complete nucleotide triphosphate forms, adenosine and guanosine triphosphate (ATP and GTP) are the recognizable units of RNA and DNA. Therefore, purines are initially formed as ribonucleotides rather than as free bases.

5.1.3. Purine Nucleotide Biosynthesis is Regulated at Several Steps

Each of the three pathways that produce IMP, ATP, and GTP is controlled separately. This is important to control the total amount of purine nucleotides available for nucleic acid synthesis, avoid wasting energy and nitrogen, and reduce the amount of purine waste product, uric acid, which is toxic to cells. The pathophysiological basis of gout is excessive uric acid production, which results in its deposition in joints, causing pain and redness.

Which leads to either AMP or GMP, is where the second level of regulation takes place. Because each of these final products inhibits IMP competitively, their excessive build-up is avoided. The human body can produce enough purine through biosynthesis to meet its metabolic needs. Insufficient purine synthesis or aberrant biosynthetic pathways may result in excruciating clinical symptoms (Mendz et al., 1994, 1996, 1997).

5.1.4. Pyrimidine Ribonucleotides

Synthetic pyrimidine ribonucleotide in *H. pylori* was studied by looking at the activities of the enzymes involved in their biosynthesis pathways as well as the incorporation of preformed pyrimidines and pyrimidine ring precursors.

5.1.4.1. De Novo Pyrimidine Nucleotide Synthesis

Nine enzymes make up the pathway that produces UTP and CTP, two RNA precursors, from scratch (Figure 5.5).

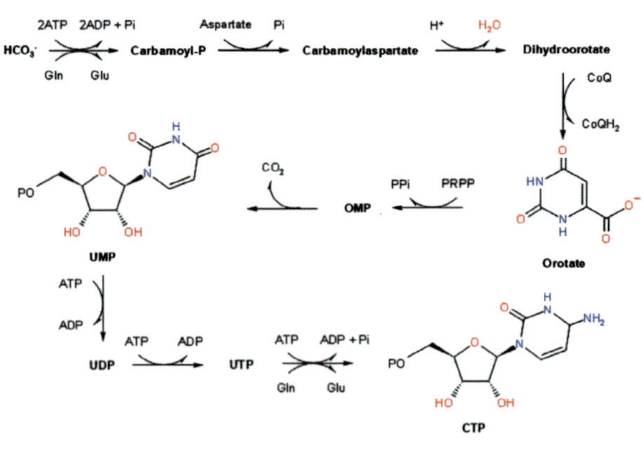

Figure 5.5. *De novo pyrimidine biosynthesis pathway.*

Source: https://zh.wikipedia.org/zh-cn/File:Pyrimidine_Synthesis.png.

Note: The enzymes encoded by the pyr genes are pyrA, carbamoyl phosphate synthase; pyrB, aspartate carbamoyl transferase; pyrC, dihydroorotase; pyrD, dihydroorotase dehydrogenase; pyrE, orotate phosphoribosyltransferase; pyrF, orotidylate decarboxylase; pyrH, UMP kinase; ndk, nucleoside diphosphokinase; and pyrG, CTP synthetase. The nucleotides are OMP, orotidine monophosphate; UMP, uridine monophosphate; UDP, uridine diphosphate; UTP, uridine triphosphate; and CTP, cytosine triphosphate.

The capability of its ability to proliferate in a specified medium devoid of preformed pyrimidines was investigated through successive passage experiments. The findings show that the bacterium can only proliferate when pyrimidine nucleotides are synthesized from scratch. The *H. pylori* de novo pyrimidine synthesis pathway's enzyme activities, have been identified in situ, and the genome of the organism may contain putative genes encoding for the corresponding enzymes.

5.1.5. Carbamoyl Phosphate Synthetase

Carbamoyl phosphate serves as the initial reactant in the metabolic pathway, being synthesized through the action of a catalyzing enzyme known as a synthetase, which utilizes glutamine, ammonium, bicarbonate, and ATP as its components. Carbamoyl phosphate serves as a pivotal point in metabolism, linking the arginine and de novo pyrimidine pathways. The introduction of radioactive carbon atoms from bicarbonate into nucleic acids in *H. pylori* may indicate the existence of a de novo pyrimidine nucleotide synthesis pathway in this organism. However, it is plausible that this incorporation occurs via the de novo purine biosynthesis process as well. Examination of the *H. pylori* genome revealed the genes HP1237 and HP0919, which share similarities with the pyrAa gene from *Salmonella enterica* serovar Cholerasuis and the pyrAb gene from *Bacillus caldolyticus*. These genes encode carbamoyl phosphate synthetases, lending credence to the interpretation of the experimental findings (Mendz et al., 1994, 1996, 1997).

5.1.5.1. Aspartate Carbamoyl Transferase

Aspartate carbamoyl transferase (ACTase) catalyzes the condensation of carbamoyl phosphate with the amino group of aspartate in the first reaction devoted exclusively to pyrimidine biosynthesis, resulting in carbamoyl aspartate. This enzyme is a dodecamer in *Escherichia coli*, consisting of three regulatory dimers (r2) and two catalytic trimers (c3). An operon transcribed from B to I contains the genes for the catalytic chain (pyrB) and the regulatory chain (pyrI). The human HP1084 gene. The pylori genome resembles the Bacillus subtilis pyrB gene, however it does not appear to contain any orthologous pyrI genes.

This enzyme's activity was expressed in *H. pylori* lysates using an analysis of radioactive tracer. The properties of H that are kinetic and regulatory, pylori ACTase were examined using radioactive tracer analysis, microtiter colorimetric assays, and nuclear magnetic resonance spectroscopy. For carbamoyl phosphate and aspartate, the apparent Kms were 0 and 11.6 mM, respectively, and at higher concentrations of either substrate, there is no indication of substrate inhibition. The ideal pH is 8.0 and the ideal temperature is 45°C. With the l-isomer of aspartate, but not the d-isomer, activity is seen. In comparison to aspartate, succinate and maleate competitively inhibit the activity of enzymes. In comparison to carbamoyl phosphate, the carbamoyl phosphate analogs acetyl phosphate and phosphonoacetic acid inhibit activity in a competitive way. At millimolar concentrations, tripolyphosphate, pyrophosphate, orthophosphate, and limiting carbamoyl phosphate inhibit the enzyme competitively with purine and pyrimidine nucleotides. At 10 mM concentrations, ribose and ribose-5-phosphate exhibit 20 and 35%, respectively, inhibition of enzyme activity. The most effective inhibitor found so far is N-Phosphonoacetyl-l-aspartate (PALA), which shows a 50% inhibition of enzyme activity at a concentration of 0 μM. PALA's inhibition is noncompetitive against aspartate and competitive against carbamoyl phosphate (Ki = 0.245 μM).

ACTase can be classified into three classes based on their catalytic and regulatory characteristics: class A ACTase is inhibited by all nucleotides, class B enzymes exhibit inhibition by CTP and activation by ATP, and class C ACTase is not affected by any nucleotides. Since ACTase inhibitors have been used as antiproliferative medications in both prokaryotes and eukaryotes, it is clear how important this enzyme is to cell survival in general. Moreover, ACTase is typically a major point of regulatory control of the de novo biosynthetic pathway in bacteria. A few characteristics of the H. The activity of the pylori enzyme is different from other ACTases, especially the well-studied *E. coli* enzyme. The substrate saturation curves for *H. pylori* ACTase exhibit hyperbolic kinetics; however, the *E. coli* curves show signs of sigmoidal behavior. Both Paracoccus abyssi and Streptococcus faecalis exhibit sigmoidal substrate kinetics. In both E, substrate inhibition is seen at higher aspartate levels, coli and S. faecalis; nevertheless, the H study shows no such effect, pylori ACTase. This study's substrate kinetics of enzyme activity are comparable to those of Pseudomonas fluorescens, a bacteria whose ACTase likewise lacks substrate inhibition and sigmoidal kinetics. The classification of ACTase is based on the kinetic and regulatory characteristics of enzyme activity; the ACTase of H. based on variations in kinetics and nucleotide regulation, pylori belongs to the class A enzyme group, despite the fact that some aspects of its regulation seem special. The H is unique among known class A enzymes in a number of catalytic properties. ACTase activity of *H. pylori*. The hyperbolic carbamoyl phosphate saturation curve of the P. was discovered. When CTP or UTP is added, fluorescens class A ACTase becomes sigmoidal; however, the outcomes for *H. pylori* show that in the presence of these nucleotides, there is no departure from the hyperbolic kinetics. The effects of substrate analogs are another distinction that was previously discussed. Vibrio costicola's class A enzyme exhibits mild activation in the presence of succinate at low aspartate concentrations, whereas the H. No such activation is seen in the pylori enzyme. Consequently, a few features of the substrate saturation curves for *H. pylori* ACTase is not the same as the ACTase found in this group.

The findings of this study imply that the availability and concentrations of *H. pylori* determine how much ACTase activity is regulated, and consequently the amount of de novo pyrimidine biosynthesis, pylori cells in their native habitat of specific metabolic regulators, and that the initial committed step in the de novo biosynthesis pathway in *H. pylori* cell-free extracts. To look into how ACTase contributes to its survival, pylori, the gene encoding for resistance to chloramphenicol was mutated to create defined isogenic pyrB mutants lacking in this enzyme activity, to make it easier to construct a pyrB-disrupted copy in *E. coli*, the whole structural gene encoding H. ACTase. The strain RU1 was cloned through pyrB auxotrophic mutant complementation. The 924 bp *H. pylori* pyrB gene encodes a hypothetical protein with a Mr of 34,278 that is consistent with H's pyrB, strains of *H. pylori* 26,695 and J99. According to the theory, the 34-kDa species is present in all bacterial and eukaryotic ACTase enzymes because the predicted protein size was extremely similar to that of the catalytic chain of those enzymes. The *H. pylori* gene exhibits strong homology with other bacterial pyrB genes, and its phylogenetic arrangement with various species aligns with the functional properties of the ACTase. Sequences that conform to the consensus recognition site linked to attenuation control are not found in the region upstream of the pyrB gene, indicating that the *H. pylori* gene does not seem to be controlled by the pyrimidine limitation mechanisms for *Bacillus* species and *E. coli*. Further evidence that regulation at the level of enzyme synthesis of the pyrB-encoded protein may not occur in *H. pylori* comes from the observation that levels of recombinant ACTase activity are unaffected by the addition of uracil, through established pathways, as observed in multiple Pseudomonas species.

The location where H transcription starts, pylori pyrB-mRNA was mapped using a ligation-anchored PCR protocol, which also revealed possible promoter regions. In several iterations of *H. pylori* cells, no pyrB mutants resistant to chloramphenicol were found. Effective mutagenesis of additional *H. pylori* genes and PCR amplification of the recombined gene revealed that allelic exchange, which entailed concurrent replacement of the pyrB gene with the chloramphenicol-pyrB-disturbed copy, was the method used to create the

ACTase-negative mutants. This study revealed that the ACTase enzyme is critical for *H. pylori* survival, expanding our knowledge of pyrB beyond the examination of primary structure data. From a biomedical standpoint, the fundamental characteristics of *H. pylori* ACTase showed promise in therapeutic intervention by focusing on the pyrimidine pathway. If the distinct qualities of the *H. pylori* enzyme could be used, it might make it possible to rationally design substances that could permanently interfere with the bacterium's biosynthetic process. Though an earlier study revealed that *H. pylori* can detoxify certain medications that target ACTase, there is a lot of potential to use them alone or in combination to modify the nature of these compounds selectively.

5.1.6. Dihydroorotase and Dihydroorotate Dehydrogenase

Dihydroorotase (DHOase) and dihydroorotate dehydrogenase (DHODHase) work together to form orotate. The DHOase of Pseudomonas putida is encoded by the gene pyrC, which is orthologous to the genes HP0266 and HP0581. The DHODHase of Haemophilus influenzae is encoded by the gene HP1011, which is homologous to pyrD. The actions of H. The in-situ measurement of *H. pylori* DHOase was conducted using 1H-nuclear magnetic resonance (NMR) spectroscopy. Oxantel, an inhibitor of H, and primaquine, lawsone, and juglone—known inhibitors of DHOase in other organisms—inhibit this enzyme's activity, fumarate reductase pylori. In both liquid and solid cultures, the effects of these substances on cell growth and viability were assessed, and at different concentrations, they exhibit bactericidal properties.

The conclusion that the de novo biosynthesis of pyrimidine nucleotides is necessary for the growth and viability of the bacterium is supported by the lethal effects of blocking this pathway on cells. Lately, *H. pylori* DHODHase has been identified and expressed, mutants of *E. coli* that lack pyrD. The enzyme is a family 2 DHODHase, which is also derived from mammalian sources and is present in other gram-negative bacteria.

This membrane-associated flavoprotein can donate electrons to coenzyme Q0, menaquinone, and menadione, but it also uses an exogenous coenzyme Q6 and an endogenous FMN redox cofactor as electron acceptors. Certain pyrazole-based substances are potent H inhibitors, pylori enzyme, but not the enzyme from humans.

These substances possess strong antibacterial properties against *H. pylori* cells but not against human, gram-positive, or gram-negative cells, indicating that they are H pylori-specific antibacterial agents, pylori and supporting the notion that the de novo pyrimidine biosynthesis pathway's enzymes could be useful therapeutic targets.

5.1.6.1. Orotate Phosphoribosyl Transferase, Orotidine-5′-Phosphate Decarboxylase

The nucleotide orotidylic acid is produced when orotate phosphoribosyl transferase (OPRTase) catalyzes the transfer of a ribose 5-phosphate moiety from PRPP to orotic acid. Orotidine 5′-phosphate decarboxylase (ODCase) then decarboxylates the nucleotide to UMP. These two enzymes, whose activities in cells and lysates have been assessed using radio-tracer analysis and 31P-NMR spectroscopy, are encoded by the *H. pylori* virus. Within the bacterium's genome, the genes HP1257 and HP0005 are homologous to the pyrE and pyrF genes, respectively, encoding for OPRTase and ODCase of *Thermus aquaticus* and *B. subtilis* respectively. The high activity observed for H is a noteworthy aspect of the data on the enzymes involved in de novo pyrimidine synthesis. ODCase pylori. While OPRTase and ODCase are distinct bacterial enzymes, they seem to combine to form multifunctional enzyme complexes in some microorganisms, including Serratia marcescens and some strains of yeast, which may control or channel the catalytic efficiency of the pathway. Ehrlich ascites cells and the protozoan parasite *Crithidia luciliae* are two eukaryotic systems for which evidence of such channeling of the OPRTase-ODCase complex has been found. Consequently, the elevated ODCase activity noted in *H. pylori* could be a sign of the existence of a multienzyme complex that serves a comparable purpose.

5.1.6.2. UMP Kinase, Nucleoside Diphosphokinase, and CTP Synthase

UMP kinase and nucleoside diphosphokinase, two enzymes whose activities have been seen in *H. pylori* sequentially phosphorylate UMP to UDP and UDP to UTP in the final stages of the synthesis of UTP. The orthologs of pyrH and ndk, which encode UMP kinase and nucleoside diphosphokinase, respectively, are the genes HP0777 and HP0198. Ultimately, UTP is changed to CTP-by-CTP synthase amination. Utilizing radiolabeled UTP, the activity of this enzyme was assessed in bacterial lysates. Gene HP0349 is homologous to the pyrG gene in *H. influenzae* which codes for CTP synthase.

5.1.7. Salvage of Preformed Pyrimidines

Salvage pathways allow cells to avoid de novo pathways by using preformed nucleic bases and nucleosides for nucleotide synthesis or by recycling bases and nucleosides generated endogenously as a result of nucleotide turnover.

In *H. pylori*, the use of preformed pyrimidine bases is restricted. Orotate is an intermediate of the de novo pyrimidine nucleotide synthesis pathway. Radioactive tracer analysis experiments using this pyrimidine demonstrate that

H. moderate rates of *H. pylori* cultures. Even though their uptake is linear at the start of the incubations, labeled precursors of pyrimidine salvage pathways like uridine and uracil are substantially less well incorporated. It appears that these pyrimidine moieties were used intact because of the linearity of the uptake of orotate, uracil, and uridine during the first 10 hours of the time courses with cells. It is important to note that *E. coli* lacks a gene called uraA, which codes for a cytoplasmic membrane protein necessary for uracil uptake, colony was discovered in the *H. pylori* genetic code. The ribosyl-phosphate moiety of PRPP is transferred to a nucleotide base by phosphoribosyl transferases, which in turn yields the corresponding nucleoside monophosphate (NMP).

This process is crucial for the salvage of free nucleobases for biosynthesis. In H, the enzyme activities for uracil (UPRTase) and orotate (OPRTase) are both seen in situ, pylori. Within *E. coli*, the first gene of a bicistronic operon is the upp gene, which codes for UPRTase, and the second gene is uraA. Within *H. pylori*, the HP1257 gene that resembles the T pyrE gene. Although aquaticus OPRTase, a component of the previously mentioned de novo pathway, was discovered, no orthologous genes to Upp were discovered.

The nupC gene in *B. subtilis*, which codes for a pyrimidine nucleoside transport protein, and the HP1180 gene are similar; nevertheless, uridine is only very minimally incorporated by *H. pylori*. There are no other orthologous genes in *E. coli* that encode the enzymes needed for pyrimidine salvage have been found in *H. pylori*, including udp and udk, which encode for uridine phosphorylase and uridine kinase, respectively.

Many gram-negative bacteria have well-understood pyrimidine base and nucleoside metabolisms. One significant quality of *E. coli* and S. enterica serovar Typhimurium is that exogenous pyrimidine bases, nucleosides, and deoxynucleosides can be incorporated into nucleic acids. Exogenous precursors of pyrimidine nucleotides can be metabolized by other gram-negative bacteria, including *Aerobacter aerogenes, Citrobacter freundii, Halobacterium cutiburum, Proteus vulgaris, Pseudomonas aeruginosa, Rhodopseudomonas sphaeroides*, and *S. marcescens*. *Neisseria meningiditis*, on the other hand, has a more limited metabolism of bases and nucleosides. Although it can synthesize pyrimidine nucleotides through the de novo pathway, the only pyrimidine nucleotide precursor that it can incorporate is uracil.

To summarize, the rates of utilization for aspartate, bicarbonate, and orotate were significantly higher than those for thymine, uracil, cytosine, or any other pyrimidine nucleoside that were examined. The bacterium can obtain pyrimidine nucleotides by salvaging precursors, as demonstrated by the incorporation of uracil and uridine and the observation of UPRTase activity. Nevertheless, H. is less common than other gram-negative bacteria, pylori demonstrated a restricted capacity to absorb preformed bases and potentially nucleosides, indicating a potential resemblance in pyrimidine metabolism to Neisseria species.

5.1.8. Biosynthesis of Deoxyribonucleotides

The ribonucleoside triphosphate (rNTP) pools receive the majority of the carbon that passes through the nucleotide biosynthesis pathways, as indicated by the ratio of RNA to DNA in cells, which ranges from 5:1 to 10:1. However, since deoxyribonucleoside triphosphates (dNTPs) are almost exclusively used in the biosynthesis of DNA, the comparatively small fraction of the cell that is dedicated to their synthesis is crucial to its survival. Ribonucleoside diphosphate reductase (rNDPase) catalyzes the first step specific for deoxyribonucleotide synthesis in most organisms. It does this by converting all four ribonucleoside diphosphates to their corresponding 2′-deoxyribonucleoside diphosphates. Within *H. pylori*, the HP0680 and HP0364 genes are similar to the *E. coli* nrdA gene carrying the α subunit of this enzyme and the *Synechocystis* sp. nrdB gene, each of which encodes the β subunit.

Thymidine triphosphate (dTTP) is biosynthesized in part from deoxycytidine nucleotides and in part from the dUDP generated by rNDPase. Within *E. coli*, four more steps are needed for dTTP synthesis to be catalyzed: dUTPase (dUTP→dUMP) encoded by dut; thymidylate synthase (dUMP → dTMP) encoded by thyA; dTMP kinase (dTMP → dTDP) encoded by tmk; and nucleoside diphosphokinase (dTDP→dTTP) encoded by ndk. Within *H. pylori*, no gene orthologous to thyA has been found, however, the genes HP0865, HP1474, and HP0198 are orthologous to Dut, *tmk*, and *ndk*, respectively (Mendz et al., 1994, 1996, 1997).

5.1.8.1. E. coli

The pathway for dTTP salvage explained in *E. coli* requires the presence of thymidine kinase, encoded by the tdk gene, deoA gene, and uridine phosphorylase or thymidine phosphorylase, respectively. In *H. pylori*no genes that were similar to these genes were discovered. As an alternative, exodeoxyribonuclease may act on DNA to release nucleotide 5′-monophosphates, which can then be transformed into triphosphates as previously mentioned, thereby producing dTTP. The lexA gene in *B. subtilis*, which codes for this enzyme, is comparable to the HP1526 gene (Figure 5.6).

Deoxyadenosine is salvaged by *Helicobacter pylori*, although at much slower rates than adenosine; any deoxyguanosine incorporation is below the detection limits of the used radioactive tracer experimental techniques. Deoxynucleosides are hydrolyzed by nucleosidases to yield deoxyribose and the corresponding base, and they are phosphorylated by nucleoside phosphotransferases to yield deoxynucleoside monophosphates.

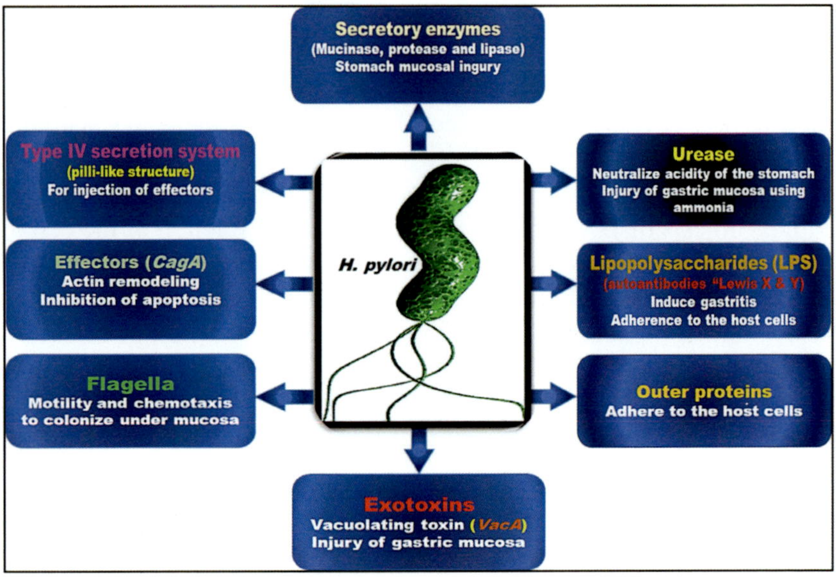

Figure 5.6. *H. pylori.*

Source: Elbehiry A, Marzouk E, Aldubaib M, Abalkhail A, Anagreyyah S, Anajirih N, Almuzaini AM, Rawway M, Alfadhel A, Draz A, et al. (2023). Helicobacter pylori infection: Current status and future prospects on diagnostic, therapeutic and control challenges. Antibiotics, 12(2), 191. https://doi.org/10.3390/antibiotics12020191.

In H, there is no detectable deoxyguanosine nucleosidase activity, but there are significant deoxyadenosine nucleosidase and phosphotransferase activities as well as deoxyguanosine phosphotransferase activity, pylori. Deoxyadenosine was incorporated by *H. pylori* and the existence of deoxyadenosine phosphotransferase indicate that the synthesis of adenosine deoxynucleotides is facilitated by the phosphorylation of the deoxynucleoside. The existence of an equally active deoxyadenosine nucleosidase, which would partially reroute the purine base moiety via some of the other biosynthetic pathways, limits the ability of this salvage mechanism to function. Despite the observed deoxyguanosine phosphotransferase activity, it appears that distinct pathways are involved in the synthesis of guanosine deoxynucleotides due to the lack of deoxyguanosine incorporation and the absence of deoxyguanosine nucleosidase activity. A portion of the deoxyribose generated by deoxyadenosine nucleosidase might be recycled and used to create guanine deoxyribonucleotides.

5.1.9. Regulation of Biosynthesis of Nucleotides

The economy of the cell depends on the rate regulation of each individual reaction in this intricate chain of biosynthetic reactions. This is achieved through repression of enzyme synthesis and negative feedback inhibition of enzyme activity. In feedback inhibition, the final product inhibits the activity of the pathway's or a branch's first enzyme. In feedback repression, a product's derivative prevents one or more of the pathway's enzymes from forming.

5.1.9.1. Regulation of Mammalian Nucleotide Metabolism and Biosynthesis

It is now known that a significant portion of the genome is transcribed into a variety of RNAs, the roles of which are still unknown. There is a significant turnover of RNA involved in cell upkeep, repair, and regulation even in quiescent cells. As a necessary part of cell division, proliferating cells must up-regulate RNA and DNA biosynthesis, which can, at least partially, regulate the rate of the entire cell cycle. This necessitates higher nucleotide synthesis-related gene expression in the late G1 phase.

Many essential transcription factors regulate the synthesis of nucleotides; two in particular, MYC and Rb/E2F, are linked to transformation and unchecked proliferation that results in cancer when they are mutated or overexpressed. In addition to coordinating the biosynthesis of RNA and proteins, MYC directly controls the expression of genes encoding the enzymes involved in the feeder pathways and nucleotide biosynthesis pathways. Additionally, MYC affects the expression of particular microRNAs that control the activity of enzymes necessary for cell division.

5.2. PURINE RIBONUCLEOSIDE OVERPRODUCTION BY *BACILLUS SUBTILIS*

When the control mechanisms are removed by different manipulations, *Bacillus subtilis* has a great capacity to overproduce purine ribonucleosides. There are two main categories of manipulations. In the first, growth-limiting concentrations of the end-product requirement are fed to auxotrophic mutants in order to decrease the intracellular levels of feedback inhibitors and repressors. In the second approach, resistance to toxic purine analogs is used to select constitutive mutants, which are no longer subject to feedback repression, or desensitized mutants, which are resistant to feedback inhibition.

5.2.1. Purine Nucleotide Overproduction by Corynebacterium glutamicum

The ability of *Corynebacterium glutamicum* and other glutamic acid bacteria to excrete ribonucleotides following the removal of feedback controls sets them apart from the majority of microorganisms. Therefore, auxotrophs that require adenine. When glutamicum grows in an environment with limited adenine, it excretes IMP. Since intracellular accumulation of both AMP and GMP can be limited, further mutational elimination of the ability to synthesize IMP dehydrogenase results in adenine-xanthine double auxotrophs that exhibit increased IMP formation.

XMP is accumulated by guanine-requiring mutants or adenine-guanine double auxotrophs. The concentrations of XMP in culture filtrate are approximately 7 g/liter. It makes sense that too much guanine would prevent the synthesis of XMP after the base is converted to GMP, which represses the synthesis of IMP dehydrogenase and prevents it from working. It has proven more challenging to develop a direct fermentation to produce GMP than it has been for the intermediates IMP and XMP. The GMP's control over IMP dehydrogenase activity is one of the primary challenges. Many years ago, a mutation of an adenine-requiring auxotroph of *C. glutamicum* led to the creation of a mutant that accumulated a low concentration (1 g/1) of GMP to adenine-xanthine double auxotrophy, which is subsequently reversed to only require adenine. It is likely that this double mutation in the IMP dehydrogenase gene made the enzyme less sensitive to feedback inhibition. Development of the guanosine fermentation in B after that. Sub tilts indicate that the following characteristics of a bacterium are crucial for its use in a cost-effective GMP fermentation: (a) nucleotide excretion, as in *C. glutamicum*; (b) loss of adenylosuccinate synthetase synthesis ability (i.e., (c) successive resistance mutations to 8-azaguanine and/or 8-azaxanthine to desensitize IMP dehydrogenase and eliminate repression of synthesis of the common-path enzymes and IMP dehydrogenase; (d) genetic deletion of the gene coding for GMP reductase. adenine auxotrophy) to eliminate feedback inhibition of the activity of PRPP amidotransferase by AMP. There are no reports of this type of organism as of yet.

Anion-exchange resins and dioxane additions to the medium are two examples of environmental manipulations that can be used to circumvent the feedback controls that GMP imposes on its own biosynthesis. Dioxane works by quantitatively forming a complex with water and GMP. We were able to extract 4 g of IMP, 2 g of XMP, and 1 g of GMP from a one-liter culture of an azaguanine-resistant mutant of Streptomyces sp. by using the resin technique.

5.3. SALVAGE SYNTHESIS OF PURINE NUCLEOTIDES

Only hypoxanthine was excreted in the early stages of the fermentation, which led investigators to believe that *Brevibacterium ammoniagenes* was conducting a direct IMP fermentation. Then, extracellular IMP was found, and the expelled purine vanished. Research revealed that the phenomenon involved the following steps: adenine auxotrophy-induced intracellular accumulation of IMP; degradation to inosine and subsequently to hypoxanthine; excretion; and, lastly, an extraordinary extracellular' salvage syntheses of IMP from the released hypoxanthine.

Later research revealed that intact wild-type B cells were capable of converting a number of bases to their corresponding ribonucleotides. ammonia genes provided specific conditions were fulfilled. These included the addition of thiamin and pantothenate, high concentrations of inorganic phosphate and Mg^{2+}, and a growth-limiting concentration of Mn^{2+}. In a medium with a lower phosphate concentration, all of the aforementioned requirements are not required for growth under normal circumstances. However, growth is not possible without the addition of Mn^{2+}, thiamin, pantothenate, and a high concentration of Mg^{2+} when the high concentration of phosphate needed as a nucleotide precursor is used. The concentration of Mn^{2+} is essential for extracellular salvage synthesis under these aberrant growth conditions. It has to be minimal. If this is the case, it causes nucleotide formation by changing the cellular lipid composition and speeding up intracellular compound exit. Inhibition of DNA synthesis, a drop in cell viability, an increase in cell mass, and a change in cellular morphology from typical small rods and ellipsoidal cells to elongated and swollen forms occur in tandem with the permeability change. Ribose 5-phosphate, PRPP, 5-phosphoribose pyrophosphokinase, multiple purine nucleotide pyrophosphorylases, and probably ATP are among the substances that leak into the medium from the cells. Consequently, the following extracellular reactions are made possible by the addition of a purine:

$$\text{ribose 5-phosphate} + \text{ATP} \xrightarrow[\text{5-phosphoribose pyrophosphokinase}]{Mg^{2+},\ PO_4^{3-}} \text{PRPP} + \text{AMP}$$

$$\text{PRPP} + \text{added purine} \xrightarrow[\text{purine nucleotide pyrophosphorylase}]{Mg^{2+},\ PO_4^{3-}} \text{5}'\text{-purine ribonucleotide} + \text{PPi}$$

This organism converts hypoxanthine (96% of the hypoxan thine used) to IMP (15 g IMP/litre). Adenine is converted to AMP, ADP, and ATP, and guanine to GMP, GDP, and GTP. A further leakage of nucleotide phosphotransferases is responsible for the production of the di- and triphosphates of guanosine and adenosine. ATP, GTP, CTP, and PRPP are examples of phosphate donors. The addition of Hg^{2+}, Zn^{2+}, and Ag^{+}, which block the activity of nucleotide phosphotransferases, and the addition of xylene or other solvents, which increase the excretion of these enzymes and consequently increase the production of the di- and triphosphates, can regulate the ratio of mono-, di-, and triphosphates. By adding nicotinic acid or nicotinamide, *Brevibacterium ammoniagenes* can also be used to prepare nicotinic acid mononucleotide; adenine is also added to form nicotinamide adenine dinucleotide (NAD^{+}). The production of nicotinamide adenine dinucleotide occurs at 2 g/1 aL concentrations.

The addition of antibiotics or cationic surface-active agents to log-phase cells can also yield the beneficial effect of Mn^{2+} limitation in extracellular salvage synthesis. Mutants that generate nucleotides even in the presence of elevated Mn^{2+} concentrations have been identified.

It is unknown what biochemical alteration a mutation causing Mn^{2+} insensitivity causes. To cause B to bioconvert XMP to GMP, permeability modification and mutation have been combined. Ammonia genes. Mn^{2+} limitation boosted permeability, allowing XMP to enter the cells. The XMP was converted to GMP by mutation to resistance to decoyinine, an XMP aminase inhibitor, most likely by derepressing XMP aminase. GDP, GTP, and guanine were also added together. Without degrading into guanine, XMP was converted to guanine nucleotides by a related mutant with decreased 5'-nucleotidase activity. By combining this strain with an XMP-accumulating mutant in a mixed culture, guanosine monophosphate was also directly produced.

5.3.1. Purine Nucleotides

Throughout the biological world, similar pathways carry out the de novo synthesis of purine nucleotides. However, many organisms use preformed purine compounds to meet their nucleotide needs by utilizing salvage pathways that take up available purine nucleobases and nucleosides. The biosynthesis of purine nucleotides in H is poorly understood, pylori, despite research on the bacterium's capacity for survival through de novo purine synthesis, its absorption of preformed purine bases and nucleosides, and the functions of enzymes involved in purine salvage pathways.

5.3.1.1. De Novo Purine Nucleotide Biosynthesis

An investigation into the fundamental nutrients that *H. pylori* discovered that the absence of adenine or other preformed purines in specific media containing a blend of amino acids, salts, and vitamins decreased bacterial growth by 50%; additionally, the bacterium grew more effectively in the absence of uracil, guanine, and xanthine. These results suggest that the bacterium needs purine nucleotide precursors to be salvaged in order to grow and proliferate.

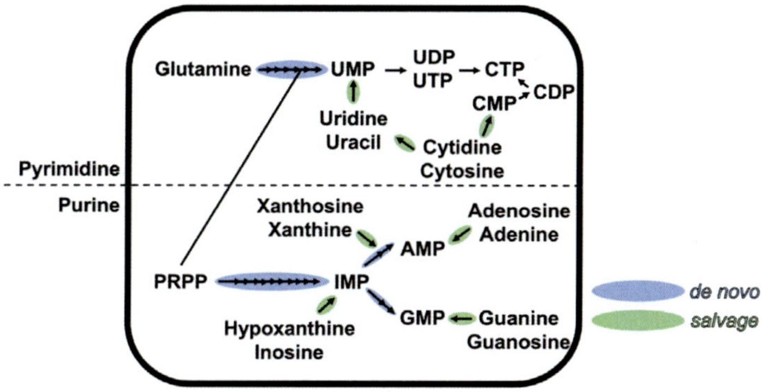

Figure 5.7. *De novo purine biosynthesis pathway.*

Note: The enzymes encoded by the pur genes are purF, glutamine: PRPP-amidotransferase; purD, phosphoribosyl glycinamide (GAR) synthetase; purN, GAR transformylase N; purT, GAR transformylase T; purL, phosphoribosyl-N-formylglycinamide synthetase; purM, phosphoribosyl-aminoimidazole synthetase; purK, phosphoribosyl-carboxyaminoimidazole synthetase; purE, phosphoribosyl-carboxyaminoimidazole mutase; purC, phosphoribosyl-N-succino-carboxamide-aminoimidazole; purB, adenylosuccinate lyase; purH, phosphoribosyl-carboxamide-aminoimidazole transformylase; and purH, inosinic acid cyclohydrolase. The compounds are Gln, glutamine; PPi, pyrophosphate; Glu, glutamate; ATP, adenosine triphosphate; ADP, adenosine diphosphate; Pi, phosphate; N-Formyl-THF, N^{10}-formyl-tetrahydrofolate; THF, tetrahydrofolate; CO_2, carbon dioxide; Asp, aspartate; and H_2O, water.

Source: Ayoub N, Gedeon A and Munier-Lehmann H (2024) A journey into the regulatory secrets of the de novo purine nucleotide biosynthesis. Front. Pharmacol., 15, 1329011. doi: 10.3389/fphar.2024.1329011.

The purine ribonucleotide generated de novo in the pathway is called inosinate (IMP). In 10 enzymatic reactions, functional groups are added to PRPP stepwise to form IMP (Figure 5.7). Within the *H. pylori* genome, the only genes that are similar to those encoding the enzymes needed for IMP synthesis are HP1218 and HP1112. First, glycinamide ribonucleotide synthetase, which is orthologous to purD, and second, adenylosuccinate lyase, which is bifunctional and involved in the synthesis of AMP from IMP, are orthologous to purB. These genome analysis results corroborate the preliminary results indicating that adenine was necessary for H survival, pylori. Though the genome of *H. pylori* lacks identification for eight of the enzymes involved in de novo purine synthesis. pylori, a study using low-passage isolates and laboratory-adapted strains showed that the bacterium was still viable and growing after several passages in chemically defined media devoid of purine bases, nucleosides, and nucleotides. In the presence of bovine serum albumin and catalase, the bacterium grew and multiplied continuously for five passages in defined media devoid of preformed purines, suggesting that growth is not due to the uptake of leftover preformed purines that may have been brought into the media by bacteria that were harvested from plates or made available by lysis of a portion of the cell populations. These findings suggest that to satisfy its needs for purine nucleotides, *H. pylori* has the ability to synthesize purines from scratch.

To comprehend how adenine contributes to *H. pylori*, cells were cultivated in various conditions using chemically defined media containing or lacking adenine. If exogenous catalase is present, the viability of bacterial cells stays constant, but it decreases with each passage if both catalase and adenine are missing from the media. Culture media for *H. pylori* growth are used when blood or serum are not available. catalase is added in order to lower the levels of reactive oxygen species. Adenine has been shown to support *H. pylori*. By introducing adenine to the cultures and stopping the decrease in the quantity of colony-forming units, pylori viability in the absence of catalase can be restored. As a scavenger of hydroxyl radicals (OH·), adenine may play a role in protecting cells from hydroxyl radicals. This idea is supported by cell growth experiments conducted with this purine base and several other scavengers of OH. A folate analog called methotrexate (MTX) stops the growth of bacteria that need folate by blocking the enzyme dihydrofolate reductase. Dose-response curves demonstrate that MTX has potent bactericidal effects on *H. pylori*. Any biosynthetic reaction in which tetrahydrofolate serves as a one-carbon unit carrier, such as the biosynthesis of glycine and methionine, thymidylate, and purines, may be affected by inhibition of tetrahydrofolate production. The mechanism by which MTX inhibits cell growth and how the purine de novo biosynthesis pathway contributes to H. By cultivating the bacterium in the presence of MTX, bases such as adenine or hypoxanthine, or aminoimidazole carboxamide, pylori were studied. The MTX-induced growth inhibition is reversed in the presence of purine metabolites, indicating that the antifolate's bactericidal effects on *H. pylori* work by preventing the purine de novo biosynthesis pathway.

Despite the lack of a glutamine: PRPP-amidotransferase gene ortholog for purF, the first committed step in de novo purine biosynthesis (Figure 5.7), which is found in the *H. pylori* genome, 1H- and 31P-NMR spectroscopy has been used to show that the bacterium possesses this enzyme activity (Mendz, unpublished results). This amidotransferase activity is present in the bacterium because *H. pylori* lysates incubated with the substrates glutamine and PRPP yield as products glutamate and PPi (the third product of this enzymic reaction, 5-phosphoribosylamine, decomposes rapidly). As structural analogs of glutamine, azaserine and 6-diazo-5-oxonorleucine permanently block glutamine amidotransferases in a variety of organisms. The glutamine:P RPP-amidotransferase activity is inhibited in *H. pylori* lysates by these substances demonstrated that the bacterium contained this enzyme (Mendz, unpublished results).

Each branch of the pathway needs four different enzymes to convert IMP to GTP or ATP, and *H. pylori* appears to carry the genes encoding for these enzymes. IMP dehydrogenase, which is encoded by guaB orthologous to HP0829, GMP synthetase, which is encoded by guaA orthologous to HP0409, and 5′-guanylate kinase, which is encoded by gmk orthologous to HP0321, are all involved in GDP synthesis. Adenylosuccinate synthetase, encoded by purA orthologous to HP0255, adenylosuccinate lyase, encoded by purB orthologous to HP1112, and adenylate kinase, encoded by adk orthologous to HP0618, are the enzymes involved in the synthesis of ADP. Nucleoside diphosphate kinase uses ATP as the phosphate donor to convert GDP and the majority of nucleoside diphosphates to triphosphates. The HP0198 gene is located in the *H. pylori* genome bears a strong resemblance to the *Myxococcus xanthus* ndk gene, which codes for this enzyme. Given that both pathway genes are present, it is likely that the bacteria uses them to convert inosinic acid into ATP and GTP. However, bacterial lysates show no signs of guanylate kinase activity. Using GTP or ITP as phosphate donors, *H. pylori* adenylate kinase can phosphorylate AMP and produce GDP and IDP, respectively (Mendz et al., 1994, 1996, 1997).

Table 5.1. *Some Flavodoxin-Containing Bacteria* [a]

Microorganism	Protein Name	Gene Name	Seq. Length	Long/short Chain	Phylum	Gram Stain
Anabaena (Nostoc) sp.	Flavodoxin	*isiB*	170	Long	Cyanobacteria	Negative
Aquifex aeolicus	Flavodoxin	*fldA*	185	Long	Aquificae	Negative
Azotobacter vinelandii	Flavodoxin 1	*Avin45950*	174	Long	Proteobacteria	Negative
Azotobacter vinelandii	Flavodoxin 2	*nifF*	180	Long	Proteobacteria	Negative

Azotobacter chroococcum	Flavodoxin B	*nifF*	180	Long	Proteobacteria	Negative
Bacillus cereus	Flavodoxin	*BC_1376*	148	Short	Firmicutes	Positive
Bacillus cereus	Flavodoxin	*BC_3541*	154	Short	Firmicutes	Positive
Bacillus subtilis	Probable flavodoxin 2	*ykuP*	151	Short	Firmicutes	Positive
Bacillus subtilis	Probable flavodoxin 1	*ykuN*	158	Short	Firmicutes	Positive
Bacteroides uniformis	Flavodoxin	*BACUNI_04544*	178	Long	Bacteroidetes	Negative
Buchnera aphidicola	Flavodoxin	*fldA BUsg_289_*	154	Long [b]	Proteobacteria	Negative
Buchnera aphidicola	Flavodoxin	*fldA BU299*	171	Long	Proteobacteria	Negative
Buchnera aphidicola	Flavodoxin	*fldA bbp_277*	174	Long	Proteobacteria	Negative
Campylobacter jejuni	Flavodoxin	*fldA*	163	Long	Proteobacteria	Negative
Clostridium beijerinckii[c]	Flavodoxin		138	Short	Firmicutes	Positive
Clostridium pasteurianum	Flavodoxin	*CLPA_c13840*[d]	140	Short	Firmicutes	Positive
Clostridium saccharobutylicum	Flavodoxin	*floX*	160	Long	Firmicutes	Positive
Desulfovibrio desulfuricans	Flavodoxin	*Ddes_1951*	148	Short	Proteobacteria	Negative

Desulfovibrio gigas[c]	Flavodoxin		146	Short	Proteobacteria	Negative
Desulfovibrio gigas[c]	Flavodoxin		147	Short	Proteobacteria	Negative
Desulfovibrio salexigens	Flavodoxin	*Desal_0805*	146	Short	Proteobacteria	Negative
Desulfovibrio vulgaris	Flavodoxin	*DVU_2680/ DvMF_1143*	148	Short	Proteobacteria	Negative
Escherichia coli	Protein MioC	*mioC*	147	Short	Proteobacteria	Negative
Escherichia coli	Uncharacterized protein YqcA	*yqcA*	149	Short	Proteobacteria	Negative
Escherichia coli	Flavodoxin 2	*fldB*	173	Long	Proteobacteria	Negative
Escherichia coli	Flavodoxin 1	*fldA*	176	Long	Proteobacteria	Negative
Fusobacterium nucleatum	Flavodoxin	*FN0724*	167	Long	Fusobacteria	Negative
Haemophilus influenzae	Protein MioC homolog	*mioC*	146	Short	Proteobacteria	Negative
Haemophilus influenzae	Flavodoxin	*fldA*	174	Long	Proteobacteria	Negative
Helicobacter pylori	Flavodoxin	*fldA*	164	Long	Proteobacteria	Negative
Klebsiella pneumoniae	Flavodoxin	*fldA/nifF*	176	Long	Proteobacteria	Negative

Lactobacillus reuteri	Flavodoxin/ nitric oxide synthase	*Lreu_1727*	149	Short	Firmicutes	Positive
Listeria monocytogenes	Lmo 2153 protein	*lmo2153*	145	Short	Firmicutes	Positive
Megasphaera elsdenii (*Peptostreptococcus elsdenii*) [c]	Flavodoxin		137	Short	Firmicutes	Negative
Pantoea agglomerans (*Enterobacter agglomerans*)	Flavodoxin	*nifF*	177	Long	Proteobacteria	Negative
Pasteurella multocida	Protein mioC homolog	*mioC*	147	Short	Proteobacteria	Negative
Pectobacterium carotovorum[c]	Exoenzyme regulation regulon ORF2		151	Short	Proteobacteria	Negative
Pseudomonas aeruginosa	Uncharacterized protein PA3435	*PA3435*	150	Short	Proteobacteria	Negative

Pseudomonas aeruginosa	Flavodoxin FldP	*fldP*	184	Long	Proteobacteria	Negative
Pseudomonas putida	Flavodoxin	*mioC*	151	Short	Proteobacteria	Negative
Rhodobacter capsulatus	Flavodoxin	*nifF*	182	Long	Proteobacteria	Negative
Salmonella Typhi	Flavodoxin 2	*fldB*	173	Long	Proteobacteria	Negative
Salmonella Typhi	Flavodoxin	*fldA*	176	Long	Proteobacteria	Negative
Salmonella Typhimurium	Flavodoxin 2	*fldB*	173	Long	Proteobacteria	Negative
Salmonella Typhimurium	Flavodoxin 1	*fldA*	176	Long	Proteobacteria	Negative
Shewanella oneidensis	Flavodoxin Protein MioC	*mioC*	146	Short	Proteobacteria	Negative
Shewanella oneidensis	tRNA pseudouridine synthase C-associated flavoprotein YqcA	*yqcA*	154	Short	Proteobacteria	Negative
Shewanella oneidensis	Flavodoxin	*fldA*	175	Long	Proteobacteria	Negative

Shigella flexneri	Uncharacterized protein YqcA	*yqcA*	149	Short	Proteobacteria	Negative
Shigella flexneri	Flavodoxin 1	*fldA*	176	Long	Proteobacteria	Negative
Streptococcus agalactiae [e]	Flavodoxin	*mioC*	147	Short	Firmicutes	Positive
Streptococcus pneumoniae	Flavodoxin	*flaV*	147	Short	Firmicutes	Positive
Synechococcus sp.	Flavodoxin	*isiB*	170	Long	Cyanobacteria	Negative
Synechocystis sp.	Flavodoxin	*isiB*	170	Long	Cyanobacteria	Negative
Treponema pallidum	Flavodoxin	*fldA*	146	Short	Spirochaetes	Negative [f]
Trichodesmium erythraeum	Flavodoxin	*fld*	171	Long	Cyanobacteria	Negative
Vibrio cholerae	Protein MioC homolog	*mioC*	144	Short	Proteobacteria	Negative
Vibrio cholerae	Flavodoxin	*fld1*	175	Long	Proteobacteria	Negative
Vibrio cholerae	Flavodoxin	*fld2*	198	Long	Proteobacteria	Negative
Wolinella succinogenes	Flavodoxin	*fldA*	171	Long	Proteobacteria	Negative

[a] *Extracted from Uniprot by searching for "flavodoxin" and refining by "reviewed," from NCBI by searching for "flavodoxin" in the "Protein" tab and refining by "Bacteria" (in the species tag), "PDB and UniProtKB/Swiss-Prot" (in the source*

databases' tag) and "from 130 to 199 residues" (in the sequence length's tag) and from References.

Source: Salillas S, Sancho J., (2020). Flavodoxins as novel therapeutic targets against Helicobacter pylori and other gastric pathogens. International Journal of Molecular Sciences, 21(5), 1881. https://doi.org/10.3390/ijms21051881.

The common precursor IMP allows for the interconversion of guanine and adenine nucleotides. When the bacterium is able to salvage nucleobases in its environment, these conversions are crucial in maintaining the balance of both types of nucleotides in the cellular pool. There are two routes by which adenine nucleotides can be changed into guanine nucleotides, neither of which has orthologous genes in *H. pylori* genetic material. GMP reductase converts GMP to IMP, pylori DNA using the guaC orthologous gene HP0854.

Research on the biosynthesis of purines from scratch in *H. pylori* emphasize the necessity of comparing the results of genomic analysis with metabolic data in order to comprehend the physiology of the bacterium and suggest that this pathway is crucial to the bacterium.

Table 5.2. *Targets for Hp Infection Treatment*

Target	Pathway
Type II 3-dehydroquinase dehydratase (DHQ2)	Shikimate pathway
Shikimate 5-dehydrogenase (SDHase)	Shikimate pathway
Shikimate kinase (SK)	Shikimate pathway
Chorismate synthase	Shikimate pathway
Phosphopantetheine adenylyltransferase (PPAT)	Coenzyme A biosynthesis
Carbon starvation protein A	Starvation response, utilization of peptides, and host–pathogen interactions
Methylthiotransferase (MiaB)	Protein synthesis
Ribosomal RNA small subunit methyltrans-ferase E	Protein synthesis
Ribosomal protein L11 methyltransferase	Protein synthesis
Tetrapyrrole (Corrin-Porphyrin) methylase family protein	Protein synthesis
Peptide chain release factor 1	Protein synthesis

Fumarate reductase (FrdA, FrdB, and FrdC)	Krebs cycle and anaerobic respiration
Glu-tRNA^{Gln} amidotransferase, subunits A (GatA), B (GatB), and C (GatC)	Protein synthesis
Helicase-nuclease DNA Repair Enzymes (AddAB)	DNA damage reparation
Cytochrome C-type biogenesis protein CcdA	Cytochrome C synthesis
Cytochrome C oxidase, subunits CcoN, CcoO, CcoP and CcoQ	ATP synthesis
Flavodoxin (Fld)	Oxidative decarboxylation of pyruvate
Pyruvate:ferredoxin oxidoreductase (POR), subunit α (porA), β (porB), (porC or porG) and δ (porD)	Oxidative decarboxylation of pyruvate
Flavodoxin:quinone reductase (FqrB)	Oxidative decarboxylation of pyruvate
2-oxoglutarate:acceptor oxidoreductase, subunits A (OorA), B (OorB), C (OorC) and D (OorD)	Succinyl-CoA production
N-succinyl-L,L-diaminopimelic acid desuccinylase, SDAP-deacylase (DapE)	Succinylase pathway (lysine biosynthesis)
Glutamate racemase MurI	Peptidoglycan biosynthesis
Multi-drug resistance protein MsbA	Lipopolysaccharide biosynthesis
UDP-galactose 4-epimerase (GalE)	Lipopolysaccharide biosynthesis
Urease, subunits α (UreA), and β (UreB)	Acclimation to low pH
Nickel-responsive regulator (NikR)	Urease expression and nickel uptake regulator
Nickel–cobalt transporter (NixA)	Nickel/cobalt transport
Urease accessory protein UreE	Urease maturation
Urease accessory protein UreF	Urease maturation
Urease accessory protein UreG	Urease maturation
Urease accessory protein UreH	Urease maturation

Hydrogenase/urease maturation factor (HypA)	Urease maturation
Hydrogenase/urease maturation factor (HypB)	Urease maturation
Heat Shock Protein A (HspA)	Nickel homeostasis
Hpn	Nickel homeostasis and storage
Acid-activated urea channel (UreI)	Urea permeability
α-carbonic anhydrase	Acclimation to low pH
β-carbonic anhydrase	Acclimation to low pH
Spore coat polysaccharide biosynthesis protein C (PseC)	Pseudaminic acid biosynthesis pathway (Pse): flagellin glycosylation
Heat-inducible transcription repressor (HrcA)	Flagella biosynthesis
Transcriptional repressor of DnaK operon (HspR)	Flagella biosynthesis
Major flagellin FlaA	Flagellar filament composition
Minor flagellin FlaB	Flagellar filament composition
Flagellar hook-associated protein 2 (FliD)	Flagellum assembly (filament capping)
Flagellar hook-associated protein 1 (FlgK)	Flagellum assembly (hook-filament junction formation)
ATP-binding protein (YlxH)	Flagella biosynthesis
Flagellar basal body L-ring protein (FlgH)	Flagellum assembly (L-ring composition)
Flagellar basal body P-ring protein (FlgI)	Flagellum assembly (P-ring composition)
Flagellar basal body M-ring protein (FliF)	Flagellum assembly (MS ring composition)
Flagellar biosynthetic protein (FliP)	Flagellum assembly (Flagellar export component)

Flagellar biosynthetic protein (FliQ)	Flagellum assembly (Flagellar export component)
Flagellar motor switch protein (FliY)	Flagellum assembly (C-ring composition; Flagellar export component)
Flagellum-specific ATP synthase (FliI)	Flagellum assembly (Flagellar export component)
Flagella-specific σ factor (FliA)	Flagellum assembly (regulatory protein)
FlgM (putative antagonist of FliA)	Flagellum assembly (regulatory protein)
Cytotoxin-associated gene A (CagA)	*cag* pathogenicity island (host cell metabolism modulation, inflammation, metaplasia and precancerous transformations)
cag-Type IV secretion system (T4SS)	*cag* pathogenicity island (translocation of bacterial factors (e.g., Cag A and peptidoglycan) into host cells)
HopQ adhesin (outer membrane protein)	Adhesion to host cells and translocation of CagA into host cells
Vacuolating cytotoxin (VacA)	Cellular vacuolation, apoptosis and inhibition of cell cycle progression and host immune response
Blood group antigen binding adhesin (BabA)	Adhesion to host cells
High temperature requirement A (HtrA)	Chaperone and proteolytic activities (intercellular adhesion cleavage)
Sialic acid-binding adhesin (SabA) (outer membrane protein)	Bacterial migration to epithelium surface
HopZ adhesin (outer membrane protein)	Adhesion to host cells

OipA adhesin (outer membrane protein)	Adhesion to host cells
AlpA/B adhesin (outer membrane protein)	Adhesion to host cells
Cation efflux system protein CusA	Efflux of cobalt/zinc/cadmium
Cobalt/Zinc/Cadmium efflux system membrane fusion protein	Efflux of cobalt/zinc/cadmium
Cobalt/Zinc/Cadmium resistance protein (CzcA, CzcB and CzcC)	Efflux of cobalt/zinc/cadmium
CznABC metal efflux pump	Efflux of cadmium/zinc/nickel
Ferrix siderophore transport system TonB periplasmic binding protein	Iron transport
Ferric siderophore transport system ExbB biopolymer transport protein	Iron transport
Haemin uptake system ATP binding protein	Iron transport
Glutathionyl spermidine synthetase	Intracellular thiol redox balance regulation
Iron-binding ferritin-like antioxidant protein	Prevention of toxic reactive species formation
DNA-binding protein Dps	DNA breaking protection
Superoxide dismutase	Superoxide dismutation
Thioredoxin reductase	Prevention of toxic reactive species formation
RNA polymerase σ^{54} factor	Survival under stress conditions
Multi-drug resistance protein MsbA	Efflux of hydrophobic drugs
Exodeoxyribonuclease (LexA)	SOS response activation
Homeostatic stress regulator (HsrA)	Regulation of gene expression

Source: Salillas S, Sancho J., (2020). Flavodoxins as novel therapeutic targets against helicobacter pylori and other gastric pathogens. International Journal of Molecular Sciences, 21(5):1881. https://doi.org/10.3390/ijms21051881.

5.3.2. Salvage of Preformed Purines

Numerous organisms possessing the ability to synthesize purines from scratch can also scavenge a broad variety of purines, transform them into alternative purines, and integrate them into nucleic acids. Purine nucleotide synthesis salvage routes in *H. pylori* were examined by monitoring salvage enzyme activity and adenine and guanine incorporation (Figure 5.9). Adenine and guanine with radiolabeled labels were incorporated at 64.5 and 27.5 pmol h^{-1} (106 cells)$^{-1}$, respectively. Significant adenine-, guanine-, and hypoxanthine phosphoribosyl transferase activities are seen in H. luteinus. Phosphoribosyl transferases are essential enzymes in the salvage of purines, pylori lysates. The orthologous HP0572 genes in *E. coli* encode adenine phosphoribosyl transferase, similar to apt and HP0735, an ortholog of gpt that codes for a xanthine/guanine phosphoribosyl transferase (XGPRTase) in *H. influenzae* were found. The *H. pylori* -measured rate, HPRTase activity in converting hypoxanthine to IMP was much lower than that of APRTase and GPRTase, and the *H. pylori* strain did not have an orthologous gene for any particular HPRTase. The GPRTase and HPRTase, respectively, are encoded by the genes gpt and hpt in *Escherichia coli* and S. enterica serovar Typhimurium. But in *H influenza*, the substrates that XGPRTase uses to synthesize IMP, XMP, and GMP are hypoxanthine, xanthine, and guanine; and in *B. subtilis*, guanine or hypoxanthine are used by the same transferase. Therefore, it's possible that the *H. pylori* enzyme that the gpt gene encodes has the capacity to use hypoxanthine and guanine. Interestingly, adenine, guanine, and hypoxanthine uptake are correlated with the activities of the corresponding phosphoribosyl transferases in *H. pylori*, indicating that uptake processes—which resemble those in *E. coli*—are subject to the same metabolic restrictions as the salvage biosynthesis pathways, *B. subtilis*, and *S. enterica* serovar Typhimurium. The substantial concentrations of purine bases that *H. pylori* and the comparatively high transferase activities found in the measurements suggest that the bacterium is capable of effectively salvaging purines through this route (Mendz et al., 1994, 1996, 1997).

Phosphorylation of the corresponding nucleoside is one way to synthesize IMP or GMP from inosine or guanosine, respectively. Within *E. coli* and S. enterica serovar Typhimurium, guanosine kinase, which is encoded by gsk, is the enzyme that catalyzes this step. This pathway is absent in *H. pylori*, as evidenced by the lack of guanosine kinase activity and the corresponding gene. Within *E. coli* and *S. enterica serovar Typhimurium*, a key mechanism for the recovery of purine nucleosides involves their breakdown to the corresponding base by a purine-nucleoside phosphorylase encoded by deoD, which is then followed by the relevant phosphoribosyl transferases' conversion to a nucleotide monophosphate.

H. pylori did not exhibit any purine-nucleoside phosphorylase activity, although its genome contained the orthologous gene HP1178 to deoD. The comparatively high adenine and guanine nucleosidase activities seen in H may help to explain this discrepancy, pylori cell extracts, which can act as a stand-in for purine-nucleoside phosphorylase in the nucleoside hydrolysis process by hydrolyzing the nucleosides to ribose and the corresponding base. Since the phosphorylase enzyme is measured concurrently with the nucleosidase activity, it is possible that the phosphorylase activity was obscured by the nucleosidase activity. Alternatively, the bacterial growth conditions used in that study might prevent deoD from being expressed. Purine nucleosides in the growth medium stimulate the synthesis of purine-nucleoside phosphorylase in *E. coli* and *S. typhimurium serovar Typhimurium* while simultaneously suppressing the contributions of the de novo pathway to the purine nucleotide pool. The fact that *H. pylori* had significant purine nucleoside phosphotransferase activity was noteworthy, extracts from cells. These enzymes may represent a different nucleoside salvage pathway not present in *E. coli* adenosine or guanosine, respectively, and phosphorylate them to AMP or GMP, *B. subtilis*, or *S. enterica serovar Typhimurium*, subtilis. There are no known genes in the H that encode these enzymes, pylori genetic code.

The main pathway for adenosine utilization in *H. pylori* is suggested by the significant activities of adenosine nucleosidase and adenosine phosphoribosyl transferases, the lower activity of adenosine phosphotransferase, and the absence of adenosine kinase activity, *H. pylori* is through the purine ring being salvaged after adenosine is hydrolyzed to adenine, and phosphorylation by adenine phosphoribosyl transferases; however, adenosine phosphotransferase can also directly phosphorylate adenosine to produce AMP in certain amounts. When it comes to guanosine utilization, similar mechanisms are in place. Consequently, ribonucleoside and deoxyribonucleoside catabolism in *H. pylori* seem to take comparable routes as in *S. enterica serovar Typhimurium* and *E. coli*. Instead of being immediately phosphorylated in these bacteria, the nucleosides are first broken down into free bases, which then combine with PRPP to produce the corresponding nucleoside monophosphate derivative.The presence of adenylate kinase in *H. pylori* is suggested by the high activities of this enzyme and its capacity to phosphorylate AMP using ATP, GTP, or ITP is a useful method for assisting in maintaining the equilibrium of purine nucleotide pools inside cells. The lack of discernible AMP hydrolysis by *H. pylori* lysates show very little, if any, activity of AMP hydrolase and AMP nucleosidase. One significant metabolic function of purine nucleotides is the synthesis of coenzymes with the adenylate moiety, which explains the transfer of phosphate groups from GTP or ITP to AMP in *H. pylori* could be a strategy employed by the bacteria to maintain the required levels of coenzyme A, nicotinamide nucleotides, and flavin nucleotides (Mendz et al., 1994, 1996, 1997).

REFERENCES

1. Adair, L. B., & Jones, M. E. (1972). Purification and characteristics of aspartate transcarbamylase from *Pseudomonas fluorescens*. *Journal of Biological Chemistry*, *247*, 2308–2315.
2. Ahonkhai, I., Kamekura, M., & Kushner, D. J. (1989). Effects of salts on the aspartate transcarbamylase of a halophilic bacterium, *Vibrio costicola*. *Biochemistry and Cell Biology*, *67*, 666–669.
3. Alm, R. A., Ling, L. S., Moir, D. T., King, B. L., Brown, E. D., Doig, P. C., Smith, D. R., Noonan, B., Guild, B. C., de Jonge, B. L., Carmel, G., Tummino, P. J., Caruso, A., Uria-Nickelsen, M., Mills, D. M., Ives, C., Gibson, R., Merberg, D., Mills, S. D., Jiang, Q., Taylor, D. E., Vovis, G. F., & Trust, T. J. (1999). Genomic-sequence comparison of two unrelated isolates of the human gastric pathogen *Helicobacter pylori*. *Nature*, *397*, 176–180. https://doi.org/10.1038/16710.
4. Bergh, S. T., & Evans, D. R. (1993). Subunit structure of class A aspartate transcarbamylase from *Pseudomonas fluorescens*. *Proceedings of the National Academy of Sciences USA*, *90*, 9818–9822.
5. Bethell, M. R., & Jones, M. E. (1969). Molecular size and feedback regulation of bacterial aspartate transcarbamylase. *Archives of Biochemistry and Biophysics*, *134*, 352–365.
6. Burns, B. P., Hazell, S. L., & Mendz, G. L. (1997). *In situ* properties of aspartate carbamoyltransferase activity in *Helicobacter pylori*. *Archives of Biochemistry and Biophysics*, *347*, 119–125.
7. Burns, B. P., Hazell, S. L., & Mendz, G. L. (1998). A novel mechanism for resistance to the antimetabolite N-phosphonoacetyl-l-aspartate by *Helicobacter pylori*. *Journal of Bacteriology*, *180*, 5574–5579.
8. Burns, B. P., Hazell, S. L., Mendz, G. L., Kolesnikow, T., Tillett, D., & Neilan, B. A. (2000). The *Helicobacter pylori* pyrB gene encoding aspartate carbamoyltransferase is essential for survival. *Archives of Biochemistry and Biophysics*, *380*, 78–84. https://doi.org/10.1006/abbi.2000.2003.
9. Chang, T. Y., & Jones, M. E. (1974). Aspartate transcarbamylase from *Streptococcus faecalis*. Purification, properties, and nature of an allosteric activator site. *Biochemistry*, *13*, 629–638. https://doi.org/10.1021/bi00701a017.
10. Chu, C., & West, T. P. (1990). Pyrimidine biosynthetic pathway of *Pseudomonas fluorescens*. *Journal of General Microbiology*, *136*, 875–880. https://doi.org/10.1099/00221287-136-5-875.
11. Copeland, R. A., Marcinkeviciene, J., Haque, T. S., Kopcho, L. M., Jiang,

W., Wang, K., Ecret, L. D., Sizemore, C., Amsler, K. A., Foster, L., Tadesse, S., Combs, A. P., Stern, A. M., Trainor, G. L., Slee, A., Rogers, M. J., & Hobbs, F. (2000). *Helicobacter pylori*-selective antibacterials based on inhibition of pyrimidine biosynthesis. *Journal of Biological Chemistry, 275*, 33373–33378. https://doi.org/10.1074/jbc.M004367200.

12. Grem, J. L., King, S. A., O'Dwyer, P. J., & Leyland-Jones, B. (1988). Biochemistry and clinical activity of N-(phosphonacetyl)-l-aspartate: A review. *Cancer Research*, *48*, 4441–4454.
13. Guy, H. I., & Evans, D. R. (1994). Cloning and expression of the mammalian multifunctional protein CAD in *Escherichia coli*. Characterization of the recombinant protein and a deletion mutant lacking the major interdomain linker. *Journal of Biological Chemistry*, *269*, 23808–23816. https://doi.org/10.1016/S0021-9258(19)46949-5.
14. Jyssum, S., & Jyssum, K. (1979). Metabolism of pyrimidine bases and nucleosides in *Neisseria meningiditis*. *Journal of Bacteriology*, *138*, 320–323. https://doi.org/10.1128/JB.138.2.320-323.1979.
15. Kaneko, T., Tanako, A., Sato, S., Kotani, H., Sazuka, T., Miyajima, N., Sugiura, M., & Tabata, S. (1997). Sequence analysis of the genome of the unicellular cyanobacterium *Synechocystis* sp. strain PCC 6803. I. Sequence features in 1 Mb region from map position 64% to 92% of the genome. *DNA Research*, *2*, 153–166. https://doi.org/10.1093/dnares/2.3.153.
16. Kenny, M. J., McPhail, D., & Shepherdson, M. (1996). A reappraisal of the diversity and class distribution of aspartate transcarbamylases in gram-negative bacteria. *Microbiology*, *142*, 1873–1879. https://doi.org/10.1099/00221287-142-7-1873.
17. Mendz, G. L., Jimenez, B. M., Hazell, S. L., Gero, A. M., & O'Sullivan, W. J. (1994). De novo synthesis of pyrimidine nucleotides by *Helicobacter pylori*. *Journal of Applied Bacteriology*, *77*, 1–8. https://doi.org/10.1111/j.1365-2672.1994.tb01939.x.
18. Mendz, G. L., Jimenez, B. M., Hazell, S. L., Gero, A. M., & O'Sullivan, W. J. (1994). Salvage synthesis of purine nucleotides by *Helicobacter pylori*. *Journal of Applied Bacteriology*, *77*, 674–681. https://doi.org/10.1111/j.1365-2672.1994.tb01957.x.
19. Mendz, G. L., Shepley, A. J., Hazell, S. L. (1996). Survival of *Helicobacter pylori* by *de novo* synthesis of pyrimidine nucleotides. *Gut Suppl.*, *39*, A73.
20. Mendz, G. L., Shepley, A. J., Hazell, S. L., & Smith, M. A. (1997). Purine metabolism and the microaerophily of *Helicobacter pylori*. *Archives of Microbiology*, *168*, 448–456. https://doi.org/10.1007/s002030050425.

21. Misawa, M., Nara, T., Udagawa, K., Abe, S., & Kinoshita, S. (1964). *Agricultural and Biological Chemistry, 28*, 688.

22. Nakayama, K., Sato, Z., Tanaka, H., & Kinoshita, S. (1968). *Agricultural and Biological Chemistry, 32*, 1331.

23. Nara, T., Misawa, M., & Kinoshita, S. (1968a). *Agricultural and Biological Chemistry, 32*, 561.

24. Neuhard, J., & Kelln, R. A. (1996). Biosynthesis and conversion of pyrimidines. In F. C. Neidhart (Ed.), *Escherichia coli and Salmonella: Cellular and Molecular Biology* (2nd ed., pp. 580–599). American Society for Microbiology.

25. Neumann, J., & Jones, M. E. (1964). End-product inhibition of aspartate transcarbamylase in various species. *Archives of Biochemistry and Biophysics, 104*, 438–447.

26. Nygaard, P. (1993). Purine and pyrimidine salvage pathways. In A. L. Sonenshein (Ed.), *Bacillus subtilis and Other Gram-Positive Bacteria: Biochemistry, Physiology and Molecular Genetics* (pp. 359–378). American Society for Microbiology.

27. O'Donovan, G. A., & Neuhard, J. (1970). Pyrimidine metabolism in microorganisms. *Bacteriological Reviews, 34*, 278–343.

28. Pragobpol, S., Gero, A. M., Lee, C. S., & O'Sullivan, W. J. (1984). Orotate phosphorybosyltransferase and orotidylate decarboxylase from *Crithidia luciliae*: Subcellular location of the enzymes and evidence for substrate channeling. *Archives of Biochemistry and Biophysics, 230*, 285–293.

29. Purcarea, C., Erauso, G., Prieur, D., & Hervé, G. (1994). Aspartate transcarbamylase from the deep-sea hyperthermophile archeon *Pyrococcus abyssi*: Genetic organisation, structure, and expression in *Escherichia coli. Microbiology, 140*, 1967–1975.

30. Quinn, C. L., Stephenson, B. T., & Switzer, R. L. (1991). Functional organisation and nucleotide sequence of the *Bacillus subtilis* pyrimidine biosynthetic operon. *Journal of Biological Chemistry, 266*, 9113–9127.

31. Reynolds, D. J., & Penn, C. W. (1994). Characteristics of *Helicobacter pylori* growth in a defined medium and determination of its amino acid requirements. *Microbiology, 140*, 2649–2656.

32. Roland, K. L., Powell, F. E., & Turnbough, C. L. (1985). Role of translation and attenuation in the control of pyrBI expression in *Escherichia coli. Journal of Bacteriology, 163*, 991–999.

33. Roth, B. (1983). Selective inhibitors of bacterial dihydrofolate reductase: Structure-activity relationships. In G. H. Hitchings (Ed.), *Inhibition of*

Folate Metabolism in Chemotherapy (pp. 107–127). Springer-Verlag.

34. Shepley, A. J., Mendz, G. L., & Hazell, S. L. (1995). The essential role of de novo pyrimidine nucleotide synthesis in *Helicobacter pylori*, abstr. P-58. In *Proc. 7th FAOBMB Congress*. Australian Society for Biochemistry and Molecular Biology.
35. Swyrid, E. A., Seaver, S. S., & Stark, G. R. (1974). N-(phosphonacetyl)-l-aspartate, a potent transition state analog inhibitor of aspartate transcarbamylase, blocks proliferation of mammalian cells in culture. *Journal of Biological Chemistry*, *249*, 6945–6950.
36. Tomb, J. F., White, O., Kerlavage, A. R., Clayton, R. A., Sutton, G. G., Fleishmann, R. D., Ketchum, K. A., Kienk, H. P., Gill, S., Dougherty, B. A., Nelson, K., Quackenbuch, J., Zhou, L., Kirkness, E. F., Peterson, S., Loftus, B., Richardson, D., Dodson, R., Khalk, H. G., Glodek, A., McKenney, K., Fitzgerald, L. M., Lee, M., Adams, M. D., Hickey, E. K., Berg, D. E., Gocayne, I. D., Fujii, C., Bowman, C., Watthey, L., Wallin, E., Hayes, W. S., Borodovsky, M., Karp, P. D., Smith, H. O., Fraser, C. M., & Venter, J. C. (1997). The complete genome sequence of the gastric pathogen *Helicobacter pylori*. *Nature*, *388*, 539–547. https://doi.org/10.1038/41303.
37. Traut, T. W., & Jones, M. E. (1977). Inhibitors of orotate phosphoribosyltransferase and orotidine-5′-decarboxylase from mouse Ehrlich ascites cells: A procedure for analyzing the inhibition of a multi-enzyme complex. *Biochemical Pharmacology*, *26*, 2291–2296.
38. Turner, R. J., Lu, Y., & Switzer, R. L. (1994). Regulation of the *Bacillus subtilis* pyrimidine biosynthetic (pyr) gene cluster by an autogenous transcriptional attenuation mechanism. *Journal of Bacteriology*, *176*, 3708–3722.
39. Umezu, K., Amaya, T., Yoshimoto, A., & Tomita, K. (1971). Purification and properties of orotidine-5′-phosphate pyrophosphorylase and orotidine-5′-phosphate decarboxylase from baker's yeast. *Journal of Biochemistry (Tokyo)*, *70*, 249–262.
40. Wheeler, P. R. (1990). Recent research into the physiology of *Mycobacterium leprae*. *Advances in Microbial Physiology*, *31*, 71–124.
41. Wild, J. R., & Belser, W. L. (1977). Pyrimidine biosynthesis in *Serratia marcescens*: A possible role for nonsequential enzyme interactions in mimicking coordinate gene expression. *Biochemical Genetics*, *15*, 157–172.
42. Wild, J. R., & Belser, W. L. (1977). Pyrimidine biosynthesis in *Serratia marcescens*: Polypeptide interactions of three nonsequential enzymes. *Biochemical Genetics*, *15*, 173–180.

Chapter **6**

Amino Acids

Contents

INTRODUCTION

Amino acids are the building blocks of proteins and peptides, they are known as the building blocks of life. The nitrogen content of amino acids is relatively high (~16%), setting them apart from fats and carbohydrates. Amino acids are divided into two groups: nonessential amino acids, which the body synthesizes, and essential amino acids, which are not synthesized. Because they are essential for food, feed, personal care, and pharmaceutical industries as nutrients, additives, rejuvenators, and medications, amino acids are in high demand and have many uses. They have distinct and important roles in a variety of applications and have the potential to be therapeutic. A few essential amino acids, like methionine and lysine, have seen a sharp rise in demand over the past 20 years due to their widespread use in the food, feed, and pharmaceutical industries. These amino acids are added to animal feed to enhance the quality and quantity of meat produced while also maximizing animal growth. The availability of these animal products has undoubtedly been very helpful in overcoming the essential amino acid deficiency that is quite noticeable in the staple foods of underdeveloped and densely populated regions of the world. Due to its widespread use in the food industry, the demand for non-essential amino acids, such as glutamic acid, has also grown dramatically in recent years. The demand for different amino acids is expected to rise over the next few decades, so scientists will need to concentrate more on creating sophisticated manufacturing processes that make use of biotechnological tools.

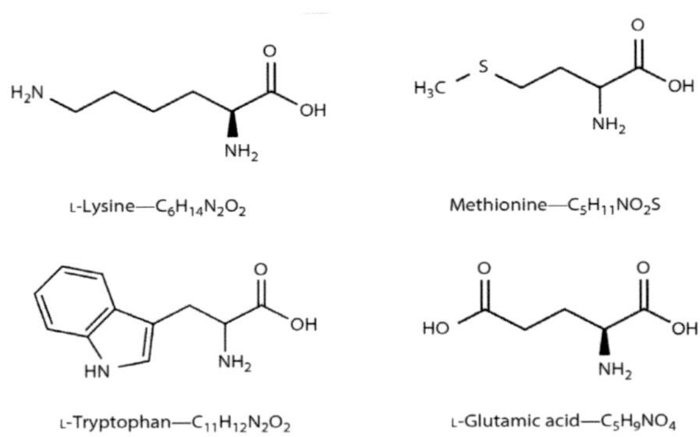

Figure 6.1. *Structure and molecular formula of lysine, methionine, tryptophan, and glutamic acid.*

Source: https://www.pnpcollege.edu.in/upload/lib_teaching_material/science/botany/2015-Micro-bialamino-acids-production.pdf.

This chapter's objective is to provide a thorough overview of several of the aforementioned amino acids, including glutamic acid, lysine, methionine, and tryptophan (Figure 6.1), as well as their historical contexts, current issues, and the use of biotechnological techniques and instruments to boost bacterial strains' production capacities.

The use of microorganisms in the production of goods for sale is thought to have begun in prehistoric times. As a result, fermented foods and drinks have been produced, but they lack the authentic meaning of fermentation. Pasteur was the one who first used the term "fermentation" and disproved the notion of spontaneous generation. Many researchers have used bacteria and filamentous fungi in their subsequent studies along these lines. However, the perception and comprehension of microbial physiology and metabolic processes were not fully realized until the scientific interpretation of fermentation began.

A number of metabolites, including antibiotics and amino acids, were discovered as a result of the biotransformation of organic molecules into beneficial products, which introduced a novel idea to the history of fermentation. Numerous pharmaceutical and biochemical products have been discovered as a result of the ongoing research on the topic. Simultaneously, developments in fermentation engineering greatly aided in the commercial production of these beneficial products. With the advancement of knowledge in molecular biology, genetic engineering, and microbial physiology, the fermentation industry has expanded significantly in terms of introducing novel products with excellent industrial and medical applications. These developments have added significantly to our understanding of these microorganisms, their isolation, modification, and growth conditions. They have also opened the door to the possibility of increasing the productivity of the desired product by adjusting specific nutrients or optimizing the process.

Modern biotechnology began with early advancements in strain mutation and the application of microbial physiology to create appropriate media composition for increased yield of desired products. The assessment of novel biological commodities with potential applications in pharmaceuticals, nutraceuticals, or cosmeceuticals has advanced remarkably, with fermentation engineering and industrial microbiology playing opposing roles. The increased commercial exploitation of biological processes suggests that biotechnology will become a rapidly expanding industry in the near future, one that will have a profound impact on people's lives and well-being worldwide.

6.1. L-GLUTAMIC ACID

6.1.1. Historical Background and New Challenges

One of the most commercially significant amino acids is *L*-glutamic acid. Glutamates are the carboxylate ions and salts of glutamic acid. This amino acid was first discovered in wheat gluten treated with sulfuric acid in 1866 by German scientist Karl Heinrich Leopold Ritthausen (Plimmer and Hopkins, 1912). At Tokyo Imperial University in 1907, Kikunae Ikeda produced the crystalline glutamic acid salt, monosodium glutamate (MSG), by developing and patenting a production process. This marked the first successful commercial production of glutamic acid. The Japanese company Ajinomoto Corporation subsequently filed for a patent on MSG in 1909. The discovery of *Micrococcus glutamicus*, which ferments glucose medium to produce 30 g/L of *L*-glutamic acid, was a turning point in the history of glutamic acid production (Kinoshita et al., 1957b). The increased demand for MSG, used as a flavor-enhancing agent, led to interest in large-scale *L*-glutamic acid fermentation.

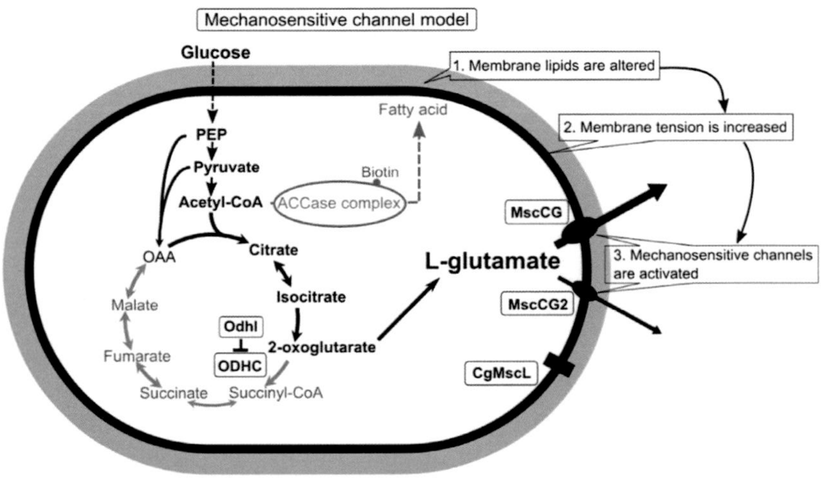

Figure 6.2. *Biochemistry of C. glutamicum.*

Source: Nakayama Y. (2021). Corynebacterium glutamicum mechanosensing: From osmoregulation to l-glutamate secretion for the avian microbiota-gut-brain axis. Microorganisms, 9(1), 201. https://doi.org/10.3390/microorganisms9010201.

Manufacturers face challenges as market demand continues to rise. They can choose to increase the capacity of their current plants or use cutting-edge biotechnological tools to enhance production through strain improvement and fermentation medium modification. The biosynthesis of *L*-glutamic acid using

Corynebacterium glutamicum involves relatively simple metabolic pathways.

Glycolytic pathways are primarily involved in the metabolism of glucose. The synthesis of oxaloacetate and α-ketoglutarate depends on pyruvate and phosphoenol pyruvate. When glutamate dehydrogenase is present, α-ketoglutarate is transformed into l-glutamic acid (Figure 6.2). Because the genes governing this conversion process are situated next to the genes governing cell wall synthesis and cell division, the phosphorylation stage is highly significant. Consequently, disruptions in cell wall synthesis or cell division processes may prevent α-ketoglutarate dehydrogenase activity from occurring. Consequently, there could be an impact on the process of α-ketoglutarate conversion to glutamic acid.

6.1.2. Biotechnological Approach and Production

A great deal of biotechnological tools have been used by various researchers with an understanding of microbial physiology to modify the strains for increased production of glutamic acid, especially utilizing *C. glutamicum*. The progress, with respect to the capabilities of the different strains of *C. glutamicum* to produce glutamic acid, have been reported from time to time over the last few decades. The biology, biosynthetic pathways, nutritional requirements, mutations, and potential mechanosensitive channels of this organism were all thoroughly understood before doing this. Numerous researchers have examined the microbial synthesis of l-glutamic acid in great detail, employing various Coryneform species strains. The method for producing l-glutamic acid using these strains has even been patented. Among the most well-known Coryneform species are *C. glutamicum*, *Brevibacterium lactofermentum*, *Brevibacterium ammoniagenes*, *Brevibacterium thiogenetalis*, *Brevibacterium saccharoliticum*, *Corynebacterium lilium*, *Corynebacterium herculis*, *Brevibacterium flavum*, and *Brevibacterium divarticum* (Kinoshita 1999). *Bacillus cereus*, *Bacillus megaterium*, *Bacillus circulans*, *Escherichia coli*, and *Sarcina lutea* are additional organisms that produce glutamic acid. In the industrial setting, genetically modified strains of *Corynebacterium* or *Brevibacterium* are used in batch/fed-batch submerged fermentation processes to produce glutamic acid. Large fermenters furnished with every amenity needed to regulate various aspects and maximize yield—including cooling, pH and dissolved oxygen measurement, and others—are used for commercial production. A typical example of the submerged fermentation process on a laboratory scale for the production of l-glutamic acid utilizing a mutant strain of *C. glutamicum* has been summarized below.

Organisms are usually maintained on Hottinger slant agar containing beef trypsin digestion, sodium chloride, dipotassium phosphate, and agar at low temperature. After subculturing, a 24-h-old culture is used to inoculate 25 mL of sterile seed medium in 250 mL Erlenmeyer flasks in triplicate. The following seed medium composition can be used: glucose (8%), NH_4Cl (0.5%), corn steep liquor (0.3%), K_2HPO_4 (0.5%), KH_2PO_4 (0.5%), $MgSO_4{\cdot}7H_2O$ (0.03%), $CaCO_3$ (1.0%), and deionized water to make 100%. The pH of the medium can be adjusted to 7.2 using NaOH. The inoculated flasks are grown in an orbital shaker incubator maintained at 30°C and 230 rpm for 15 h. The entire contents of the one flask is then transferred to a 2.0 L capacity Eyla Mini Jar fermenter with 500 mL of sterile nutrient medium containing molasses (20%), KH_2PO_4 (0.5%), KH_2PO_4 (0.5%), $MgSO_4{\cdot}7H_2O$ (0.3%), urea (0.8%), $CaCO_3$ (1.0%), and deionized water to make 100%. In most cases, the optimum pH of the medium was recorded as 7.0. The fermentation is usually initiated with continuous agitation and aeration for 48 h at 30°C. At the end of the incubation period, the culture broth is separated from the cells by centrifugation at 10,000 rpm for 5 min at 4°C and the supernatant is diluted 50-fold with 7% (v/v) of glacial acetic acid. The diluted sample is then further centrifuged at 10,000 rpm for 5 min at 4°C. The supernatant is then finally filtered through a nylon membrane of 0.22 µm pore size and then collected for further analysis and crystallization (Pasha et al., 2011; Khan et al., 2013a,b).

Since the discovery of the original strain of *C. glutamicum*, a large number of strains, which are capable of producing significant quantities of glutamic acid, have been developed through the process of genetic engineering. Yields have now exceeded 100 g/L and even more with modified strains compared with original wild strains, which produced only 10 g/L. Most original glutamic acid-producing strains of *C. glutamicum* were observed as biotin auxotrophs. Thus, growing such strains in a medium deficient in biotin was found to "trigger" glutamic acid production.

The biotin-deficient medium has the property of altering the cell membrane of the organism because of suboptimal fatty acid biosynthesis and is thus responsible for increased production. In view of this hypothesis, it is believed that some other physicochemical factors such as fermentation at higher temperatures, the addition of surfactants (e.g., Tween 40) or antibiotics (e.g., penicillin) in the culture medium can also trigger the excretion of glutamic acid. These factors have been linked to a triggering mechanism resulting in a decrease or repression of the enzyme α-ketoglutarate dehydrogenase. Thus, finally causing redistribution of metabolites at the branch point in the tricarboxylic acid (TCA) cycle leading from α-ketoglutarate to succinyl-CoA or glutamate (Figure 6.2). However, this mechanism has limitations, therefore, a further increase or overproduction of glutamate can be achieved through optimization of the metabolic flux by tuning the glutamate dehydrogenase activity. l-Glutamic acid production was also achieved through immobilized cells of *C. glutamicum*. Using more than 93 g/L, good yields were achieved through batch fermentation but with low productivity (3.8 g/L/h), whereas 73 g/L of glutamic acid was recovered through continuous fermentation with high productivity of approximately 29.1 g/L/h. Several other studies have also been published, documenting the role of immobilized cells in the production of l-glutamic acid.

It has been reported that the immobilization of microbial cells in biological processes can occur either as a natural phenomenon or through artificial processes. Although attached cells in their natural habitat exhibit significant growth, the artificially immobilized cells also allow restructuring of growth. The authors reported obtaining 48.5 g/L of l-glutamic acid through the immobilized cells of a mutant strain of *C. glutamicum*. The use of immobilized cells for the production of l-glutamic acid and some other amino acids is progressing rapidly and is expected to have some more valuable and useful data for commercial exploitation in the near future.

6.1.3. Metabolic Pathways Involved in the Biosynthesis of Glutamic Acid from Glucose

Figure 6.3 illustrates the glucose-to-glutamate biosynthesis route. The main path, indicated by the heavy arrows, requires a minimum of 16 enzyme steps. Reductive amination is the process that turns alpha-oxoglutarate into glutamate. Glutamate dehydrogenase that is specific to $NADP^+$ is the enzyme that is causing this conversion.

This enzyme must exist in order for glutamate to form. α-oxoglutarate instead of glutamate accumulates when glucose is incubated with resting cells without the presence of ammonia. The previous isocitrate dehydrogenase reaction provides the NADPH needed for the glutamate dehydrogenase reaction. In turn, glutamate dehydrogenase supplies the $NADP^+$ needed by isocitrate dehydrogenase.

The low content of α-oxoglutarate dehydrogenase favors glutamate production. The importance of the lack of this enzyme has been demonstrated with E. colt which does not require biotin and is not a glutamate excreter. Even without the biotin requirement, a mutant which lacks ce-oxoglutarate dehydrogenase was found to excrete 2.3 g of glutamate per liter while its parent excreted none.

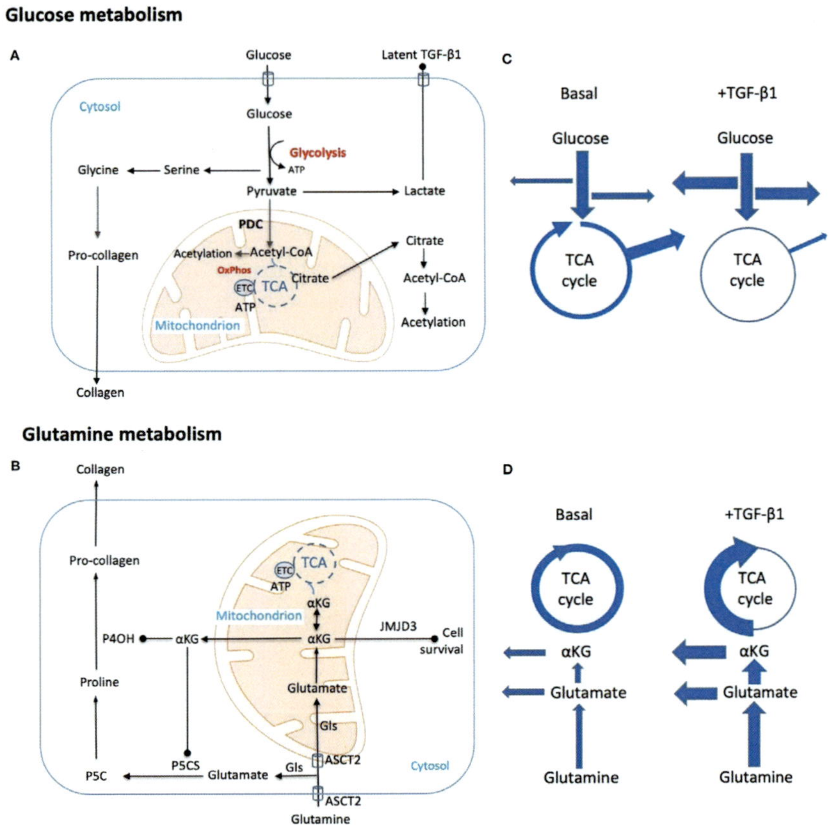

Figure 6.3. *Metabolic pathways involved in biosynthesis of glutamic acid from glucose.*

Source: Hewitson TD and Smith ER (2021) A Metabolic reprogramming of glycolysis and glutamine metabolism is a requisite for renal fibrogenesis—Why and how? Front. Physiol., 12, 645857. doi: 10.3389/fphys.2021.645857.

In addition to the EMP pathway, glutamic acid-producing bacteria also use the HMP pathway to convert glucose to three-carbon and two-carbon compounds. These compounds can then feed into the TCA cycle. Estimates of the relative utilization of glucose by the two pathways have shown the predominance of the EMP pathway under fermentation conditions.

When a glutamate fermentation is carried out in the presence of $^{14}CO_2$, the radioactivity is fixed into the α-carboxyl group of glutamate. Two enzymes have been found in glutamate excretion which participate in the fixation of carbon dioxide, namely oxaloacetate carboxylase and the $NADP^+$ -linked malic enzyme which catalyzes the fixation of carbon dioxide to pyruvate to yield malate.

Malate is oxidized to oxaloacetate by malate dehydrogenase. The oxaloacetate is converted to citrate. Because of a deficiency in the TCA cycle, glutamic acid-producing bacteria use the glyoxylate pathway shown by thin lines in Figure 6.3. Therefore, the two competing reactions of isocitrate are important. The isocitrate reaction is needed during the growth phase for energy production and to produce intermediates for biosynthetic reactions. However, after the growth phase, glutamate production would be best without the operation of the isocitrate reaction. This would indicate that optimal conditions for the growth phase and the glutamate production phase should be different. After growth, the ideal fermentation would proceed by the following reaction:

$$C_6H_{12}O_6 + NH_3 + 1.5O_2 \rightarrow C_5H_9O_4N + CO_2 + 3H_2O$$

This represents a 100% molar conversion of 81.7% weight conversion of sugar to glutamic acid. On the other hand, the poorest fermentation would be represented by complete oxidation of glucose as occurs with growth in media containing an optimal concentration of biotin and which results in no conversion. What is actually found lies between these limits, i.e., 50–75% molar conversion by resting cells. This indicates that carbon is being lost as carbon dioxide by reversal of action of the malic enzyme and of oxaloacetate carboxylase during operation of the glyoxylate bypass. Therefore, assimilation of ammonium ions in this organism is almost totally dependent upon action of $NADP^+$ -linked glutamate dehydrogenase.

6.1.4. Alteration of Permeability and Glutamic Acid Production

Alteration of permeability in relation to glutamic-acid production was reviewed by Demain and Birnbaum (1968). The permeability barrier is altered in bacteria to produce a significant amount of glutamic acid. Penicillin added to the growth medium, fatty acid derivative treatment, oleic acid deficiency in an oleic acid-requiring auxotroph, glycerol deficiency in a glycerol-requiring auxotroph, and biotin deficiency can all cause increased permeability in glutamic-acid bacteria. A deviation in the typical.

It was discovered that auxotrophs that require oleic acid and biotin cannot be used to produce L-glutamic acid from n-paraffins. When penicillin or cephalosporin was added, the excretion of L-glutamic acid was followed by the excretion of N-acetylglucosamine and phospholipid, which are known to be parts of the cell wall and membrane, respectively. This observation led to the discovery of a Corynebacterium alkanolyticum auxotroph that requires glycerol, and in this auxotroph, the amount of glycerol supplied controls the synthesis of phospholipids in the cells. In the culture, the mutant generated approximately 40 g of L-glutamic acid per liter from n-paraffins when glycerol was present at 0.01% but penicillin was not. Research utilizing the auxotroph indicates that the phospholipid content regulates the membrane permeability to L-glutamic acid and that the cellular content of unsaturated fatty-acyl residues does not always control the permeability of L-glutamic acid. The addition of penicillin to log-phase cultures of *C. glutamicum* results in a rapid (97–99.5%) decrease in viability accompanied by a rapid rate of glutamate excretion. Cell mass (measured by optical density) and total cell count remain fairly constant after a slight initial increase. The packed cell volume, however, decreases by 60–80% during the phase of decreasing viability indicating a change in the surface properties of the cells. The cells continue to produce glutamate for 40–50 h after penicillin addition with no lysis throughout the entire fermentation. Apparently, the lack of an overabundance of potent mucopeptidases allows the cells to retain their form and to metabolize as permeable 'resting' entities. During growth in a medium containing a high concentration of biotin, *C. glutamicum* synthesizes glutamate until the cell becomes saturated at 25–35 μg per mg of dry weight. Cessation of production was assumed to be due to some feedback control of glutamate on its own synthesis. When the cells become more permeable, glutamate passed from the cells into the medium, thus relieving the feedback effect and allowing further synthesis.

6.1.5. Industrial Application and Therapeutic Role

The food industry uses glutamic acid and its salt most frequently to enhance flavor. A significant amount of free glutamic acid, also known as its salt MSG, can be found in a variety of food items, such as soy sauces and cheese. Furthermore, glutamate serves as a special brain fuel and carries out a number of other crucial tasks like detoxifying ammonia, acting as a hepatoprotective agent (glutamine), assisting in the healing of peptic ulcers, and more because it is an essential component of cellular metabolism (Chaitow, 1985; Zareian et al., 2012).

As a neurotransmitter, glutamic acid plays a major role in pharmaceuticals. The central nervous system (CNS) contains large amounts of glutamic acid, which is thought to function as a neurotransmitter. Additionally, it might stimulate the cerebral cortex's metabolism, which would enhance mental clarity and memory. Glutamic acid has a greater concentration in the central nervous system than any other neurotransmitter that is widely known. Furthermore, it appears from the biochemical and electrophysiological data that glutamic acid or its analogue functions as an excitatory neurotransmitter in the central nervous system. The hippocampus has a high concentration of N-methyl-d-aspartate (NMDA) receptors. Blocking NMDA receptors can have a significant impact on a person's memory and general mental function. In order to increase cation conductance, depolarize the cell membrane, and unblock NMDA receptors, glutamic acid and aspartic acid can combine with NMDA receptors. These biochemical characteristics ultimately result in enhanced mental and memory function (McEntee and Crook, 1993).

6.2. L-LYSINE

6.2.1. Historical Background and New Challenges

Among the list of essential amino acids, l-lysine is one of the most popular and widely used amino acids. It plays a significant role in the nutrition of both humans and animals. The possibility of producing amino acids by microorganisms was first thoroughly investigated in the late 1940s, and in the latter half of the 1950s, some encouraging findings were published (Mitchell and Houlahan, 1948; Windsor, 1951).

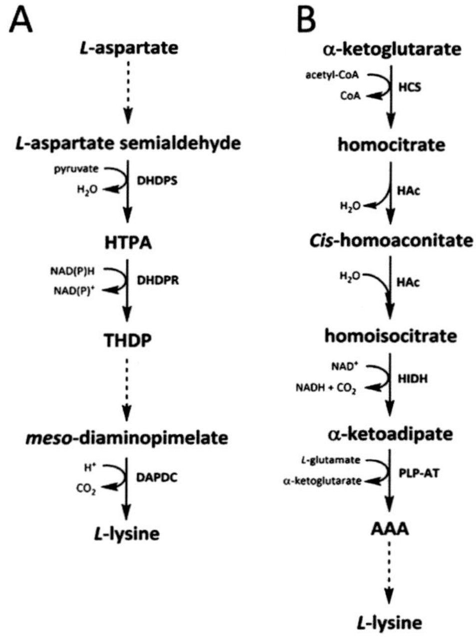

Figure 6.4. *Production of l-lysine through the conversion of its immediate precursor DAP (two-step process).*

Source: By Tatiana soaresdacosta – Own work, CC BY-SA 4.0, https://commons.wikimedia.org/w/index.php?curid=65237872.

Moreover, several researchers reported that yeast and Neurospora mold can produce l-lysine from α-aminoadipic acid, while *E. coli* can synthesize it from diaminopimelic acid (DAP). However, Chas Pfizer and Company Inc.'s microbial synthesis of l-lysine via DAP decarboxylation was recognized as the first commercially scaled production in the United States. Then, simultaneously, a modified process (Figure 6.4) was found and reported that used a single organism to produce DAP and convert it to l-lysine.

The introduction of l-glutamic acid fermentation using *C. glutamicum* had a great economic effect in the field of l-lysine fermentation. Further research in this direction resulted in the development of an efficient mutant of *C. glutamicum* for the commercial production of l-lysine (Kinoshita et al., 1957b, 1958). This modified method was also successfully used to produce some other amino acids (Mahmood 2010).

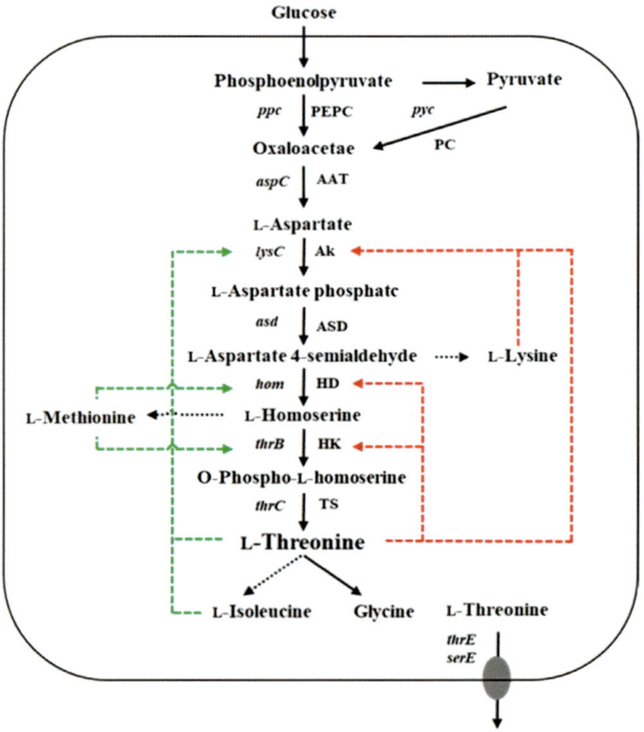

Figure 6.5. *Regulatory pathways of l-lysine biosynthetic pathways in C. glutamicum.*

Source: Duan M, Chen S, Liu X, Liu J, Zhu D. (2023). The application of corynebacterium glutamicum in L-threonine biosynthesis. Fermentation, 9(9):822. https://doi.org/10.3390/fermentation9090822.

Modifications in the biosynthetic pathway extended one of the biggest advantages by producing some other amino acids along with l-lysine using *C. glutamicum*, *B. flavum*, and *B. lactofermentum*. The isolation of auxotrophs can be carried out by using penicillin selection techniques followed by ultraviolet radiation. The identification of growth factors (such as vitamins and amino acids) of the auxotrophs can be made auxanographically. Further advancements in the production of l-lysine were achieved by introducing analogue-resistant *C. glutamicum C. glutamicum* and *B. flavum*. An l-lysine analogue, S-(2-aminoethyl)-l-cysteine (AEC), is the best example of developing such analogue-resistant strains. The development of an AEC-resistant strain is primarily based on the treatment of a suitable homoserine auxotroph in a suitably complete culture medium containing meat extract, peptone, yeast extract, sodium chloride, and agar. The medium is supplemented with N-methyl-N-nitro-N-nitrosoguanidine in phosphate buffer. The treatment time is short (~30 min) and at a very low temperature (0°C).

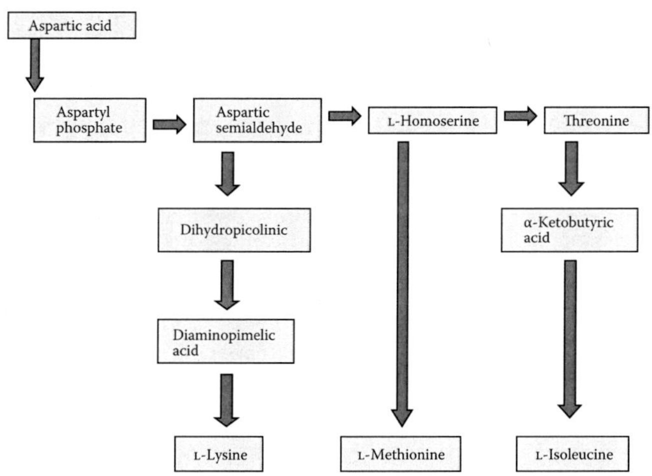

Figure 6.6. *Production of l-lysine by the auxotrophic mutants (one-step process).*

Source: https://www.pnpcollege.edu.in/upload/lib_teaching_material/science/botany/2015-Micro-bialamino-acids-production.pdf.

This is followed by the treatment of cells with AEC after washing with saline in a minimal medium supplemented with glucose as a carbon source, inorganic salts, and amino acids, for example, l-threonine and dl-methionine. It is now evident that auxotrophy and resistance to feedback inhibition when genetically combined into a single strain (i.e., mutants resistant to AEC), results in increased production of l-lysine. The overproduction of l-lysine might be due to the starvation of threonine, which decreases the feedback inhibition of aspartate kinase, the first key enzyme in l-lysine biosynthesis (Figure 6.5). The genetic induction of an auxotropic or regulatory mutant that can overproduce l-lysine is of course extremely important. However, at the same time, process optimization (especially the medium composition and the addition of certain growth factors and culture conditions) is also indispensable for the commercial production of l-lysine (Figure 6.6).

The increased demand for l-lysine around the world has made it difficult for researchers to develop new genetically modified organisms that can produce exceptional amounts of l-lysine or to further improve the strains that currently exist. Global Biochem Technology (China), COFCO Biochemical (China), Changchun Dacheng (China), Cheil Jedang (with facilities in China, Indonesia, and Brazil), Ajinomoto (with manufacturing facilities in Japan, France, Italy, United States, Brazil, China, and Thailand), and Archer Daniels Midlands (United States) are the world's leading producers of l-lysine. Together with other businesses, these significant producers generated 1,755,300 tons of l-lysine in 2011. For 2013, that amount is projected to reach 2,376,300 tons, and by the end of 2018, that amount is expected to have grown to 2,518,000 tons, or approximately US$5.9 billion. China is the world's largest producer of l-lysine and will likely remain so for some time to come, making up about 65% of the global capacity for lysine production (Wippler 2011; IHS Chemical 2012).

Meeting the production demand with cost-effective raw materials and recovery processes is quite challenging and will be crucial to the success of the project given the steadily rising demand for l-lysine. All of the raw materials used in the fermentative synthesis of l-lysine are substances that are found in nature or the body. Fermentation of l-lysine has not been found to produce any toxic byproducts. Conversely, valuable components are still present in the spent broth, allowing for the recovery of numerous byproducts. The spent broth still has a number of beneficial ingredients that could be added to animal feed or used as fertilizer, such as potassium salts, phosphorus compounds, and organic and inorganic nitrogen compounds. It has also been reported that feed-grade l-lysine hydrochloride can be successfully produced by a recent process that includes spray-drying the entire fermented broth at the conclusion of the incubation period (Eggeling and Sahm, 2011). This will undoubtedly lower the cost of waste material disposal as well as the production of feed-grade l-lysine.

6.2.2. Biotechnological Approach and Production

Even though l-lysine production has advanced significantly, scientists are still working to create an economical fermentation process with new biotechnological techniques. In addition, biotechnology firms are always looking for new research directions and attempting to attain market leadership in the production of l-lysine and other amino acids through the use of diverse management models and business strategies. In addition to strains with high productivity, using economical raw materials and recovery techniques has become crucial for various manufacturers to compete and survive in the market. Introduced for the purpose of producing l-lysine, genetically modified strains have the capacity to produce up to 170 g/L. Metabolic engineering techniques can be employed to achieve more sophisticated and focused advancements in the l-lysine biosynthetic pathway and its secretion. Maintaining control over these dynamics will undoubtedly guarantee efficient delivery of precursors and meet the cells' energy needs. Researcher interest in engineering new targets for l-lysine overproduction is growing for energy metabolism and osmoregulation, in addition to the genetic engineering of transport systems, biosynthetic pathways, and central metabolic pathways. Furthermore, strains other than *C. glutamicum*, such as *Bacillus methanolicus* (methylotropic bacteria) are also gaining interest in the production of l-lysine using methanol and lignocellulose as an alternate raw material (Brautaset and Ellingsen, 2011).

A certain amount of attention has been drawn to ensuring an ideal carbon and energy flow within the bacterial cell central metabolism (CCM) to achieve optimized metabolic production, thanks to advancements in molecular biology and genetic engineering. A specific example is that of *C. glutamicum* in which the CCM involves glycolysis, the pentose phosphate pathway, and the TCA along with the anaplerotic and gluconeogenic reactions. It is anticipated that a thorough understanding of CCM and the unique enzyme activities of the pathways will undoubtedly present a chance to establish cell factories that are perfect for producing not only l-lysine but also a few other industrially significant metabolites (Papagianni, 2012).

The selection of bacterial strains, media composition, process optimization, and end product recovery all play a major role in the overall yield and economics of the industrial production of l-lysine. It is important to balance the sources of carbon and nitrogen, as well as the strain's nutritional needs and the addition of specific precursors. Additionally, a higher yield of l-lysine depends critically on having an adequate supply of oxygen to meet the cell's oxygen needs. Submerged fermentation is the most widely used method worldwide and is very effective when applied on an industrial scale. The main raw ingredients used in the fermentation of l-lysine, such as starch, glucose, molasses, and others, are widely available and used in other fermentation-related industries. The most commonly used organism for l-lysine fermentation is *C. glutamicum*. Therefore, to understand its production, a specific example has been mentioned below using a genetically modified strain of *C. glutamicum*.

On nutrient agar slants, organisms are typically kept at a low temperature of 6 to 8°C according to a monthly transfer schedule. For seed, a 24-hour culture generated by subculturing from stock culture is typically utilized, succeeded by fermentation. It is possible to use a natural or synthetic medium that can meet the needs of the specific strain. Glucose and molasses are the two main carbon sources used in commercial starches. Care must be taken to create a balance between carbon and nitrogen sources such as corn steep liquor, soybean cake acid hydrolysate, yeast extract, peptone, and the like, along with the addition of inorganic salts such as KH_2PO_4, K_2HPO_4, $MgSO_4{\cdot}7H_2O$, $FeSO_4{\cdot}7H_2O$, $ZnSO_4{\cdot}7H_2O$, $MnSO_4{\cdot}7H_2O$, $(NH_4)_2\ SO_4$, and others. Other necessary nutrients are biotin and vitamin B1. In some cases, the production of l-lysine can be enhanced by using the leucine fermentation liquor (0.2%–15%) in the production medium. In most cases, the optimum pH of the medium has been recorded as 7.2 and temperature at 30°C. The seed stage cultivation requires around 24 h, whereas the fermentation stage is complete by approximately 96 h. After this, harvesting is done and the product, l-lysine, is recovered using some suitable and economical method.

- **Example: Production of L-Lysine on Glucose and Molasses Medium Supplemented with Fish Meal Under Submerged Condition in Mini Jar Fermenter**

Numerous researchers have emphasized the significance of l-lysine due to its use in the food, feed, and pharmaceutical industries. In children, l-lysine deficiency can have a significant impact on bone growth and development; in adults, l-lysine aids in calcium absorption and preserves nitrogen balance. As a result, it might also aid in preventing the osteoporosis-related bone loss. Additionally, it aids in the production of antibodies, hormones, enzymes, and collagen as well as strengthening the body's defense mechanisms, which are particularly effective against cold sores and herpesviruses. Its advantages in tissue repair are also indisputable.

The demand for l-lysine is thus continuously increasing, asking investigators to meet the challenge by increasing the production capacity using high-yield bacterial strains and the best-possible economical source of raw material, especially carbon and nitrogen for its production. This case study is also part of this program. In the current experiment, l-lysine production was evaluated with two different types of media using a homoserine auxotrophic mutant of *C. glutamicum* FRL no. 61989, which is also resistant to AEC. The auxotrophic and regulatory mutants are good options for increased l-lysine production. Isolation of the auxotrophic mutant is based on the treatment of the wild strain of *C. glutamicum* with ultraviolet radiation followed by a penicillin selection method, whereas AEC-resistant strains were developed by treating the homoserine auxotrophic mutant with N-methyl-N-nitro-N-nitrosoguanidine, followed by growing the cells in a medium supplemented with AEC. Colonies that appeared on the surface of the agar plate during 2 to 7 days of incubation were picked up as AEC-resistant mutants.

6.2.2.1. Cultivation of C. glutamicum FRL No. 61989

The organism was maintained on a monthly transfer schedule on nutrient agar slants. A seed medium of following composition in grams per liter was used: glucose (30.0), peptone (10.0), yeast extract (5.0), sodium chloride (3.0), KH_2PO_4 (0.5), K_2HPO_4 (1.5), $MgSO_4$ (0.5), NH_4SO_4 (0.5), sodium acetate (0.1), and 25.5 µg/L of biotin. The pH was adjusted to 7.2. Erlenmeyer flasks of 250 mL capacity containing 30 mL of the sterilized medium were inoculated with a loopful of culture grown on nutrient agar slant for 24 h. The flasks were shaken at 200 rpm at 30°C for 24 h in an orbital shaker-incubator to prepare the seed culture.

Fermentation media I and II with the following composition in grams per liter were used for comparative studies in terms of yield and cost affectivity. Medium I contains glucose (110.0), yeast extract (5.0), peptone (10.0), and meat extract (5.0). Medium II contains molasses (400.0), fish meal (25.0), and corn steep liquor (5.0). Both media were supplemented with the following ingredients in grams per liter: KH_2PO_4 (0.5), K_2HPO_4 (1.5), $MgSO_4{\cdot}7H_2O$ (0.5), $(NH_4)_2SO_4$ (15.0), calcium carbonate (10.0), $MnSO_4{\cdot}4H_2O$ (0.01), ferrous sulfate $4H_2O$ (0.01), sodium acetate (2.0), and vitamin B1 (0.001). The concentration of biotin was 125.0 and 25.0 µg/L in media I and II, respectively. The pH was adjusted to 7.2 for both media, and 1.0 mL of 20% silicon RD in deionized water was added as antifoam.

One liter of media I and II was transferred separately to 2.0 L capacity mini jar fermenters and serialized by autoclaving for 15 min at 15 psi. The respective fermenters containing media I and II were inoculated with the contents of one seed flask and stirred at 400 rpm with an air flow of 1.0 vvm at 30°C. The fermenters were harvested after 96 h.

6.2.2.2. Isolation, Extraction, and Purification of L-Lysine

Following the incubation period, the fermenters' temperature was raised to 60 °C for 15 minutes, and the liquid culture was then transferred to separate Erlenmeyer flasks with a capacity of 3.0 L. After cells and other precipitated materials were separated from the broths using a high-speed centrifuge set to 10,000 rpm for 20 minutes, the broths were purified and separated using the ion exchange method. The supernatants from both fermenters were adjusted to pH

2.0 using 6 N HCl and then run through a separate column, measuring 30 cm in length and 6 cm in diameter, filled with 250 g of the potent cation exchange resin Amberlite IR-120. To adsorb the *L*-lysine, the supernatant flow rate through the column was set at approximately 100 drops per minute. Following this, 500 mL of diluted ammonia was eluted using 250 mL of deionized water that had been passed through both columns twice. The entire liquid collected from each column was then run through the weak cation exchange resin Amberlite IRC-50, keeping the resin quantity, liquid flow rate, and column diameter constant. After two rounds of washing with 250 mL of deionized water, the solution was adjusted to pH 5 and treated for six hours with 10 g/L of activated carbon. To produce *L*-lysine crystals, the lysine and carbon slurry was vacuum filtered, concentrated to approximately 30% under reduced pressure, and then left overnight at 10 °C. The crystalline material was dried in a vacuum dryer after being filtered and rinsed with cold deionized water.

The purity of the material was checked by performing thin-layer chromatography in a mixture of n-propanol and strong ammonia solution (67:33), using ninhydrin as a spray reagent as well as by comparing the infrared absorption spectrum with standard l-lysine. The purity was recorded as 98.9% and 98.2%, respectively. An assay for l-lysine was also performed and the endpoint was determined potentiometrically. The average of three experiments produced 67.80 and 65.20 g/L of l-lysine monohydrochloride by glucose and molasses medium with a corresponding yield ($Y_{p/s}$) of 0.6116 gg^{-1} (61.6% conversion efficiency) and 0.592 gg^{-1} (59.2% conversion efficiency), respectively.

6.2.2.3. Effect of Media Composition and Fermentation Parameters on L-Lysine Production

The fermentation of l-lysine is observed to be a complex microbiological process that is influenced by many biochemical and physical parameters. Therefore, to achieve high product yields through the optimization of fermentation parameters and the selection of the most appropriate carbon and nitrogen sources, a thorough understanding of bacterial physiology and biochemistry is imperative. In the present case study, after several sideline studies, the nutritional requirement and fermentation parameters were established for a strain of *C. glutamicum* FRL no. 61989 and accordingly applied for a final comparative study to observe the production capacity on glucose and molasses medium. The strain FRL no. 61989 was developed by treating a homoserine auxotroph with AEC.

The selection of raw materials, particularly the source of carbon and nitrogen, is primarily determined by economic factors in the industrial production of l-lysine. Determining the ideal amount of carbon and nitrogen source based on the nutritional properties of the production culture is therefore crucial. Shake flask experiments with media containing various sugar concentrations supplemented with an adequate amount of nitrogen source to meet the demand imposed by raising the carbon level were employed in the current study as a method. The information required to determine the percentage yield of product is obtained by analyzing the amount of l-lysine generated and the amount of sugar left over after the process. One can consider the sugar concentration that yields the highest percentage of l-lysine to be the ideal concentration at which to proceed (Daoust 1976).

Our conclusion is that, on both glucose and molasses medium, the yield of l-lysine is relatively similar due to the current case study. Therefore, comparing the cost of glucose and molasses, the composition of medium II justified the use of molasses as a carbon source for the production of l-lysine. Furthermore, by lowering the concentration of biotin in the molasses medium, the cost of the product can be further decreased because molasses contains a sufficient amount of biotin. Similarly, considering the high cost of peptone, yeast extract, meat extract, and soybean protein acid hydrolysate, the use of fish meal as a nitrogen source was also justified. Therefore, with such reasonably priced materials, an l-lysine fermentation can be performed at a significantly lower cost.

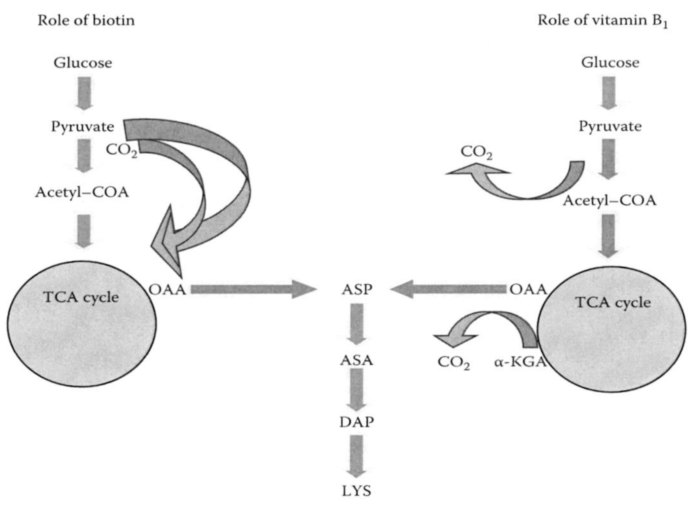

Figure 6.7. *Effect of biotin and vitamin B_1 on l-lysine production.*

Source: https://www.pnpcollege.edu.in/upload/lib_teaching_material/science/botany/2015-Micro-bialamino-acids-production.pdf.

Vitamin B_1 and biotin concentrations were set at 1.0 mg/L and 125 µg/L, respectively. It is well known that biotin and vitamin B_1 play important roles in the fermentation of amino acids. According to reports, biotin is involved in cell permeability, protein synthesis, glucose oxidation, and carbon dioxide transport of covalent bonds. Additionally, aspartic acid, a crucial intermediate that helps with the synthesis of l-lysine, is produced by enzymes that contain biotin. The oxidative decarboxylation of pyruvate and α-ketoglutaric acid in the bacterial system is said to be facilitated by vitamin B_1, which increases the production of acetyl-CoA and oxaloacetic acid, which in turn contributes to the increased formation of l-lysine. The impact of vitamin B_1 and biotin is emphasized in Figure 6.7. Additionally, Figure 6.8 presents the entire laboratory-scale l-lysine production mechanism.

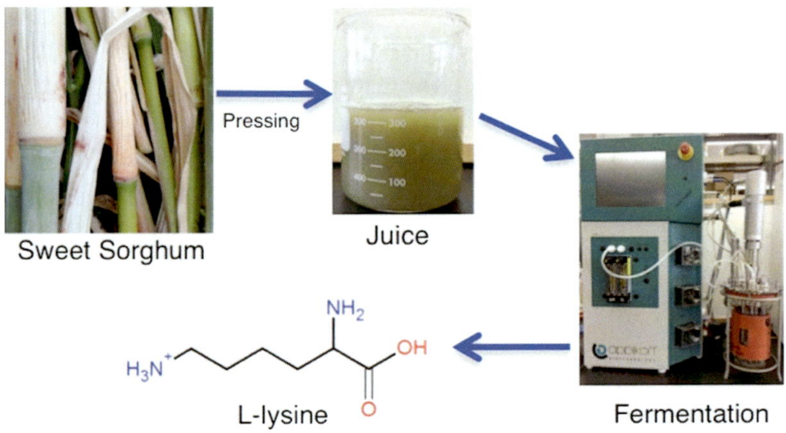

Figure 6.8. *An example of the laboratory-scale production of l-lysine.*

Source: Johnston DB, Nghiem N. (2018). Evaluation of Sweet Sorghum Juice for the Production of Lysine Using Corynebacterium glutamicum. Fermentation, 4(2):29. https://doi.org/10.3390/fermentation4020029.

6.2.3. Industrial Application and Therapeutic Role

One of the most varied uses and functions in the food, feed, and pharmaceutical industries is that of l-lysine. It is thought to be the first limiting amino acid for pigs and the second limiting amino acid for poultry. It is the most significant amino acid for animals that are monogastric. Consequently, it is a crucial supplement to animal feed in order to maximize the growth of hens and pigs. Figure 6.9 emphasizes l-lysine's metabolic uses.

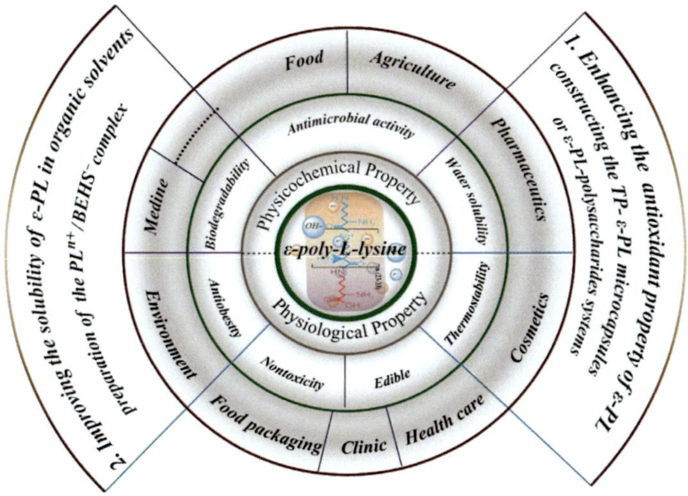

Figure 6.9. *Metabolic fate: application and role of l-lysine.*

Source: Li, S., Mao, Y., Zhang, L. et al. (2022). Recent advances in microbial ε-poly-L-lysine fermentation and its diverse applications. Biotechnol. Biofuels, 15, 65. https://doi.org/10.1186/s13068-022-02166-2.

In the food industry, l-lysine is a common ingredient in many dietary or nutritional supplements used by bodybuilders, weight lifters, athletes, and even some individuals to increase energy and prevent muscle breakdown. High-l-lysine nutritional supplements come in a variety of dosage forms, including syrups, film-coated tablets, capsules, and powder for instant beverages. The body needs l-lysine in order to produce l-carnitine, a chemical needed to convert fatty acids into energy. In addition to aiding in the formation of collagen and calcium absorption, l-lysine is crucial for the health of bones and muscles. In addition, it aids in the synthesis of antibodies, some hormones, and enzymes, either directly or indirectly.

Additionally, l-lysine is also recommended for the treatment of some viral infections, for example, herpes simplex, cold sores, shingles, and human papillomavirus infections such as genital warts and genital herpes. It has reportedly also been used to treat various other forms of pain and inflammation, including migraines. A lack of it can lead to serious health issues with growth and development, kidney stones, low thyroid hormone production, asthma, and persistent viral infections. Nausea, exhaustion, vertigo, anemia, appetite and energy loss, difficulty concentrating, irritability, bloodshot eyes, hair loss, growth retardation (particularly in children), and reproductive issues are among the symptoms. 500–1500 mg of l-lysine, divided into two doses and combined with a low-arginine diet, is the recommended therapeutic dosage for maintaining an antiherpes effect. In active stages, this dosage can reach up to 3,000 mg. The daily requirements of the human body for l-lysine are as follows: 103 mg for infants, 1,600 mg for children, 800 mg for adult males, and 500 mg for adult females.

6.3. METHIONINE

6.3.1. Historical Background and New Challenges

The discovery of glutamic acid-producing bacterium, *C. glutamicum*, eventually led to fermentation processes for producing various amino acids. Nevertheless, the fermentative synthesis of methionine received little attention in the early stages. Scientific publications about the synthesis of methionine by microorganisms are extremely rare. There weren't many references available on the topic until 1967. During this time, it was reported that a *Pseudomonas* sp. strain and a leucine-requiring strain of *Ustilago maydis* accumulated 6 g/L of methionine on a synthetic medium. Methionine accumulation in G-132-13 was 13.2 g/L.

Many researchers made valuable contributions to the commercial production of methionine between 1990 and the beginning of 2001. Before being used on a big scale, the fermentative process still requires more thorough research to determine its reproducibility and higher yield. Nowadays, there are two ways to make methionine: chemical synthesis or protein hydrolysis by enzymes. Both procedures come at a high cost. Whereas enzymatic hydrolysis of proteins yields a complex mixture from which methionine must be separated through an expensive recovery process, chemical synthesis yields a mixture of dextro and Levo rotatory methionine. Using continuous flow immobilized enzyme bioreactors, methionine isomers generated through chemical reactions can be resolved with the aid of the fungal enzyme aminoacylase.

The issues surrounding chemical synthesis are extremely difficult for manufacturers and researchers to solve. They must investigate the introduction of fermentative production utilizing genetically modified organisms on a commercial scale and come up with the best solution for these issues. Methionine is produced in bacteria from aspartate, which is derived from oxaloacetate, just like other essential amino acids. Methionine is another common amino acid found in animal feed, in addition to l-lysine. It does, however, represent nearly half of the market for l-lysine. According to reports from 2011, the global market for animal feed methionine was valued at US$285 billion, with 850,000 tons sold. To produce biomethionine, two major corporations, Cheil Jedang (South

Korea) and Arkema SA (France), are building a sizable plant in Malaysia. It is anticipated that the operation will begin in late 2013 and yield 80,000 tons annually (Saidak, 2012).

6.3.2. Biotechnological Approach and Production

Methionine has been produced using biotechnological methods using auxotrophic, regulatory, or auxotrophic regulatory mutants. Since methionine (branched pathway amino acids) does not inhibit or repress its own production, auxotrophic regulatory mutants are the most appropriate since auxotrophic mutants are generally less feasible for methionine. Higher concentrations of methionine (up to 25 g/L) can be accumulated by microorganisms like Figure 6.10 illustrates the regulation of methionine biosynthesis in *Corynebacterium* and *Brevibacterium*.

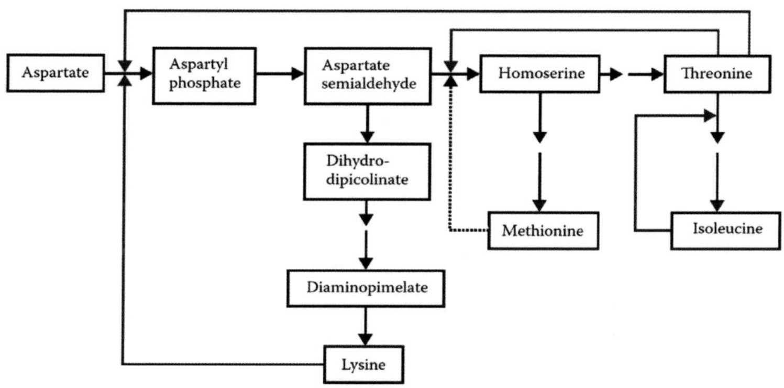

Figure 6.10. Regulation of methionine biosynthesis in C. glutamicum. The blue arrows show inhibition and dotted arrows represent repression.

Source: https://www.pnpcollege.edu.in/upload/lib_teaching_material/science/botany/2015-Micro-bialamino-acids-production.pdf.

Regulatory mutants are also used for producing methionine. Recently, a new approach to screening for methionine-producing bacteria was reported using methionine auxotrophs of *E. coli*, whereas another study relating to the production of l-methionine using B. cereus isolated from soil samples was also reported (Dike and Ekwealor, 2012). The regulation of methionine biosynthesis in *E. coli* is shown in Figure 6.11.

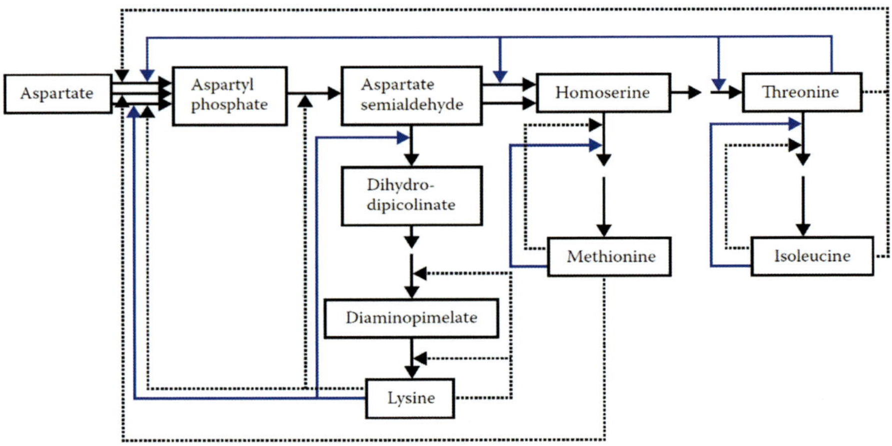

Figure 6.11. *Regulation of methionine biosynthesis in E. coli. The blue arrows show inhibition and dotted arrows show repression.*

Source: https://www.pnpcollege.edu.in/upload/lib_teaching_material/science/botany/2015-Micro-bialamino-acids-production.pdf.

In *Corynebacterium* and *Brevibacterium*, the biosynthesis mechanism for methionine production is quite simple as compared with the biosynthesis mechanism in *E. coli* (Figure 6.11) in which all enzymes are inhibited or repressed by the end product.

For the fermentative production of l-methionine, the majority of researchers reported using a carbon source that was either 10% glucose or 5% maltose, supplemented with inorganic salts and varying amounts of biotin and vitamin B_1. The product must be treated with ion exchange resins in order to be recovered. The process of concentrating under vacuum, treating with absolute alcohol, and drying at a low temperature of about 4°C for 24 hours will yield crystalline methionine.

6.4. TRYPTOPHAN

6.4.1. Historical Background and New Challenges

There were no published scientific studies on the direct microbial production of tryptophan until 1967. During this time, researchers focused more on the possibility of producing tryptophan by converting anthranilic acid, coupling indole with serine, transaminating 3-indolepyruvic acid, and converting β-indolyllactic acid. The first method employed for large-scale tryptophan production was chemical synthesis, followed by production via fermentation and enzymatic reactions. A variety of microorganisms, such as *Bacillus subtilis*, *Pseudomonas aeruginosa*, and others containing tryptophan synthase or tryptophanase, can be used for enzymatic reactions. *Escherichia coli* (*E. coli*) contains both tryptophan synthase and tryptophanase. With the introduction of efficient strains of *Corynebacterium* and *E. coli*, tryptophan is now largely produced by fermentation. The market for tryptophan remains relatively small. In 2012, global production was approximately 8,400 tons, with major companies such as Ajinomoto, Cheil Jedang, and Evonik leading the industry. However, China is expected to dominate the market.

6.4.2. Biotechnological Approach and Production

Chemical synthesis served as the foundation for the original commercial production process.

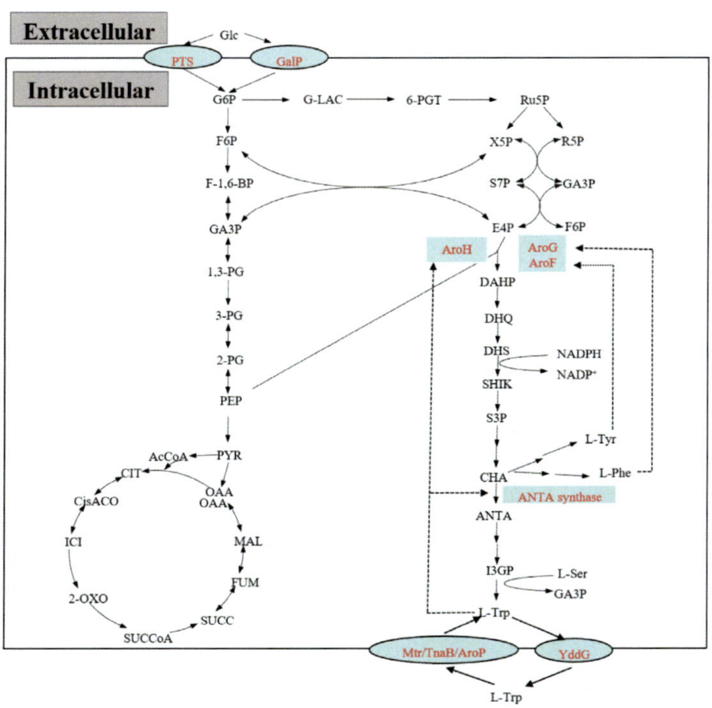

Figure 6.12. *Tryptophan biosynthetic pathway in E. coli.*

Source: *Liu L, Bilal M, Luo H, Zhao Y, Iqbal HMN. (2019). Metabolic engineering and fermentation process strategies for L-tryptophan production by Escherichia coli. Processes, 7(4), 213. https://doi.org/10.3390/pr7040213.*

However, due to the rising demand for tryptophan, particularly from the feed and pharmaceutical industries, additional techniques including fermentation and enzymatic reaction were also included by the end of 1980. As a result, ongoing efforts are currently being made to meet the intended production demand, particularly by applying the biotechnological approach. Among the various options, metabolic engineering has received particular attention as a means of increasing the production of l-tryptophan. The results were encouraging when *E. coli* and *Corynebacterium* were used. The tryptophan biosynthetic pathway in *E. coli* is shown in Figure 6.12. Overproducing strains of *E. coli* and *C. glutamicum* containing overexpressed genes have been developed, which can produce 40.2 g/L of tryptophan on suitable medium containing glucose as a carbon source in 40 h. A genetically engineered strain of *E. coli* was also reported to produce approximately 10.15 g/L of l-tryptophan in 48 h. It is believed that further research using more advanced biotechnological tools will be able to increase the productivity of tryptophan.

To understand the phenomenon of tryptophan fermentation, a brief method has been highlighted, which reflects the use of a genetically modified strain of *C. glutamicum* that is capable of producing tryptophan. Organisms are usually maintained on a monthly transfer schedule on nutrient agar slants supplemented with 0.5% sodium chloride. A seed medium containing glucose (20.0), peptone (5.0), yeast extract (1.0), sodium chloride (0.5), KH_2PO_4 (0.5), K_2HPO_4 (1.5), $MgSO_4$ (0.5), NH_4SO_4 (5.0), and calcium carbonate, (5.0 g/L) can be used. Optimum pH was observed at 6.8. Cultivation can be done in Erlenmeyer flasks of 250 mL capacity containing 30 mL of the above sterilized medium. Inoculation can be done with a loopful of culture grown on nutrient agar slant for 24 h. The flask should be shaken at 200 rpm at 30°C for 24 h in an orbital shaker-incubator to prepare the seed culture.

Fermentation medium may be prepared from molasses (30%), corn-steep liquor (0.7%), KH_2PO_4 (0.05%), K_2HPO_4 (0.15%), $MgSO_4{\cdot}7H_2O$ (0.025%), $(NH_4)_2SO_4$ (1.5%), and calcium carbonate (1%). A mixture of the following ingredients should be added in grams per liter: vitamin B1 (1000.0), biotin (50.0), l-phenylalanine (200.0), and l-tyrosine (175.0). The pH of the seed and fermentation media remain unchanged, that is, adjusted to 6.8, whereas 1.0 mL of 20% silicon RD in deionized water is added as antifoam. For the laboratory scale, mini jar fermenters of 2 L capacity containing 1 L of the medium can be used. Serialization can be affected by autoclaving for 15 min at 15 psi. After cooling, the fermenter is inoculated by one seed flask and stirred at 400 rpm with an air flow of 1.0 vvm (air volume/liquid volume per minute) at 30°C. The fermenters can be harvested after 72 h. Product recovery is usually done using ultracentrifugation at around 10,000 rpm, followed by treatment with cation exchange resin and decolorization with activated carbon. After further centrifugation, the mixture can be subjected to drying under a vacuum dryer. The product yield usually ranges between 10 and 25 g/L.

6.5. L-ISOLUECINE

Biosynthesis of branched amino acids in bacteria is controlled by the following regulatory mechanisms: (i) feedback inhibition of L-threonine dehydratase by isoleucine; (ii) inhibition of α-acetolactate (ALA) synthetase by valine; (iii) multivalent repression of isoleucine-valine biosynthetic enzymes by all four end-products of this sequence, i.e., valine, leucine, isoleucine and pantothenic acid; (iv) feedback inhibition of α-isopropylmalate (IPM) synthetase by leucine; and (v) regulation of intracellular levels of threonine, a precursor of isoleucine (Umbarger, 1969).

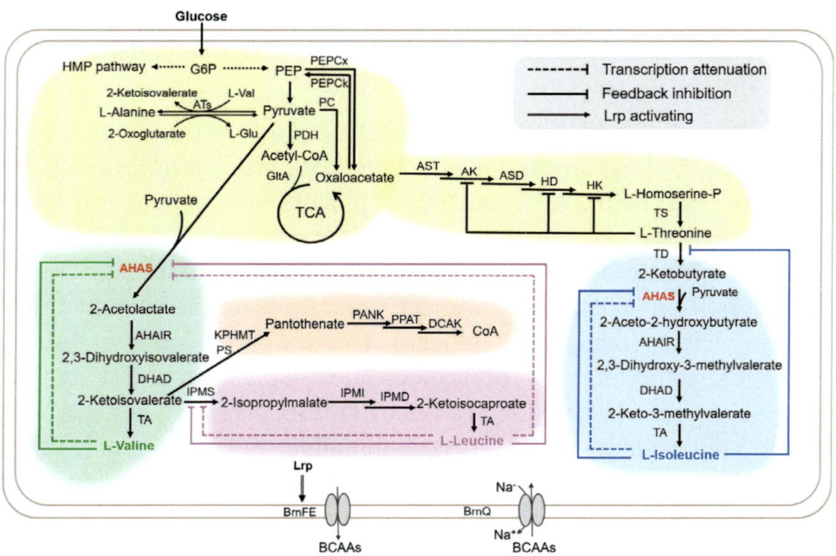

Figure 6.13. *Biosynthetic pathway and regulations of BCAAs in C. glutamicum. The green areas represent L-valine metabolic pathway, the pink areas represent L-leucine metabolic pathway, the blue areas represent L-isoleucine metabolic pathway, the yellow areas represent central metabolic pathway and L-threonine metabolic pathway, the orange areas represent pantothenate metabolic pathway. The terminating symbol of dotted lines indicate transcription attenuation, the terminating symbol of solid lines indicate feedback inhibition, and the double-line arrow indicates Lrp activation.*

Source: *Yu, S., Zheng, B., Chen, Z, et al., (2021). Metabolic engineering of corynebacterium glutamicum for producing branched chain amino acids. Microb. Cell Fact., 20, 230. https://doi.org/10.1186/s12934-021-01721-0.*

Note: PEPCx: PEP carboxylase; PEPCk: PEP carboxykinase; PC: pyruvate carboxylase; ATs: aminotransferase; PDH: pyruvate dehydrogenase; GltA: citrate synthase; AST: Aspartate aminotransferase; AK: aspartate kinase; ASD: semialdehyde dehydrogenase; HD: homoserine dehydrogenase; HK: homoserine kinase; TS: threonine synthase; TD: threonine dehydratase; AHAS: acetohydroxy acid synthase; AHAIR: acetohydroxy acid isomerase; DHAD: hydroxy acid dehydratase; TA: branched-chain amino acid aminotransferase; IPMS: isopropyl malate synthase; IPMI: α-isopropylmalate isomerase; IPMD: α-isopropylmalate dehydrogenase; Lrp: leucine-responsive protein; KPHMT: ketopantoate hydroxymethyltransferase; PS: pantothenate synthetase; PANK: pantothenate kinase; PPAT: phosphopantetheine adenylyltransferase; DCAK: dephospho-CoA kinase.

Figure 6.13 illustrates the regulation. Thus wild-type strains of bacteria do not excrete appreciable amounts of isoleucine in a salts-sugar medium. Microbial production of isoleucine was, therefore, first established by addition of precursors such as α-aminobutyrate(αAB), D-threonine and α-hydroxybutyric acid.

Fermentation methods using the precursors have been established using various bacteria by several groups and have been utilized industrially. DL-β-Bromobutyric acid could also be utilized as a precursor for isoleucine production. In addition to its role as a precursor by de-amination to α-oxobutyrate, αAB stimulates isoleucine production by releasing the feedback inhibition of threonine dehydratase by isoleucine, and acts as a regulatory factor on acetohydroxyacid (AHA) synthetase, stimulating synthesis of acetohydroxybutyrate (isoleucine precursor) and inhibiting that of acetolactate (valine precursor) in Bacillus subtilis. D-Threonine induced synthesis of D-threonine dehydratase which is resistant to end-product inhibition by L-threonine in contrast to L-threonine dehydratase in species of Serratia and Pseudomonas. Hence, D-threonine supplies α-oxobutyrate continuously for isoleucine biosynthesis, avoiding feedback control by isoleucine. In addition, αAB formed from an excess of α-oxobutyrate exerts a stimulating effect on the enzyme systems in these bacteria in favor of isoleucine biosynthesis. In Serratia marcescens no. 1, D-threonine derepresses synthesis of D-threonine dehydratase and AHA synthetase. In an αAB-resistant mutant derived from S. marcescens no. 1, D-threonine dehydratase and AHA synthetase are derepressed. This strain produced 15–16 g of L-isoleucine per liter in the presence of D-threonine.

Recently a process for L-isoleucine production directly from carbohydrates has been established. Isoleucine hydroxamate (IH) shows false feedback inhibition on L-threonine dehydratase activity. A mutant of Serratia marcescens which is resistant to IH and is to desensitized L-threonine dehydratases, and αAB-resistant mutants which are genetically derepressed for the isoleucine-valine biosynthetic enzyme, produce no isoleucine in a salts-sugar medium in the absence of isoleucine precursors. A double mutant (IHA 818), resistant to both IH and αAB and isolated from the IH-resistant strain, produced 6–7 g of L-isoleucine per liter in a medium lacking threonine. The double mutant lacked both feedback inhibition and repression of isoleucine biosynthesis. Increasing the αAB resistance led strain GIHVLA 2795 to produce 12 g of L-isoleucine per liter in a glucose medium.

In *Brevibacterium flavum*, growth inhibition by α-amino-ßhydroxyvaleric acid (AHV), an analog of threonine, was reversed not only by L-threonine but also partially by L-isoleucine. Threonine-producing mutants, which were isolated as analogue-resistant mutants, produced a small amount of L-isoleucine. Strain ARI-129, which is resistant to AHV and was isolated on a medium supplemented with 2 mg of AHV per ml, produced 11 g of L-isoleucine per liter. Homoserine dehydrogenase from strain ARI-129 was insensitive to feedback inhibition by L-isoleucine and threonine, while no difference was observed in the activity of threonine dehydratase between strains ARI-129 and 2 24 7A, a wild-type parent strain. Another AHV-resistant mutant (ARI-199), which was selected on a plate containing 3 mg of AHV per liter, produced 15.1 g of L-isoleucine per liter. The medium had the following composition: 10% glucose, 5% $(NH_4)_2SO_4$, 0.15% KH_2PO_4, 0.05% $MgSO_4 \cdot 7H_2O$, 2p.p.m. Fe^{2+}, 2p.p.m. Mn^{2+}, 200 µg biotin per liter, 400 µg thiamin HC1 per liter, 4 mg Mieki per liter and 5% $CaCO_3$. Maximum production of L-isoleucine was obtainable after 66 h. Strain ARI-129 grew slightly better than ARI-199, and maximum production of L-isoleucine was obtained after 44 h. The productivity of strain ARI-129 was more stable than that of ARI-199. The difference between the isoleucineand L-threonine-producing mutants was attributed to differences in permeability of L-threonine. In isoleucine-producing mutants, intracellular accumulation of threonine overcomes feedback inhibition of threonine dehydratase by isoleucine because inhibition of threonine dehydratase by isoleucine is competitive with respect to L-threonine. An O-methylthreonine-resistant mutant (AORI-126), which was derived from ARI-129, produced 14.5 g of L-isoleucine per liter. The specific activity of threonine dehydratase from strain AORI-126 increased about two-fold over that of strains no. 2247A and ARI-129, whereas the degree of inhibition of the enzyme by L-isoleucine was the same.

An ethionine (20 mg per ml)-resistant mutant of Brev. flavum was isolated after mutagenic treatment of strain BB-69, an AHV-resistant mutant which produces L-threonine in a yield of 13% from glucose. Mutant strain no. 168 produced L-isoleucine in a yield of 10%. Another ethionine (15 mg per ml)-resistant mutant was derived from Brev. flavum FAB-3-1, which was resistant to both thialysine, a lysine analogue, and AHV and produced L-threonine in

11% yield. The mutant produced L-isoleucine in 12% yield. Production of L-isoleucine by the process using acetic acid as the carbon source reached 33.5 g per liter, a 10% yield (Papagianni, 2012).

REFERENCES

1. Abdenacer, M., Kahina, B. I., Aicha, N., Nabil, N., Jean-Louis, G., & Joseph, B. (2012). Sequential optimization approach for enhanced production of glutamic acid from *Corynebacterium glutamicum* 2262 using date juice. *Biotechnology and Bioprocess Engineering, 17*, 795–803.
2. Adelberg, E. A., & Myers, J. W. (1953). Modification of the penicillin technique for the selection of auxotrophic bacteria. *Journal of Bacteriology, 65*(3), 348–353.
3. Amin, G. A., & Al-Talhi, A. (2007). Production of L-glutamic acid by immobilized cell reactor of the bacterium *Corynebacterium glutamicum* entrapped into carrageenan gel beads. *World Applied Sciences Journal, 2*(1), 62–67.
4. Asakura, Y., Kimura, E., Usuda, Y., Kawahara, Y., Matsui, K., Osumi, T., & Nakamatsu, T. (2007). Altered metabolic flux due to deletion of odhA causes L-glutamate overproduction in *Corynebacterium glutamicum*. *Applied and Environmental Microbiology, 73*(4), 1308–1319.
5. Austin, N. W., & Esmond, E. S. (1965). Formation and interrelationships of tryptophanase and tryptophan synthetases in *Escherichia coli*. *Journal of Bacteriology, 89*(2), 355–363.
6. Balch, J. F., & Balch, P. A. (1990). *Prescription for Nutritional Healing* (pp. 27–31). Avery Publishing Group Inc.
7. Banik, A. K., & Majumdar, S. K. (1975). Effect of minerals on production of methionine by *Micrococcus glutamicus*. *Indian Journal of Experimental Biology, 13*, 510–520.
8. Bloom, F. E. (2001). Neurotransmission and the central nervous system. In J. G. Hardman, L. E. Limbird, & A. G. Gilman (Eds.), *Goodman and Gilman's The Pharmacological Basis of Therapeutics* (pp. 293–335). McGraw-Hill.
9. Brautaset, T., & Ellingsen, T. E. (2011). Lysine: Industrial uses and production. In M. Moo-Young (Ed.), *Comprehensive Biotechnology* (pp. 541–554). Elsevier.
10. Chaitow, L. (1985). *Amino Acids in Therapy* (pp. 43–106). Thorsons Publishers Limited.
11. Daoust, D. R. (1976). Microbial synthesis of amino acids. In B. M. Miller & W. Litsky (Eds.), *Industrial Microbiology* (pp. 106–127). McGraw-Hill Book Company.
12. Davis, B. D. (1952). Biosynthesis interrelations of lysine, diaminopimelic acid, and threonine in mutants of *Escherichia coli*. *Nature, 169*, 534–536.

13. Dike, K. S., & Ekwealor, I. A. (2012). Production of L-methionine by *Bacillus cereus* isolated from different soil ecovars in Owerri, South East Nigeria. *European Journal of Experimental Biology, 2*(2), 311–314.
14. Dulaney, E. L. (1967). Microbial production of amino acids. In H. J. Peppler (Ed.), *Microbial Technology* (pp. 308–343). Reinhold Publishing Corporation.
15. Eggeling, L., & Sahm, H. (2011). Amino acid production. In M. Moo-Young (Ed.), *Comprehensive Biotechnology* (pp. 625–649). Elsevier.
16. Fong, C. V., Goldgraben, G. R., Konz, J., Walker, P., & Zank, N. S. (1981). Condensation process for DL-methionine production. In A. S. Goldfrab (Ed.), *Organic Chemicals Manufacturing Hazards* (pp. 115–194). Ann Arbor Science Publisher.
17. Funfstuck, R., Straube, E., Schildbach, O., & Tietz, U. (1997). Prevention of reinfection by L-methionine in patients with recurrent urinary tract infection. *Med. Klin., 92*, 574–581.
18. Gu, P., Yang, F., Kang, J., Wang, Q., & Qi, Q. (2012). One-step of tryptophan attenuator inactivation and promoter swapping to improve the production of L-tryptophan in *Escherichia coli. Microbial Cell Factories, 11*, 30–36.
19. Hamilton, B. K., Hasiao, H., Swann, W. E., Andersen, D. M., & Delente, J. J. (1985). Manufacture of amino acids with bioreactors. *Trends in Biotechnology, 3*, 64–68.
20. Ikeda, M., & Takeno, S. (2013). Amino acid production by *Corynebacterium glutamicum. Microbiological Monographs, 23*, 107–147.
21. Kase, H., & Nakayama, K. (1975). L-methionine production by methionine analog-resistant mutants of *Corynebacterium glutamicum. Agricultural and Biological Chemistry, 39*, 153–160.
22. Khan, N. S., Singh, R. P., & Prasad, B. (2013a). Studies on substrate inhibition in the microbial production of L-glutamic acid. *International Journal of Engineering Research and Technology, 2*(1), 1–7.
23. Khan, N. S., Singh, R. P., & Prasad, B. (2013b). Modelling the fermentative production of L-glutamic acid by *Corynebacterium glutamicum* in a batch bioreactor. *International Journal of Engineering Research and Technology, 5*(1), 192–199.
24. Kinoshita, S. (1999). Taxonomic position of glutamic acid producing bacteria. In M. C. Flickinger & S. W. Drew (Eds.), *Encyclopaedia of Bioprocess Technology: Fermentation, Biocatalysis, and Bioseparation* (pp. 1330–1336). John Wiley & Sons.
25. Kinoshita, S., Nakayama, K., & Kitada, S. (1958). L-lysine production

using microbial auxotroph. *Journal of General and Applied Microbiology, 4*(2), 128–129.

26. Kinoshita, S., Tanaka, T., Udaka, S., & Akita, S. (1957b). Glutamic acid fermentation. *Proceedings of the Symposium on Enzyme Chemistry, 2*, 464–468.
27. Kinoshita, S., Udaka, S., & Shimono, M. (1957a). Studies on amino acid fermentation: Part I. Production of L-glutamic acid by various microorganisms. *Journal of General and Applied Microbiology, 3*(3), 193–205.
28. Kita, D. A., Heights, J., & Huang, H. T. (1958). Fermentation process for the production of L-lysine. U.S. Patent 2,841,532, 1–2.
29. Koike, M., & Reed, L. J. (1960). A-keto acid dehydrogenation complex. II. The role of protein-bound lipoic acid and flavin adenine dinucleotide. *Journal of Biological Chemistry, 235*, 1931–1938.
30. Kumar, D., & Gomes, J. (2005). Methionine production by fermentation. *Biotechnology Advances, 23*, 41–61.
31. Kumar, D., Bisaria, V. S., Sreekrishan, T. R., & Gomes, J. (2003). Production of methionine by multianalogues resistant mutant of *Corynebacterium lilium*. *Process Biochemistry, 38*, 1165–1171.
32. Mahmood, Z. A. (1996). *Production of L-Lysine Through Fermentation* (PhD thesis, Department of Pharmaceutics, University of Karachi, Karachi, Pakistan).
33. Mahmood, Z. A. (2010). *L-Lysine: Production Through Fermentation.* VDM Verlag.
34. McEntee, W. J., & Crook, T. H. (1993). Glutamate: Its role in learning, memory, and the aging brain. *Psychopharmacology, 111*(4), 391–401.
35. Mitchell, H. K., & Houlahan, M. B. (1948). An intermediate in the biosynthesis of lysine in *Neurospora*. *Journal of Biological Chemistry, 174*, 883–887.
36. Mondal, S., & Chatterjee, S. P. (1994a). Enhancement of methionine production by methionine analogue resistant mutants of *Brevibacterium heali*. *Acta Biotechnologica, 14*, 199–204.
37. Mondal, S., Das, Y. B., & Chatterjee, S. P. (1994b). L-methionine production by double auxotrophic mutants of an ethionine resistant strain of *Brevibacterium heali*. *Acta Biotechnologica, 14*, 61–66.
38. Moss, J., & Lane, M. D. (1971). The biotin dependent enzymes. In A. Miester (Ed.), *Advances in Enzymology* (Vol. 35, pp. 321–332). John Wiley & Sons Inc.

39. Nakayama, K., & Kinoshita, S. (1961). Studies on lysine fermentation II. α, ε-diaminopimelic acid and its decarboxylase in lysine producing strain and parent strain. *Journal of General and Applied Microbiology, 7*(3), 155–160.

40. Neuvonen, P. J., Tokola, O., Toivonen, M. L., & Simell, O. (1985). Methionine in paracetamol tablets, a tool to reduce paracetamol toxicity. *International Journal of Clinical Pharmacology, Therapy and Toxicology, 23*, 497–500.

41. Odunfa, S. A., Adeniran, S. A., Teniola, O. D., & Nordstrom, J. (2001). Evaluation of lysine and methionine production in some lactobacilli and yeast from Ogi. *International Journal of Food Microbiology, 63*, 159–163.

42. Ozulu, U. S., Nwanah, O. U., Ekwealor, C. C., Dike, S. K., Nwikpo, C. L., & Ekwealor, I. A. (2012). A new approach to screening for methionine producing bacteria. *British Microbiology Research Journal, 2*(1), 36–39.

43. Papagianni, M. (2012). Recent advances in engineering the central carbon metabolism of industrially important bacteria. *Microbial Cell Factories, 11*, 50–63.

44. Parcell, S. (2002). Sulfur in human nutrition and application in medicine. *Alternative Medicine Review, 7*, 22–44.

45. Pasha, S. Y., Ali, M. N., Tabassum, H., & Mohammad, M. K. (2011). Comparative studies on production of glutamic acid using wild type, mutants, immobilized cells, and immobilized mutants of *Corynebacterium glutamicum*. *International Journal of Engineering Science and Technology, 3*(5), 3941–3949.

46. Pham, C. B., Galvez, C. F., & Padolina, W. G. (1992). Methionine fermentation by batch fermentation from various carbohydrates. *Asian Food Journal, 7*, 34–37.

47. Plimmer, R. H. A., & Hopkins, F. G. (1912). *The Chemical Composition of the Proteins. Monographs on Biochemistry. Part 1. Analysis* (2nd ed.). Longmans.

48. Pontecorvo, G. (1949). Auxanographic technique in biochemical genetics. *Journal of General Microbiology, 3*(1), 122–126.

49. Porter, R. J., Mulder, R. T., Joyce, P. R., & Luty, S. E. (2005). Tryptophan and tyrosine availability and response to antidepressant treatment in major depression. *Journal of Affective Disorders, 86*(2, 3), 129–134.

50. Potera, C. (1991). Researchers identify tryptophan contaminant. *Genetic Engineering News, 11*(2), 1–24.

51. Prasad, M. P., Gupta, N., Gaudani, H., Gupta, M., Gupta, G., Krishna,

V. K., Trivedi, S., & Londhe, M. (2009). Production of glutamic acid using whole and immobilized cells of *Corynebacterium glutamicum*. *International Journal of Microbiological Research, 1*(1), 8–13.

52. Shen, T., Liu, Q., Xie, X., Xu, Q., & Chen, N. (2012). Improved production of tryptophan in genetically engineered *E. coli* with TktA and PpsA overexpression. *Journal of Biomedicine and Biotechnology, 2012*, 1–8.
53. Sidransky, H. (2001). *Tryptophan: Biochemical and Health Implications*. CRS Press.
54. Tosa, T., Mori, T., Fuse, N., & Chibata, I. (1967). Studies on continuous enzyme reactions for preparation of a DEAE-sephadex-aminoacylase column and continuous optical resolution of acyl-DL-amino acids. *Biotechnology and Bioengineering, 9*, 603–608.
55. Tosaka, O., & Takinami, K. (1978). Pathway and regulation of lysine biosynthesis in *Brevibacterium lactofermentum*. *Agricultural and Biological Chemistry, 42*(1), 95–100.
56. Tosaka, O., Hirakawa, H., & Takinami, K. (1979). Effect of biotin levels on L-lysine formation in *Brevibacterium lactofermentum*. *Agricultural and Biological Chemistry, 43*(3), 491–495.
57. Umerie, S. C., Ekwealor, I. A., & Nawabo, I. O. (2000). Lysine production from various carbohydrate and seed meals. *Bioresource Technology, 75*, 249–252.
58. Van der Heijden, F. M., Fekkes, D., Tuinier, S., Sijben, A. E., Kahan, R. S., & Verhoeven, W. M. (2005). Amino acids in schizophrenia: Evidence for lower tryptophan availability during treatment with atypical antipsychotics. *Journal of Neural Transmission, 112*(4), 577–585.
59. Zareian, M., Ebrahimpour, A., Bakar, F. A., Mohamed, A. K. S., Forghani, B., Safuan, M., Kadir, A., & Saari, N. (2012). A glutamic acid producing lactic acid bacteria isolated from Malaysian fermented foods. *International Journal of Molecular Sciences, 13*, 5482–5497.

Chapter 7

Lipids and Fatty Acids

Contents

INTRODUCTION

The production of microbial lipids and fatty acids represents a promising frontier in biotechnology, with significant implications for sustainable energy, nutrition, and industrial applications. Microbial lipids, often referred to as single-cell oils (SCOs), are produced by oleaginous microorganisms, including certain yeasts, fungi, and bacteria, which have the capability to accumulate lipids to more than 20% of their biomass under specific conditions.

The process begins with the cultivation of these microorganisms in bioreactors, where they are provided with a nutrient-rich medium. Key factors such as carbon source, nitrogen limitation, and environmental conditions (temperature, pH, and oxygen levels) are meticulously controlled to optimize lipid accumulation. Carbon sources like glucose, glycerol, and agricultural waste are commonly used, making the process both versatile and cost-effective.

During lipid accumulation, microorganisms convert these carbon sources into various types of fatty acids and lipids, which can be harvested and refined for different purposes. The types of fatty acids produced include saturated, monounsaturated, and polyunsaturated fatty acids, each with distinct applications. For example, polyunsaturated fatty acids (PUFAs) such as omega-3 and omega-6 are highly valued for their health benefits and are used in nutritional supplements and functional foods.

Microbial lipids also have significant potential as renewable biofuels. As the world seeks alternatives to fossil fuels, microbial lipids can be converted into biodiesel, providing a sustainable energy source with lower environmental impact. Additionally, these lipids are used in the production of bioplastics, cosmetics, and pharmaceuticals, showcasing their versatility.

The advantages of microbial lipid production are manifold. It offers a sustainable alternative to traditional lipid sources, such as plant oils and animal fats, which require extensive land and water resources. Furthermore, microbial processes can utilize waste materials, thereby contributing to waste reduction and a circular economy. Advances in genetic engineering and metabolic pathway optimization continue to enhance the efficiency and yield of microbial lipid production, making it an increasingly viable industrial process.

The production of microbial lipids and fatty acids is a dynamic and rapidly evolving field with broad applications in energy, health, and industry. Through the optimization of microbial cultivation and bioprocessing techniques, this technology holds the promise of contributing to a more sustainable and efficient future.

7.1. ECONOMIC MICROBIOLOGY OF LIPIDS AND FATTY ACIDS

The study and utilization of microbial processes to generate these valuable compounds for diverse industrial sectors is the scope of the field known as economic microbiology of lipids and fatty acids. This area of study integrates economics, biotechnology, and microbiology to maximize the production and marketing of microbial fatty acids and lipids, providing affordable and sustainable substitutes for conventional sources. By combining microbiology, biotechnology, and economics, the economic microbiology of lipids and fatty acids investigates the microbial processes involved in the synthesis, refinement, and commercial application of these valuable compounds. The goal is to provide sustainable and economically viable substitutes for conventional lipid sources. This multidisciplinary field uses the ability of oleaginous microorganisms, such as certain yeasts, fungi, and bacteria, to efficiently accumulate lipids under particular conditions to address the growing global demand for biofuels, nutritional supplements, cosmetics, pharmaceuticals, and bioplastics. Microbial lipids, also called single-cell oils (SCOs), are very important to industry because of their environmental advantages and versatility.

Microbial lipids are a promising renewable source for the production of biodiesel in the biofuels industry. Microbial lipids are a more sustainable option than plant-based oils because they can be produced on a large scale without taking up agricultural land. Furthermore, utilizing waste substrates for microbial growth lowers production costs and has a positive environmental impact, making microbial lipids a competitive substitute for fossil fuels. Within the field of nutritional supplements, polyunsaturated fatty acids (PUFAs)—which are derived from microbiological sources and are crucial for human health—are being included more frequently in functional foods and dietary supplements. These PUFAs offer health benefits without depleting marine resources, which is a traditional method of extracting omega-3, supporting sustainability and human health at the same time.

Because of their biocompatibility and functional qualities, microbial lipids are also essential in the formulation of medications and cosmetics. They fulfill the increasing demand from consumers for natural and environmentally friendly products by acting as emollients, moisturizers, and carriers for active ingredients. This makes them a sustainable substitute for compounds derived from petrochemicals. Additionally, it is possible to transform microbial lipids into biodegradable plastics, which addresses the issue of plastic pollution globally and encourages the use of renewable resources in line with the circular economy's tenets. The utilization of waste materials and by-products as substrates allows for cost-effective production, which significantly reduces production costs and promotes waste valorization and resource efficiency. These are some of the economic benefits of microbial lipid production. Large-scale microbial lipid production is now possible thanks to developments in bioreactor design and process optimization. Genetic engineering, continuous fermentation techniques, and metabolic pathway optimization increase productivity and yield, making the process competitive with conventional lipid sources. The market for microbial lipids and fatty acids is driven by the rising demand for environmentally friendly and health-promoting products, which is being bolstered by a shift towards renewable resources and heightened awareness of environmental issues.

Notwithstanding noteworthy progressions, obstacles like refining microbial strains, refining lipid extraction techniques, and augmenting process efficacy persist. To get over these obstacles and produce at a reasonable cost, more research and development is required. It can be difficult to navigate regulatory frameworks and gain market acceptance, but successful commercialization of microbial lipid products depends on meeting consumer preferences and safety standards. The environmental impact of the entire production process—from sourcing substrates to disposing of finished products—must be continuously evaluated and minimized, even though microbial lipid production is intrinsically more sustainable than traditional methods. In summary, the economic microbiology of lipids and fatty acids presents enormous promise for environmentally and economically responsible industrial applications, providing valuable compounds in an efficient and responsible manner. The continuous progress in this domain holds the potential to unveil novel prospects and propel the shift towards a more eco-friendly and circular economy.

7.1.1. Industrial Significance and Applications of Microbial Lipids and Fatty Acids

1. **Biofuels**: Microbial lipids are increasingly recognized as a renewable source for biodiesel production. Unlike plant-based oils, microbial lipids can be produced on a large scale without competing for agricultural land, making them a sustainable alternative. The ability to use waste substrates for microbial growth further reduces production costs and environmental impact.
2. **Nutritional Supplements**: Polyunsaturated fatty acids (PUFAs) such as omega-3 and omega-6 fatty acids, derived from microbial sources, are essential for human health. These fatty acids are

incorporated into dietary supplements and functional foods, providing health benefits without depleting marine resources, which are traditionally used for omega-3 extraction.

3. **Cosmetics and Pharmaceuticals**: Microbial lipids are used in the formulation of cosmetics and pharmaceuticals due to their biocompatibility and functional properties. They serve as emollients, moisturizers, and carriers for active ingredients, offering a sustainable alternative to petrochemical-derived compounds.
4. **Bioplastics**: Microbial lipids can be converted into biodegradable plastics, providing an eco-friendly alternative to conventional plastics. This application not only addresses the plastic pollution problem but also promotes the use of renewable resources.

7.1.2. Economic Advantages

1. **Cost-Effective Production:** Utilizing waste materials and by-products as substrates for microbial growth significantly lowers production costs. This approach not only makes microbial lipid production economically viable but also contributes to waste valorization and a circular economy.
2. **Scalability**: Advances in bioreactor design and process optimization have made large-scale microbial lipid production feasible. Continuous fermentation processes, genetic engineering, and metabolic pathway optimization enhance yield and productivity, making the process competitive with traditional lipid sources.
3. **Market Demand**: The growing demand for sustainable and health-promoting products drives the market for microbial lipids and fatty acids. The increasing awareness of environmental issues and the shift towards renewable resources provide strong market incentives for microbial lipid production.

7.1.3. Challenges and Future Prospects

1. **Technical Challenges**: Despite significant advancements, challenges such as optimizing microbial strains, improving lipid extraction methods, and enhancing process efficiency remain. Continued research and development are essential to overcome these hurdles and achieve cost-effective production.
2. **Regulatory and Market Barriers**: Navigating regulatory frameworks and market acceptance can be challenging. Ensuring that microbial lipid products meet safety standards and consumer

preferences is crucial for successful commercialization.

3. **Sustainability and Environmental Impact**: While microbial lipid production is inherently more sustainable than traditional methods, it is essential to continually assess and minimize the environmental impact of the entire production process, from substrate sourcing to product disposal.

Lipid and fatty acid economic microbiology has enormous potential for long-term industrial uses. We can effectively and ethically create valuable compounds by utilizing microbial processes, which helps us to solve environmental and economic problems. The continuous progress in this domain holds the potential to unveil novel prospects and propel the shift towards a more eco-friendly and circular economy.

7.2. BACTERIA AS SOURCES OF (COMMERCIAL) LIPIDS

7.2.1. Commercial Utilization of Bacterial Lipids

Lipids are useful substances with a wide range of commercial uses. They can take the place of fossil fuels in many industrial processes. The use of oils as sustainable alternative feedstock to produce a wide range of fundamental oleochemicals for organic synthesis, catalysis, or biotechnology in the chemical industry has advanced remarkably (Biermann et al., 2011). However, replacing fossil fuels in the transportation sector is a major global goal, and the most common biofuel at the moment is biodiesel, which is derived from plant, animal, or microbial oils (Kosa et al., 2011).

Triacylglycerols (TAG) derived from vegetable oil are transesterified with short-chain alcohols, typically methanol or ethanol, to produce conventional biodiesel. However, because the oil content of common oil crops is typically less than 5% of their biomass, the oil yield is very low. For instance, canola yields only 1,200 L/ha, and oil palm about 6,000 L/ha. The enormous demand for arable land, which invariably clashes with the world's food supply, is another disadvantage. Certain microorganisms have oil contents that can reach up to 70% of their cell dry weight (CDW), but they only need a small amount of space and fertile land to grow. Establishing alternative oil sources is therefore necessary in order to replace fossil fuels in a way that is both more sustainable and, over time, economically viable. This might be particularly possible if the microbes made use of inexpensive industrial waste products, waste compounds, or plant material that can be grown on relatively arid land (so-called "energy crops"). Therefore, microorganisms are viewed as promising sources of high-value lipids and offer a number of benefits over conventional oils derived from plants or animals. Compared to multicellular eukaryotes, bacteria can produce a greater variety of different types of lipids and have a less complex genome, metabolism, and cell compartmentation. Furthermore, they are typically easier to genetically modify in order to produce optimized strains that raise overall process productivity and competitiveness even further. Wax esters (WE) are another type of lipid that eukaryotes typically do not accumulate in significant amounts. They are made up of long-chain alcohol and fatty acid residues. We are important components of lubricants, coatings, and medical and cosmetic goods, among other things. They are typically extracted from plants that need a lot of time and money to grow, like jojoba or carnauba wax. When it comes to composition, certain bacteria can provide an intriguing substitute source of WE for jojoba oil. Additionally, the selection of precursor substrates can be used to target the desired composition. As far as is currently known, wax ester synthase/ acyl-CoA: diacylglycerol acyltransferase (WS/DGAT) catalyzes significant bacterial TAG or WE synthesis. While some bacterial species have alternative pathways for synthesizing TAG, these typically don't explain a sizable build-up of this storage compound (Röttig et al., 2013).

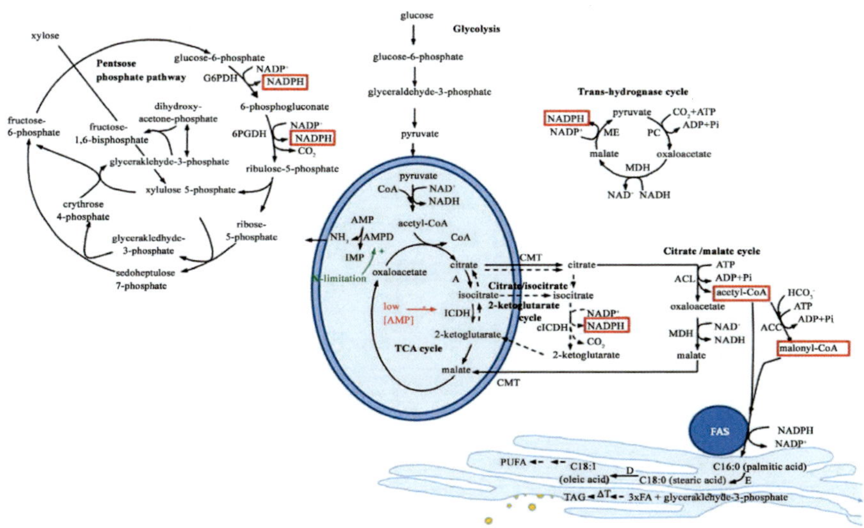

Figure 7.1. *Model chemical structures of industrially relevant lipids produced by bacteria.*

Source: Santek, M. I., Beluhan, S., & Santek, B. (2018). Production of Microbial Lipids from Lignocellulosic Biomass. InTech. doi: 10.5772/intechopen.74013.

Additional bacterial lipids, like rhamnolipids and polyesters (polyhydroxyalkanoates, or PHA), are also very important to industry as biosurfactants or bioplastics, respectively. This section, which focuses on bacteria that produce neutral and non-polymeric lipids like FFA, fatty acid ethyl esters (FAEE), WE, or TAG, does not address these topics (Figure 7.1).

A schematic overview of various methods for using bacteria to transform various feedstock types into valuable lipids for use in commercial or industrial settings is provided in Figure 7.2. While defined carbon sources might only be taken into consideration for extremely high-value end products (like those used in medicine), much less expensive substrates, such as sugars derived from (hemi)cellulose or industrial waste or by-products, are required to lower the cost of producing bulk lipids, which are needed to produce biodiesel, and to improve the process' overall sustainability. Some extremely promising methods try to use the bacterial community in activated sludge to convert waste water and CO_2 produced by cyanobacteria or microalgae to produce fats. The potential feedstock and promising bacterial production strains—either naturally able to accumulate lipids or genetically engineered to do so—will be covered in the following sections (Silva et al., 2023).

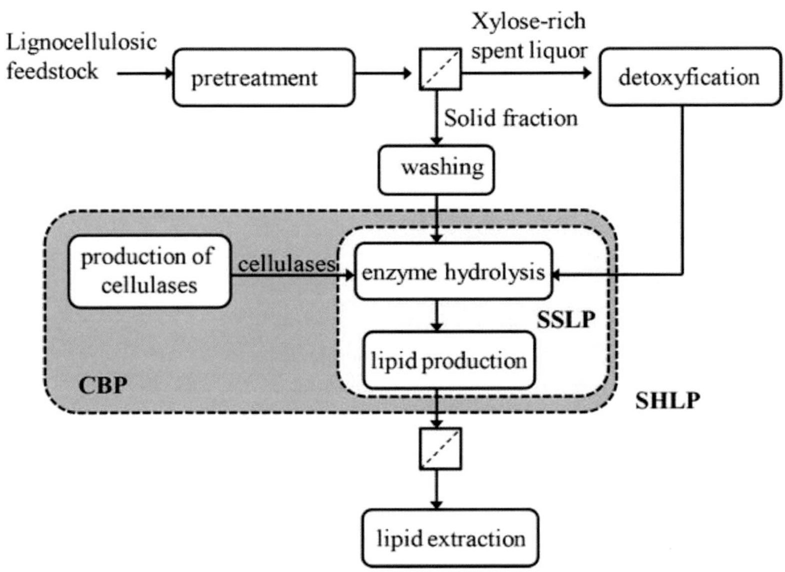

Figure 7.2. *Strategies to employ bacteria for the production of industrially relevant lipids from diverse feedstock.*

Source: Santek, M. I., Beluhan, S., & Santek, B. (2018). Production of Microbial Lipids from Lignocellulosic Biomass. InTech. doi: 10.5772/intechopen.74013.

7.2.2. Feedstock for the Production of Bacterial Lipids

It is unethical and economically problematic to use sugars derived from starch or sucrose for large-scale bacterial lipid production. Alkanes and alkanoic acids, two structurally related substrates, are typically the most productive sources of many different types of lipids. However, an economic production system with these substrates is problematic and unlikely due to their high prices and low toxicity and miscibility, respectively. The cost of the substrate has a significant impact on the price competitiveness of lipid-derived fuels when compared to conventional fuel, accounting for as much as 50% of the total production cost. With roughly 415 billion tons of different kinds of agricultural waste accumulating annually, lignocell comes from agricultural, industrial, and forest residues and is the largest and least expensive source of sugars in the world. Furthermore, using this cheap and plentiful feedstock would make lipid production geopolitically independent of oil-producing nations as well as ecologically sound. However, lignocellulose's breakdown into fermentable sugars necessitates a time- and energy-consuming pretreatment because of its complex and stable structure. To create so-called "second-generation biofuels," it would be beneficial to use bacterial strains that either degrade (hemi)cellulose or effectively use the released hexoses and pentoses, such as D-xylose, L-arabinose, D-glucose, D-mannose, and D-galactose.

Other intriguing strategies that are presently being discussed seek to develop efficient methods for obtaining microbial lipids from carbon dioxide or organic waste to produce "third-generation biofuels." Specialized anaerobic microorganisms gather lipid mixtures made of TAG, WE, or PHA during the wastewater treatment process. An estimated 50% of all biological wastewater treatment plants in the United States are thought to extract and transesterify these lipids. S. could contribute to a future energy strategy that makes the greatest use of renewable resources by meeting 0.5 percent of its annual diesel demand (Muller et al., 2014). Microorganisms that are oleaginous photosynthetic organisms are capable of effectively converting solar energy and recycling carbon dioxide into lipids. Under stressful conditions, microalgae store TAG, but certain cyanobacteria store large amounts of lipids in the form of diacylglycerol in their thylakoid membranes. These bacteria can even be genetically modified to overproduce FFA (Liu et al., 2011).

7.2.3. Bacterial Lipid Producers

Several crucial considerations should be made when assessing bacterial strains for possible use as industrial lipid producers: i) cost-prohibitive carbon sources and media supplements should not be required; ii) high growth rates, achievable cell densities, and product concentrations, particularly in fermentation conditions; and iii) easy product extraction. Continuous cultivation and lower extraction costs would be made possible by an extracellular product. If directed strain optimization is required, the bacteria also need to be non-pathogenic and genetically accessible.

7.2.3.1. Lipid Accumulation in Gram-Positive Bacteria

Bacteria that belong to the gram-positive Actinomycetales family are primarily capable of accumulating large amounts of TAG as intracellular storage deposits. E.g., a species *of Streptomyces, Dietzia, Gordonia, Nocardia, Rhodococcus*, or Arthrobacter of which, in particular *R. opacus* PD630 is considered a highly promising strain for industrial TAG production because of its fast growth, high substrate tolerance, and acceptance of a wide range of carbon sources. It is well-known for its impressive TAG contents of up to about 80% of its CDW (Hetzler et al., 2013). Despite reports from many years ago that Arthrobacter AK19 could also accumulate large amounts, this strain has not been used in any additional research. Particularly well suited for the conversion of industrial waste materials (such as molasses, wastes from various agro-industrial processes, dairy wastewater, etc.) seem to be *Rhodococcus* or Gordonia strains or lignocellulose hydrolysates, whereas a streptomyces isolate can develop up to 50% of its CDW's fatty acids using cellobiose. The unique characteristic of streptomycetes is the inclusion of branched-chain fatty acids in addition to the straight-chain C16–C18 fatty acid residues that make up the majority of the TAG synthesized by *Rhodococcus*.

It is noteworthy that in batch cultivation, lipid contents are frequently only measured; however, continuous cultivation has the potential to greatly boost the total lipid productivity. Nevertheless, a high lipid content in the cells would not guarantee that the process is economically viable, since the total lipid yield may still be relatively low (e.g., even though the lipid content following *Rhodococcus* batch culture. The productivity of *Rhodococcus opacus* cells with molasses was found to be approximately 92% of its CDW, or only about 55 mg/L (Gouda et al., 2008). On the contrary, in a 30 L fed-batch fermentation the productivity of *R. opacus*. The amount of total fatty acids was significantly elevated to 19.4 g/L or 0.38 g/L·h, respectively, corresponding to approximately 50% of its (CDW) comes from sucrose and molasses (Voss et al., 2001). Therefore, there is no guarantee that the lipid yields obtained from individual cultivations can be compared.

According to reports, the Firmicutes isolate *Catenisphaera adipataccumulans* GK12 can accumulate FFA intracellularly in amounts as high as 25% of CDW. This phenomenon has not yet been seen in nature (Katayama et al., 2014). It is doubtful, though, if the lipids were stored as FFA, which can be extremely harmful even at low concentrations. Alternatively, FFA might have been liberated from lipids that had accumulated during the cell processing. Additionally, *Streptomyces* sp. When NP10 was grown on a complex medium, it displayed the rather unusual property of secreting large amounts of FFA with a complex composition (Ilic-Tomic, 2015).

A high percentage of actinomycetes are usually present in the anoxic tank foam in biological wastewater treatment plants' activated sludge. It would be easier to recover this lipid-rich biomass for the production of biodiesel if this intricate consortium of oleaginous bacteria could be easily skimmed from the surface (Muller et al., 2014).

7.2.3.2. Lipid Accumulation in Gram-Negative Bacteria

TAG storage in comparable amounts to those found in some Gram-positive bacteria has not yet been reported in Gram-negative bacteria. Alternatively, WE can be synthesized by certain Gram-negative bacteria, specifically Acinetobacter or hydrocarbonoclastic bacteria like *Alcanivorax* or *Marinobacter* species. These bacteria then frequently accumulate in mixed lipid inclusions with reduced levels of TAG. Only when these bacteria are grown with related carbon sources, such as n-alkanes or olive oil, are notable amounts typically obtained.

An intriguing isolate of Gram-negative bacteria, *Aeromonas* sp. 3010, attained an approximate total lipid content. 12%, of which up to 30% is made up of the polyunsaturated fatty acid eicosapentaenoic acid [20:5 n-3, EPA], which is very important to the food and pharmaceutical industries. Two rather unusual discoveries of lipid-producing Gram-negative bacteria have also been reported: an isolate belonging to the genus Nitroreductor (an α-proteobacterium) grows with short-chain organic acids and synthesizes an unusual mixture of lipids made of squalene, TAG, and light oils (like 2-butenoic acid methyl ester), with contents of up to 70% of the CDW (Okamura et al., 2016). This could be useful for treating wastewater. Furthermore, approximately a chemolithotrophic, CO_2-sequestering strain of Serratia was synthesized. Long-chain hydrocarbons and lipids make up 50% of its CDW, and extracellular lipids can reach up to 466 mg/L. The reports on oleaginous strains of Bacillus and Pseudomonas, however, need more verification because it seems improbable that these species can accumulate neutral lipids; instead, they are well-known for their capacity to store PHA. The taxonomic assignment of these isolates may have been incorrect. Additionally, the presence of neutral lipids in *P. aerogorgia* 44T1 in an independent investigation was not verified (Alvarez et al., 2002).

Since the storage of intracellular TAG has not yet been established in this class of microorganisms, cyanobacteria may represent yet another unique instance. No (WS/) DGAT gene has been found in cyanobacterial genomes that have been published. Their photosynthetic membranes, however, can act as sizable reservoirs for diacylglycerols and membrane lipids, which can also be used biotechnologically, e.g., for producing biodiesel through transesterification with methanol). Certain cyanobacterial strains, like *Cyanobacterium aponinum* or *Synechococcus* sp., have achieved remarkable lipid contents of over 40% of the CDW by using CO_2 as their only carbon source. However, because these lipids are complex substances, they must be hydrolyzed in order to release the necessary fatty acids.

For large-scale lipid production, e.g., Actinomycete bacteria, such as some species of *Streptomyces* or *Rhodococcus*, are particularly useful as feedstock for biodiesel because they are robust production strains that can grow on inexpensive raw materials and accumulate large amounts of TAG. While gram-negative bacteria that synthesize water emulsion (WEE) typically do not attain comparable high lipid contents and necessitate more costly carbon sources, certain WEs with appropriate compositions can be generated for relatively expensive end products like cosmetics.

While it is possible to isolate promising production strains from their natural environments, genetic engineering holds significant promise for improving these strains' lipid accumulation and substrate spectrum. Furthermore, non-oleaginous bacteria such as *E. coli*, as described below, can be genetically engineered to produce excessive amounts of FFA, TAG, or FAEE, turning them into useful production strains.

7.2.3.3. Free Fatty Acid Overproduction in Genetically Engineered Bacteria

Due to their toxicity in comparison to TAG or WE, FFAs are typically not overproduced and do not represent an intrinsic storage compound, except the aforementioned infrequent findings of an accumulation or excretion of FFA by natural isolates (Lennen et al., 2012). In the past 10 years, remarkable progress has been made in the genetic optimization of *E. coli* and cyanobacteria to over make and release free fatty acids.

Fatty acid secretion into the culture medium would significantly lower the overall cost of producing bacterial lipids by removing the need to harvest, dry, and extract lipid-containing cells using a solvent. Despite not having a lipid accumulation capacity by nature, *E. coli* lipid metabolism has been studied in greater detail than that of other prokaryotes and a number of well-established methods for genetic modification have been developed, coli is a desirable production organism. Therefore, substantial fatty acid overproduction can be accomplished by overexpressing genes involved in fatty acid de novo synthesis and/or blocking genes involved in fatty acid degradation. The acyl residue from acyl binding protein (ACP) is then released by the overexpression of (native or recombinant) thioesterases, causing FFA to build up (Lennen et al., 2012). Engineered *E. coli* has already achieved up to 4.8 g FFA/L strains using 3–8 g/L from woody biomass hydrolysate, or glycerol or glucose in fed-batch fermentation.

In order to weaken the cell wall integrity and allow the release of fatty acids into the medium, or to participate in competing processes (such as the unintended synthesis of PHB or cyanophycin instead of fatty acids), a cyanobacterium known as *Synechocystis* sp. SD277 underwent extensive modification. Once multiple thioesterase genes were overexpressed, this modified strain generated up to 200 mg FFA/L. Despite this, the yield is much lower than the numbers using *E. coli*, but even so, this method would be far more environmentally friendly because CO_2 can act as the only carbon source.

7.2.3.4. Triacylglycerol Synthesis in Genetically Engineered Bacteria

As was previously stated, *R. opacus* has a significant capacity to store TAG. Therefore, the majority of genetic engineering techniques have focused on expanding its substrate range in order to facilitate the production of lipids from glycerol, sugars derived from (hemi)cellulose, or hydrolyzed plant material in a more sustainable manner. Through focused alterations and an adaptive evolution approach, over 50% (w/w) TAG (or roughly 4. With corn stover hydrolysate as the only carbon source, a yield of 16 g/L was attained (Kurosawa et al., 2014). These yields are much higher than what wild-type R produced. opacus using various hydrolyzed plant materials (approx. 28% fatty acids), demonstrating the critical role that strain optimization plays in enabling competitive bacterial lipid production from a difficult feedstock.

Additionally, a number of studies concentrated on proving TAG accumulation in *E. coli*; for instance, by combining recombinant expression of essential WS/DGAT genes with increased fatty acid and diacylglycerol synthesis (Röttig et al., 2015). Nonetheless, a considerable improvement in the achievable yields is still required to compete with natural lipid accumulators.

7.2.3.5. Fatty Acid Ethyl Ester Synthesis in Genetically Engineered Bacteria

Nowadays, methanol from fossil fuels is used to produce the great majority of biodiesel (fatty acid methyl ester, or FAME). However, ethanol can also be used to produce biodiesel from fatty acid ethyl esters (FAEE), which have better physicochemical properties than FAME. A method to fully synthesize FAEE de novo (also known as "microdiesel") from unrelated carbon sources in E was devised in 2006. Salmonella (Kalscheuer et al., 2006). In the initial method and subsequent research, the ethanol synthesis pathway from *Zymomonas mobilis* (pdc for pyruvate decarboxylase and adhB for alcohol dehydrogenase B) was typically coupled with the overexpression of a WS/DGAT gene. More recent research has shown that tactics similar to those used to combat FFA overproduction—boosting fatty acid de novo synthesis and preventing fatty acid beta-oxidation—may raise FAEE levels even further.

The highest FAEE levels of 1.5 g/L have only been attained thus far with glucose and in small-scale cultures. One significant drawback of *E. coli*'s incapacity to break down lignocellulose, however, a preliminary demonstration of the concept could be accomplished by creating a specialized strain that makes use of switchgrass hydrolyzate (Steen et al., 2010).

Making branched-chain fatty acid alkyl esters from branched short-chain alcohols and (branched) fatty acids is another interesting strategy (e.g., branched-chain amino acid synthesis by taking advantage of the biosynthesis pathway for isoamyl alcohol or isobutanol. These esters perform better at low temperatures, where traditional biodiesel typically performs poorly (Tao et al., 2015). Still, a significant increase in the synthesis yield of branched FAEE is required. On the other hand, bacteria like Streptomyces isolate G25 that naturally incorporate branched fatty acids into TAGs may be useful for producing biodiesel that has a combination of residues from both straight and branched chains.

In summary, robust and fast-growing strains are ideal for the economically viable production of bacterial lipids. These bacteria should be able to use plentiful and inexpensive feedstocks, such as industrial waste or surplus materials, in addition to producing high lipid yields. Significant advancements have been made in the last 10 years in the genetic engineering of bacteria to expand their substrate ranges or lipid contents. These altered strains, however, were unable to produce lipid yields that were on par with some exceptional oleaginous wild-type bacteria, like *R. opacus*. To enable the synthesis of relatively artificial lipids, like FAEE or customized, wax ester-like compounds for more specialized uses, metabolic engineering seems to be a promising approach. Additionally, it might encourage a more thorough utilization of the whole range of carbon sources found in lignocellulosic hydrolysates, for example, or it might strengthen a production strain's resistance to specific inhibitors. To date, an impressive number of proof-of-principle studies have been carried out, which can serve as a basis for focused modifications aimed at optimizing the most promising production strains. These are currently *R. opacus* or *Candida* sp. G25 to produce TAG in large quantities from *Saccharomyces cerevisiae* arides derived from lignocellulose, specific cyanobacteria to produce lipids using CO_2, and recombinant *E. coli* strains in order to produce FAEE-biodiesel.

7.3. MICROBIAL LIPID-BASED BIOREFINERY CONCEPTS

7.3.1. Lignocellulosic Single-Cell Oil (SCO)

Biorefineries essentially aim to convert biomass into biofuels and/or fine chemical products. They can be defined as a promising path for an industry based on the use of biomass as an input for the production of high value-added products. These technologies fall under the emerging category. The challenge lies in finding a less expensive and more environmentally friendly method for transforming biomass into these highly versatile products that can be used in a wide range of industries. Furthermore, there are numerous applications and product opportunities available to the biorefinery, including a wide range of platform chemicals, like glycerol, as well as biofuels, like ethanol, and numerous biochemicals and biopolymers with a variety of industrial uses. Microorganisms, which include bacteria, yeasts, fungi, and microalgae, are classified as oilseeds due to their tendency to accumulate lipids at a weight percentage of greater than 20–25% (w/w). Cell storage lipids can build up in these microbes as a result of excess carbon content or nutrient deficiencies, such as those in nitrogen or phosphorus. Under these circumstances, normal cellular processes like protein synthesis and nucleic acid synthesis take place. Eventually, cell growth stops, and the oilseed species eats any carbon source that is available and uses biochemical reactions to convert it to lipids.

SCOs are essentially microbial oils derived from microbial sources. Their composition is strikingly similar to that of fats, edible plant oils, and animal oils.

Since the turn of the century, the potential of these microbially derived oils and fats has been acknowledged. Prior to 1980, some researchers concentrated on these oleaginous microbial biochemistry and metabolism in the process of synthesizing these oils. Since the products derived from SCO have been demonstrated to be vital to human health, the production of SCO and its oily compounds has emerged as one of the most prominent areas of research in this field during the past 20 years.

These microorganisms are able to produce oily products, have long-chain fatty acid profiles, are flexible enough to grow on a variety of carbon sources, and consume volatile fatty acids. Bettencourt et al. (2020) investigated four yeasts that were separated from the species mentioned above: *Pichia kudriavzevii*, *Candida tropicalis*, *Metschnikowia pulcherrima*, and *Apiotrichum brassicae*. They added seven different volatile fatty acids to the culture medium. The results showed that the yeasts grow in both a synthetic medium and the waste-derived effluent filtrate. Kudriavzevii exhibited a 40% (w/w) productivity, with the most common lipids (>40%) being stearic acid (20%) and palmitic acid (20%). With the use of these isolates and additional carbon sources, there is a lot of potential for larger-scale SCO production that combines environmental and financial benefits.

Di Fidio et al. (2021) investigated, a potential source of carbohydrates that oleaginous yeasts can convert to SCO. The ability of these yeasts to convert cellulose and hemicellulose fractions into oils is a critical step in enhancing the SCO production process's financial sustainability. These authors developed the bacterium *Lipomyces starkeyi* DSM 70.296 using two xylose hydrolysates obtained by pretreatment (microwave-assisted hydrolysis of hemicellulose catalyzed by $FeCl_3$ or Amberlyst-70) and two glucose hydrolysates obtained by enzymatic cellulose hydrolysis. The bacterium *L. starkeyi* produced a yield of 30% by weight. Hydrolysates that were either partially or completely detoxified were used to successfully grow. This study demonstrated how next generation biodiesel and high added value products could be effectively made from lignocellulosic raw materials from perennial grasses, with integrated cascade hydrolysis processes able to yield SCO.

7.3.2. Second-Generation Biorefineries

Second generation biorefineries are those that use biomass, which includes any type of harvest residue, leftovers from agroforestry, and grasses grown in arid conditions. The inputs used by first-generation biorefineries are directly extracted from starch and sugar.

A biorefinery can add value to lignocellulosic materials, which has led to a great deal of research into how to turn their cellulose, hemicellulose, and lignin constituents into highly valuable products. However, these materials require a pre-treatment to separate the cellulose/hemicellulose and lignin fractions due to their strong resistance to enzymatic and chemical hydrolysis.

As stated by Huang et al. (2013), a number of projects have focused on using low-cost materials to create SCO, which can then be used to use biomass in the production of microbial oil. "De novo" and "ex novo" lipid accumulation techniques are the most widely used techniques for lipid synthesis in oleaginous microorganisms. Growing conditions with limited nitrogen are usually required for the first method, which makes use of hydrophiles. In addition, SCO is fermented in hydrophobic substrates during the process of "ex novo" lipid synthesis. Low-cost substrates have two disadvantages, according to the same authors who were previously mentioned: they can lower the cost of manufacturing SCO and impede the industrialization process. The first is that there are problems with raw material transportation, and the second is that there are restrictions on raw material availability, both of which are detrimental to more consistent SCO production. Thus, the most accessible and sustainable source of SCO found in nature, lignocelluosic biomass, may be highly promising for the sustainable and clean production of SCO.

While the use of lignocellulosic biomass in the production of microbial oil is relatively new, it is widely recognized in the production of second-generation bioethanol. The three steps of the BM2BD (biomass for biodiesel) plan, which focuses on using lignocellulosic biomass for the production of biodiesel, are as follows: in the first stage, the biomass is transformed into fermentable sugars; in the second stage, oleaginous microorganisms may convert these sugars into microbial lipids; and in the third and final stage, the microbial lipids are transformed into biodiesel.

Liang et al. (2012) used lime at concentrations of 0, 0.05, and 0.1 per gram of bagasse for one or two hours to pre-treat sweet sorghum bagasse at 100°C. The pretreated samples' composition was analyzed, and the results showed that lignin and Xylan were easier to remove at higher lime concentrations. The combined impact of enzymatic hydrolysis and pretreatment was assessed. A 73.6% sugar yield was achieved with the following conditions: 0.1 g ATR/g bagasse, 10 mL water/g bagasse, and 2 hours of treatment. The sludge served as a growth medium for the oleaginous yeast *Candidatus curvatus*. The obtained neutral lipid yield (g neutral lipid/g sugar) was 0.19, which is in close proximity to the theoretical value. Therefore, sorghum bagasse can be converted into sufficient lipids to make biodiesel, valuing an agricultural by-product for the production of renewable biodiesel raw materials. Huang et al. (2013) state that a great deal of research has focused on identifying the optimal strains for lipid synthesis from lignocellulosic hydrolysates. It has been reported that the oleaginous microbes *Rhodotorula glutinis*, *Lipomyces starkeyi*, *Rharrowia lipolytica*, and *Rhodosporidium toruloides* can all produce SCO using lignocellulosic substrates.

A single cell oil lignocellulosic biorefinery and a second-generation biorefinery differ primarily in that the former uses biomass as a raw material and for oil production. The SCO lignocellulosic biorefinery uses microorganisms directly to produce oils, while the second-generation oil biorefinery uses biomass to produce fermentable sugars that are then converted into oils by yeast.

It is feasible to integrate these two kinds of biorefineries, which could lead to a reduction in greenhouse gas emissions and a more sustainable use of biomass by producing a wider range of renewable products. The main option for integration would be to feed the microorganisms employed by the SCO lignocellulosic biorefinery to produce oils by using lignocellulosic biomass as a source of fermentable sugars. The second-generation oil biorefinery refines the produced oils and uses them as raw materials to produce liquid biofuels. This can be achieved, for instance, by using microalgae-produced oils high in long-chain fatty acids as the starting material for the production of biodiesel. The other option is to feed the second-generation oil biorefinery with biomass obtained from the waste produced by the SCO lignocellulosic biorefinery during the production of oils. Yeasts in the second-generation oil biorefinery use these residues as a source of cellulose and hemicellulose to create fermentable sugars, which are subsequently transformed into oils (Silva et al., 2023).

7.3.3. Potential Markets for Microbial Lipids

According to Jin et al. (2015), economic factors are essential to a SCO biorefinery's performance. There is currently very little techno-economic research on the use of heterotrophic oleaginous bacteria to manufacture SCO. Ratledge and Cohen projected a price of USD $3000 per ton for microbial oil, excluding the cost of raw materials and taking into account research done in New Zealand to use Candida curvata to convert lactose into the equivalent of cocoa butter. Table 7.1 displays yeast strains that are capable of producing lipids, some of which entail the utilization of lignocellulosic hydrolysates.

Table 7.1. *Microorganisms with Potential Application for the Production of Lipids*

Microorganisms with High Potential	**Microorganisms with Low Potential**
S. occidentalis (Lamers et al., 2016)	*P. Silvicola* (Lamers et al., 2016)
M. pulcherrima (Lamers et al., 2016)	*C. intermedia* (Lamers et al., 2016)
L. elongisporus (Lamers et al., 2016)	*P. pertersonii* (Lamers et al., 2016)
W. lipofer (Lamers et al., 2016)	*C. lusitaniae* (Lamers et al., 2016)
Y. lipolytica (Gao et al., 2020)	*K. phaffii* (Lamers et al., 2016)

H. californica (Lamers et al., 2016)	*S. roseus* (Lamers et al., 2016)
P. anomala (Lamers et al., 2016)	*P. augusta* (Lamers et al., 2016)
T. delbrueckii (Lamers et al., 2016)	*C. bombicola* (Lamers et al., 2016)
H. beyerinckii (Lamers et al., 2016)	*K. apiculate* (Lamers et al., 2016)
C. tropicalis (Lamers et al., 2016)	*C. glabatra* (Lamers et al., 2016)
R. toruloides (Michou et al., 2022)	*T. elliptica* (Abomohra et al., 2017)
L. starkeyi (Bonturi et al., 2015)	*B. braunii* (Cheng et al., 2013)
C. curvatus (Patel et al., 2018)	*R. glutinis* (Maza et al., 2020)
F. oxysporum (Matsakas et al., 2017)	*Microsphaeropsis* sp. (Subhash, 2011)
R. opacus PD630 (Kumar et al., 2015)	*R. opacus DSM 1069* (Kosa, 2011)

Source: Silva JdMe, Martins LHdS, Moreira DKT, Silva LdP, Barbosa PdPM, Komesu A, Ferreira NR, Oliveira JARd. (2023). Microbial lipid based biorefinery concepts: A review of status and prospects. Foods, 12(10):2074. https://doi.org/10.3390/foods12102074.

As we've seen above, a wide variety of microorganisms, including molds, yeasts, fungi, and microalgae, can store lipids inside their cells. Lipid droplets that are formed by SCOs are found inside cells or in the chloroplasts of algae. After fatty acid analysis, it was discovered that they resembled those in vegetable oils, the majority of which are used to make first-generation biodiesel. Furthermore, the fatty acid profile from SCO contains high concentrations of polyunsaturated fatty acids (PUFAs), also referred to as essential fatty acids, which are not produced by mammals and must be obtained through diet. Thus, the population would have greater accessibility if these PUFAs were used as a dietary supplement.

7.3.4. SCO Microbes and Culture Conditions

Triacylglycerols and free fatty acids comprise the bulk of lipids produced by organisms that are oleaginous. Lipids are ideal raw materials for producing renewable fuels because of their high carbon-to-heteroatom ratios. The high cost of the fermentation process—which is mostly caused by the carbon source—to obtain SCO may restrict the industrial use of microbial lipids. The effective use of organic waste as a carbon source for microbial lipids, such as lignocellulosic biomass, has thus been the focus of a great deal of research. Doing so would undoubtedly raise the profitability of the production process and support a biologically based economy.

Oleaginous microorganisms that produce lipids, as stated by Ma et al. (2018) can be divided into three categories: bacteria, fungi (such as mold and yeast), and microalgae. Since bacteria can only synthesis certain lipids and polyunsaturated fatty acids, they are unable to produce large quantities of lipids. As a result, the primary sources of lipids used in biorefineries are fungi and microalgae. A microbe is only regarded as oleaginous, though, if it can store more than 20% of its dry body mass as lipids, or oil. In line with Ma et al. (2018) productivity, which accounts for both biomass productivity and lipid content rather than just the lipid content produced, is the best measure of the total lipid production capacity for holophytic microalgae.

Ageitos et al. (2011) report that only 5% of oleaginous yeasts can accumulate lipids in amounts greater than 25%. These yeasts can accumulate lipids in amounts ranging from 20% to 80% of their dry cell weight. Among the genera of oilseed yeast are *Yarrowia, Candida, Rhodotorula, Rhodosporidium, Cryptococcus, Trichosporon*, and *Lipomyces*. For example, *Lipomyces starkeyi, Rhodotorula glutinis*, *Candida curvata*, and *Cryptococcus albidus* produce 58, 65, 64, and 72% of lipids, respectively. Yeasts are superior to other microbial sources in a number of ways when it comes to lipid production. Their doubling times are frequently less than an hour, and their cultures may be expanded more easily than those of microalgae. Because of the variety of microorganisms and growth conditions they support, oily yeasts can also be useful for the production of triglycerides, surfactants, or polyunsaturated fatty acids.

Microalgae are unicellular photosynthetic organisms that can produce significant amounts of lipids and hydrocarbons when exposed to sunlight and carbon dioxide from flue gases. The average amount of lipid produced by algal cells is between 1 and 70%, but in certain cases, it can reach 90% of the dry weight. 25–75% of *Bradylochococcus braunii* and *Chlorella* sp. (32–28%), *Nannochloropsis* sp. (31–68%), *Nitschschia* sp. (45–47%) as well as *Schizochytrium* sp. Among other things, microalgae that produce lipids make up 50–77%.

Microalgae are more advantageous than land plants and other microorganisms because of their high lipid productivity, four different cultivation modes (autotrophic, heterotrophic, mixotrophic, and photo-heterotrophic), independence from arable land and other environmental factors, effective nutrient extraction, and rapid growth. Large-scale applications such as combined or separate wastewater treatment and biofuel generation are made possible by heterotrophic microalgae culture, which is frequently less expensive, simpler to set up, and easier to use than autotrophic crop cultivation. Lipid production from microorganisms requires increasing and lowering the cost of yields through the use of adequate carbon concentrations and alternative or renewable sources, such as lignocellulosic biomass from forestry or agricultural residues, or other substrates of organic residues. The most common renewable source in the biosphere is non-edible lignocellulosic biomass, which is composed of cellulose, hemicellulose, and lignin. Whether by physical, chemical, or enzymatic means, it has been widely used as a raw material in the production of biofuels to hydrolyze a variety of inedible lignocellulosic biomass to produce fermentable sugars.

Combining microbial lipids with oleaginous bacterial cultures is one possible way to make them commercially viable. However, to achieve an efficient bioconversion of inexpensive substrates into microbial lipids, consideration must be given to the features of both the microorganism and the substrate at the same time. Due to the extreme variability of the activity of the yeasts and the lignocellulosic raw materials utilized, as well as the parameters of the applied processes, such as time and temperature and whether or not pre-treatments are performed, it is impossible to characterize the carbon concentrations required for good productivity. On the other hand, Mu et al. (2015), the process using Chlorella sp. can be examined with a biomass concentration of 5.8 g/L and a produced lipid content of 34.0%. as a successful productivity outcome. According to Patel et al. (2021) oleaginous bacteria possess the unique ability to take up multiple carbon sources from renewable substrates, such as hydrophilic and hydrophobic ones. One of the greatest sources of raw materials for the synthesis of these microbes and lipids is sugarcane waste. Microorganisms like *Chlorella* sp. can use the hydrolysate, which is rich in glucose, xylose, and arabinose, that is produced after this material is hydrolyzed enzymatically and the lipid generation.

While de novo lipid synthesis on sugar-based substrates produces more triacylglycerols, ex novo lipid synthesis is more beneficial for growth on hydrophobic substrates. In oleaginous bacteria, these substrates accumulate during three different physiological stages: growth, lipid accumulation, and lipid turnover in the presence of a high carbon/nitrogen ratio. In the growth phase, proteins, polysaccharides, polar lipids, and biomass are produced by absorbing available nutrients, especially the carbon source. Wynn et al. (2010) describe how this lipid accumulation occurs during the stationary phase in the presence of excess carbon and nutritional restriction, which inhibits cell division and growth due to a nitrogen deficit. Storage lipids are subsequently created from the excess carbon. Temperature, pH, aeration, light, and CO_2 levels all have an impact on the formation of lipids in addition to nutrients.

According to Liu et al. (2016), it is essential to comprehend how biomass inhibitors affect microbial strain performance in order to develop strategies that can boost fermentation performance. The presence of toxic compounds in lignocellulosic hydrolysates, such as furfural, vanillin, and 5-hydroxymethylfurfural (5-HMF), is a significant factor affecting the efficiency of hydrolysate-based fermentation processes. Therefore, they call attention to this problem and propose that one possible way to obtain oleaginous microorganisms with enhanced tolerance to toxic compounds and the ability to grow in biomass hydrolysates is by employing evolved strains. As stated by Liu et al. (2016) states that the inhibitory concentrations will differ depending on the microorganism. Based on tests conducted using *R. toruloides*, values ranging from 4–8 mM for furfural, 0–5 mM for vanillin, and 1–5 mM for hydroxymethylfurfural can be found for a wide variety of microorganisms. Ngamsirisomsakul et al. (2021) compared oilseed yeast strains from five distinct genera and were able to identify the optimal strain for using sugarcane bagasse as a raw material to produce lipids. Because the hydrolysate's buffering ability improved the growth and production of lipids in all strains, demonstrating better performance in the pH range between 5.5 and 7.5, temperature of 30°C, and the use of a flask shaking at a speed of 200 rpm, it is possible that sugarcane bagasse can be used as a raw material for the production of microbial lipids if the control criteria are met.

Many microalgal strains that are grown in heterotrophic or mixotrophic environments can use organic acids, sugars, and alcohols as carbon sources. The distinction is that in the heterotrophic stage of development, microalgae are grown in the absence of light, consume organic matter from their environment to meet their energy requirements, and release carbon dioxide. In a mixotrophic culture, microalgae consume exogenous organic substances from the environment along with CO_2 by using light energy. As stated by Subramaniam et al. (2010), limitations in nitrogen, phosphorus, silicon, salinity, and iron can cause an alga to reach a high quantity of cellular lipids, which is contingent upon the stage of cell growth. The development, lipid buildup, and chrysolamine in indicated that at 3 mM of initial nitrogen concentration, the highest biomass concentration (6.21 ± 0.17 g/L) was obtained, while higher initial nitrogen concentrations, like 18 mM, were more likely to cause lipid and chrysolamine accumulation, which in turn led to a higher lipid content (43.80 ± 0.57 percent of DW) (Wang et al., 2020). Singh et al. (2020) looked into the production of lipids and beta-carotene by freshwater microalgae in the Scenedesmus genus, Tetradesmus obliquus SGM19. Under ideal growth conditions, the strain produced 2.5 grams of biomass/L, 0.67 mg/g of β-carotene in dry biomass, 1.5 grams of nitrogen per liter of $NaNO_3$, and 28.5% of lipids by weight, with 71% of C16 and C18 fatty acids in the profile. The lipid content indicated that the biodiesel's characteristics (cetane number = 48.59) were within the tolerances specified by EN 14,214 and ASTM D6751 standards. The growth rate, lipid content, and fatty acid composition of microbial cells can all be impacted by the kind and concentration of the nitrogen source. When compared to inorganic nitrogen sources like ammonium sulfate or nitrate, the use of organic nitrogen sources like yeast extract, peptone, or tryptone has typically produced higher biomass and lipid yields.

One of the primary factors that have been found to impede the intracellular accumulation of lipids in microorganisms is the carbon source. Generally speaking, carbon sources that are metabolized quickly, like glucose, can produce high biomass yields but can also lead to lower lipid accumulation. Conversely, more complex or slowly metabolized carbon sources, like xylose or cellobiose, allow for a more significant accumulation of lipids. (i) Oxygen availability, as low oxygen levels can inhibit microbial growth and decrease lipid accumulationd (ii) Temperature: Lower temperatures may cause lipids to accumulate more significantly, whereas higher temperatures may cause lipids to accumulate less significantly but grow more quicklyd (iii) Since different microorganisms have different ideal pH ranges for growth and lipid accumulation, pH has a significant impact; and (iv) The availability of nutrients, since in addition to nitrogen, the presence of phosphorus, sulfur, and trace elements can also influence the accumulation of lipids, and the growth and accumulation of lipids in microorganisms can be restricted by the inadequacy of these nutrients. Utilizing symbiotic cultures—either sequential or mixed—offers a number of benefits over pure cultures and can be a tactic to boost the synthesis of microbial lipids and carotenoids. Processing costs can be reduced when used in conjunction with inexpensive substrates such as effluent or industrial waste. Dias et al. (2019) state that certain yeast and microalgae species have extensive dietary requirements. The production of microbial lipids from renewable raw sources is one of the most promising processes for the production of oleochemicals and biodiesel. Investing in a combination of computer-aided designs, synthetic biology techniques, metabolic systems engineering, and multi-omics integration analysis is necessary to increase the production/accumulation of lipids by microorganisms in comparison to the theoretical maximum limitations. In Figure 7.3 it is possible to observe the schematic diagram of fatty acid metabolic pathways.

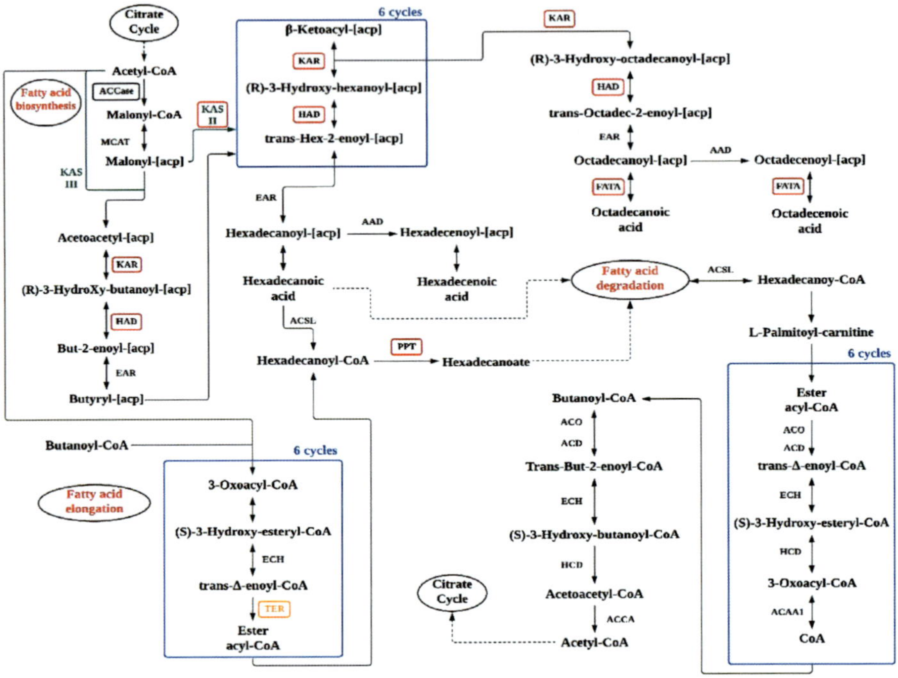

Figure 7.3. *Schematic diagram of fatty acid metabolic pathways.*

Source: Silva JdMe, Martins LHdS, Moreira DKT, Silva LdP, Barbosa PdPM, Komesu A, Ferreira NR, Oliveira JARd. (2023). Microbial lipid based biorefinery concepts: A review of status and prospects. Foods, 12(10):2074. https://doi.org/10.3390/foods12102074.

7.3.5. SCO Production from Lignocellulosic Biomass

7.3.5.1. Stages of the Bioconversion of Lignocellulosic Biomass to Lipid

Certain pretreatments, such as acid, alkaline, sequential acid-alkaline, steam explosion, and organosolv pretreatment, are necessary for the successful conversion of lignocellulosic biomass from oleaginous microorganisms to lipid. In order to supply the sugars used during the microbial lipid fermentation, pretreatment enables the disruption of the lignocellulosic matrix. Sulfuric acid is the most often used acid pretreatment, which is well-documented in the literature as the most applied technique for high sugar yield from a lignocellulosic matrix. Another common chemical technique for enhancing the enzymatic hydrolysis of various lignocellulosic matrices is alkaline pretreatment, which is less harsh on sugars and very effective at solubilizing lignin. Sequential acid-alkaline pretreatment is another effective method for lignocellulosic biomass processing. The primary benefit of this combined process is the reduction of byproducts and the efficient extraction of hemicellulose and lignin through the acid and alkaline stages, respectively. However, the length of processing, the number of steps, and the equipment cost are thought to be the primary drawbacks. Furthermore, a crucial set of pretreatments has become much more important. These are the ones that use ionic liquids, which are thought to be more eco-friendly. The process is typically carried out in milder temperatures or environments, as is the case with the use of choline acetate and 1-butyl-3-methylimidazolium chloride ([BMIM]CL).

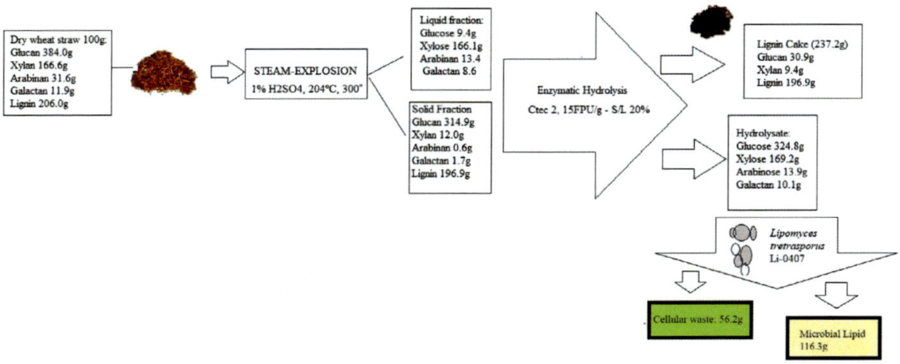

Figure 7.4. *Microbial lipid production from wheat straw using the oleaginous yeasts L. tetrasporus DSM 70314.*

Source: Silva JdMe, Martins LHdS, Moreira DKT, Silva LdP, Barbosa PdPM, Komesu A, Ferreira NR, Oliveira JARd. (2023). Microbial lipid based biorefinery concepts: a review of status and prospects. Foods, 12(10):2074. https://doi.org/10.3390/foods12102074.

They comprise a wide range of chemical components. To extract lignocellulosic biomass from hardwoods and herbs, steam explosion—a brief application of high pressure and high temperature—is a common technique. Furthermore, a number of organic solvents (e.g., lignin (liquid fraction) and cellulose (solid fraction) from lignocellulosic biomass are commonly separated using methanol, ethanol, acetic acid, formic acid, acetone, phenol, or glycerol. The pretreated biomass is combined with various enzyme kinds (e.g., cellulases and hemicellulases) for the conversion of polysaccharides into monosaccharides through enzymatic hydrolysis (e.g., glucose as well as xylose). There are two methods for combining enzymatic hydrolysis with lipid fermentation: simultaneous saccharification and fermentation (SSF) and separate hydrolysis and fermentation (SHF). Since the SSF method lowers the risk of enzyme inhibition by sugar accumulation, cross-contamination during fermentation, processing time, and financial costs, it is recommended. The temperature during the SSF process is a crucial factor to take into account because fermentation and enzymatic hydrolysis happen at distinct temperatures. Since enzymes are more sensitive to temperature changes than thermotolerant microbes, using a fermenting microorganism that is thermotolerant should be thought of as an effective method to optimize the SSF process. That is conceivable (Figure 7.4).

7.3.5.2. SCO Production from Lignocellulosic Biomass and Setbacks

Microalgae, bacteria, and fungi (including filamentous fungi and yeasts) can all produce SCO from lignocellulosic-hydrolysates biomass. The primary microbes utilized in the processes that transform lignocellulosic biomass into lipids are yeasts. Recently, a yield of 55.8 g of oil per kilogram of corncobs has been achieved when using corncobs as the substrate to produce SCO from *Rhodotorula taiwanensis* (AM2353), and more than 81.5% of the oil could be converted into biodiesel.

The native grass of the *Cyperus distans* species, which is commonly found in wetlands, abandoned areas, and roadside ditches, is an inventive source of biomass. Its application as a promising substrate for the production of SCO was recently published. In an SCO biorefinery, it was feasible to achieve roughly 53.66% of lipid accumulation in the cells by using *Yarrowia lipolytica* (MTCC 9519). Another source of biomass that is frequently used for SCO production is rice straw. The effectiveness of *Cryptococcus curvatus* (ATCC 20509) for lipid fermentation was examined in a recent study. The study's findings demonstrated that the hydrolysates could yield 0 g/g of lipid from the current substrate while producing a high total lipid of 8 g/L. Six yeast isolates were also evaluated for their potential for biodiesel production. The lignocellulosic biomass used in that study was all pretreated using steam explosion alone to prevent acetic acid, furfural, and hydroxymethylfurfural (HMF) contamination. When rice husk hydrolysate was used as the substrate, the strain *Meyrozyma guilliermondii* (G5 MK414782) produced the highest lipid content (37.99%) and the strain *Pichia kudriavzevii* (SY2 MF926445) produced the highest lipid accumulation (2.39 g/L). Corn stover is an excellent source of lignocellulosic biomass for the production of SCO. Utilizing the strain *Trichosporon dermatis* 32903—selected from 11 oleaginous yeast strains—lipid fermentation was recently used to test it. *Trichosporon dermatis* (32903) demonstrated a lipid yield of 7.46 g/L (0.104 g/g) and 6.81 g/L (0.101 g/g) from corn stover that had been pretreated with diluted acid and diluted alkali, respectively.

By getting rid of impurities or inhibitors, washing the pretreated corn stover has shown to be crucial for increasing lipid production and sugar to lipid yields. Consequently, the strain *Trichosporon dermatis* (32903) demonstrated increased lipid production from washed dilute-acid and washed dilute-alkali pretreated corn stover, exhibiting 11.43 g/L (0.156 g/g lipid yield) and 20.36 g/L (0.186 g/g lipid yield), respectively. The potential for producing SCO from sweet sorghum stalks was assessed while *Lipomyces starkeyi* (CBS 1807) was present. The maximum lipid yield of 6.40 g/L was obtained with a lipid content of 29.5% (w/w) using the hydrolysate sweet sorghum stalk juice with an initial sorghum content of 12% (w/w). It has been widely investigated whether using oleaginous bacteria species—primarily *Rhodococcus opacus*—can produce SCO from lignocellulosic biomass. *Rhodococcus opacus* was used to obtain a lipid content of 46% w/w from kraft hardwood pulp.

Another study investigated the use of black liquor's kraft lignin that had been oxygen-pretreated and then fermented with *R. opacus* (DSM 1069), whose maximum lipid yield was 0.067 mg/mL and whose main lipid products were stearic (42.7%) and palmitic (46.9%) acids. The application of ultrasound and ethanol-pretreated lignin has shown appropriate lipid production capacity (4.08%) with *R. opacus* (DSM 1069).

The same strain of bacteria, *R. opacus* (DSM 1069) was in charge of the lignocellulosic pretreatment effluent's bioconversion. About 26.99% of the cell dry weight was accumulated as lipids, which suggests a novel method for producing biodiesel and treating waste. It has been demonstrated that the use of forest biomass, specifically the species *Autechochlorella protothecoides* and Norway spruce, as a substrate for the sustainable production of SCO through oleaginous green microalgae, is advantageous. The forest biomass was processed using a hybrid organosolv-steam explosion method, and lipid production of 5.65 g/L (66% lipid content) and 5.28 g/L (63.08% lipid content) was produced when the microalgae were cultured on birch and spruce substrate hydrolysates, respectively. A few challenging challenges that directly affect the yield of SCO production are the C:N ratio of the hydrolysates, sugar content, sugar combinations, and the presence of impurities and inhibitors from the essential pretreatment stage of the process.

The main barrier to oleaginous microbe growth and SCO formation is the production of chemicals known as inhibitory chemicals, which include furfural, acetic acid, HMF, neutral and acidic phenolics, and other chemical species.

Table 7.2. *SCO Production by Different Microorganisms*

Source of SCO Production	**Microorganism Used**	**Substrate Used**	**Yield of SCO Production (g/L)**	**Advantages**	**Disadvantages**
Algae	*Chlorella vulgaris, Chlorella sorokiniana, Botryococcus braunii, A. protothecoides SAG 211-7a, Chlorella* sp.	Glucose or CO_2	10–100	Fast growth rate, high lipid-content	High production cost, contamination issues
Yeast	*Yarrowia lipolytica, Cryptococcus* sp., *Cryptococcus curvatus,*	Glucose or agro-industrial waste	30–60	Can utilize a variety of substrates, easy to manipulate	Low lipid content
Fungi	*Mortierella alpina, Mucor circinelloides, Mortierella alpina, Aspergillus, Penicillium, Fusarium and Alternaria, Cunninghamella echinulata*	Glucose or plant oil	60–80	High lipid content, can produce polyunsaturated fatty acids	Slow growth rate
Bacteria	*Rhodococcus opacus, Sterculia foetida, E. coli and Acinetobacter baylyi*	Plant oil or glucose	40–80	High lipid content, can utilize a variety of substrates	Slow growth rate, low lipid productivity

Source: Silva JdMe, Martins LHdS, Moreira DKT, Silva LdP, Barbosa PdPM, Komesu A, Ferreira NR, Oliveira JARd. (2023). Microbial lipid based biorefinery concepts: A review of status and prospects. Foods, 12(10):2074. https://doi.org/10.3390/foods12102074.

In order to overcome these obstacles, several tactics have been used, including the use of surfactants, the incorporation of detoxification procedures, and the creation of genetically modified strains that are resistant to or tolerate the inhibitor compounds.

Although certain microorganisms can accumulate lipid in their cells when inhibitors are present, detoxification is an essential step in increasing microbial lipid production, but it comes at an extra cost in terms of money and energy. *Arundo donax* L. giant reeds were used in a recent study) as a potential substrate for oleaginous yeast (*Lipomyces starkeyi* DSM 70296)-based SCO. The researchers discovered 8 g of SCO from 100 g of biomass, with a range of 15–24 weight percent in lipid yield. The SCO production potential of barley hull hydrolysate was evaluated using the *Trichosporon cutaneum* (CTM-30125) strain. To improve the accumulation of microbial lipid in the cells of *T mucosa* (CTM-30125), detoxification using overliming and adsorption treatments was necessary prior to fermentation. SCO was produced via *T. cutaneum* CX1 using a corn stover in light of the two main technical drawbacks (low C/N ratio and inhibitor contaminants).

The biodetoxification method tested by the authors was halted upon the detection of no HMF and furfural compounds. By reducing the nitrogen content, the applied method directly aids in the removal of inhibitors and increases the C/N ratio. The authors state that *T. cutaneum* CX1 bioaccumulates lipids. Using the bio-detoxified hydrolysate; attained 23.5%, which is twice as much as it did with the non-detoxified substrate. Even though the detoxification process consumes more energy and money, most recent research suggests that developing detoxification strategies targeted at high yield lipid content is crucial for the production of SCO. Table 7.2 shows a comparison of the production of SCO by various microorganisms.

7.3.5.3. Strain Development

Numerous oleaginous microorganisms, such as filamentous fungi, bacteria, yeasts, and microalgae, have been screened for their ability to produce lipids. One metabolic trait that these microbes have in common is that they are designed to convert carbon substrate into storage lipids, which build up in the cytoplasm and mostly take the form of free fatty acids (FFA) and triacylglycerides (TAG, also known as single-cell oils, or SCO). The following characteristics of oleaginous microorganisms that have been identified as promising candidates for lipid biosynthesis: (i) rapid growth rates; (ii) high intracellular lipid accumulation; and (iii) the capacity to grow in inexpensive media, such as lignocellulosic materials or agro-industrial wastes. More specifically, the ability of oleaginous bacteria to accumulate lipids is influenced by a multitude of factors, most of which are species-specific.

Table 7.3. *Main Strategies Used for the Development of Strains*

Strategies	Realization Mechanism	Example
Mutation	By creating a mutant strain with the use of physical and chemical mutagens, strain improvement is achieved.	An important commercial version of tetracycline is 6-methyl tetracycline, which is produced by a mutant strain of *Strepto-myces aureofaciens*.
Recombi-nation	The process that combines two genotypes to create a new geno-type is known as genetic recombi-nation. Effective strain improve-ment requires careful mutagenesis to maintain genetic diversity as well as the employment of sister strains, divergent strains, and ancestor crosses at predetermined intervals.	Recombination after mei-otic Return-To-Growth in a sterile polyploid hybrid yeast.
Protoplast fusion	Protoplast transformation or pro-toplast fusion is one of the most significant developments in recent years. Protoplast has a very strong negative charge on its surface, making fusion difficult.	*Sclerotium rolfsii*, a phytopathogenic fungus, protoplast isolation and fusion (sacc.)
Gene tech-nology	In vitro recombination and gene manipulation are both part of gene technology. These techniques al-low for the replication of specified DNA sequences inside prokaryote or eukaryote organisms.	Genetically modified fila-mentous fungus, bacteria, and yeast are being used in the food business to produce terpenes.

Source: Silva JdMe, Martins LHdS, Moreira DKT, Silva LdP, Barbosa PdPM, Komesu A, Ferreira NR, Oliveira JARd. (2023). Microbial lipid based biorefinery concepts: A review of status and prospects. Foods, 12(10):2074. https://doi.org/10.3390/foods12102074.

However, a common cause of oil buildup is nutritional imbalance, such as conditions of nitrogen famine (or a deficiency in sulfate, phosphate, or magnesium) in the presence of abundant carbon. Lipid biosynthesis is increased and cell proliferation is hindered by this condition. The buildup of lipids is also influenced by other physical factors, including temperature, dissolved oxygen, light intensity, and culture agitation. The biological tool of metabolic engineering strains to enhance lipid production and robust strains with other desired properties, like survival in harsh environments, substrate spectrum extension, and preference for downstream processing to lipid recovery, is metabolic regulation in oleaginous microorganisms.

Gene knockouts, overexpression, and random mutagenesis are examples of metabolic engineering techniques. In order to carry out random mutagenesis, an organism's DNA is damaged by physical or chemical treatments, which are then followed by mutant selection. Among these mutagen therapies are radiation (e.g., ultraviolet and gamma radiation), low energy nitrogen ions, and chemical agents (e.g., ethidium bromide and formaldehyde). However, multiple mutants are produced by random mutagenesis, which lengthens the time required for the mutant selection process. Gene introduction, overexpression, or deletion is addressed in the creation of a targeted genetic engineering microorganism. Target gene deletion, which is accountable for β-oxidation (e.g., acyl-CoA oxidases, as well as the lipases TGL4 and PEX10, prevent the breakdown of lipids. On the other hand, lipid accumulation is enhanced by overexpressing the genes for acetyl-CoA carboxylase (ACC1) and diacylglycerol O-acyltransferases (DGA1). In engineered Rhodotorula mucilahodotorula, the overexpression of malic enzymes increased SCO production by 18% in comparison to wild strains and improved lipid quality by acting as a NADPH cofactor for fatty acid biosynthesis. More than just enhancing lipid accumulation, genetic discoveries have made it possible to produce modified fatty acids with significant added value. *Rhodococcus opacus* and other engineered microorganisms can produce SCO at high xylose concentrations thanks to the insertion of two heterologous Streptomyces genes that encode xylose isomerase (xylA) and xylulokinase (xylB). The primary tactics for the development of strains can be confirmed in Table 7.3 (Silva et al., 2023).

7.3.5.4. Lipid Recovery

The ideal fermentative conditions are carried out in the upstream stages of SCO production in order to maximize microbial performance in lipid production, e.g., genetic modifications for strain development and fermentative parameter optimization. Critical steps that impact process efficiency are also presented in the downstream process. These steps include solvent and/or assisted extraction techniques, as well as methods for recovering lipid from oleaginous microorganisms, such as pre-treatments that disrupt cells. Because the cell walls of microorganisms are more rigid than those of other biological systems (plant and animal cells), extraction is not fully achieved when traditional solvent maceration is used alone. In order to improve lipid recovery, physical (mechanical and non-mechanical), chemical, and enzymatic methods of cell disruption have been researched. While the literature describes all of these techniques for disrupting microbial cells, physical disruption procedures have proven more applicable for large-scale applications. e.g., high-pressure homogenization, bead milling, and extraction techniques aided by ultrasound and microwave technology (Figure 7.5).

Following fermentation, enzymes are typically rendered inactive by heat treatment, or pasteurization. (e.g., lipases) and stop the microbe from consuming lipids that are stored. Following centrifugation to concentrate the biomass, cell disruption is the next stage. Because they contain polyunsaturated fatty acids, SCOs exhibit oxidation and rancidity problems that are comparable to those of plant seed oils. Therefore, to avoid lipid damage during the entire lipid recovery process, mild temperatures are needed. Cooling systems must be connected to mechanical cell disruption techniques in particular because they raise the temperature.

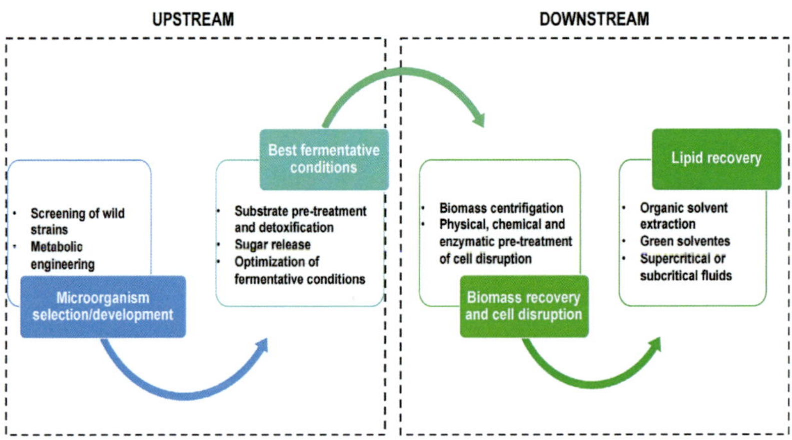

Figure 7.5. *Upstream and downstream steps of SCO production from lignocellulosic biomass.*

Source: Silva JdMe, Martins LHdS, Moreira DKT, Silva LdP, Barbosa PdPM, Komesu A, Ferreira NR, Oliveira JARd. (2023). Microbial lipid based biorefinery concepts: A review of status and prospects. Foods, 12(10):2074. https://doi.org/10.3390/foods12102074.

The bead-milling technique is a straightforward and highly effective mechanical method of cell disruption that works well with yeast, bacteria, algae, and filamentous fungus cells. This technique breaks down cells by rubbing tiny glass or stainless-steel beads against them, which grinds up oleaginous microorganisms and releases lipids. With the exception of bacteria, where tiny beads (0.1 mm) increase the effectiveness of cell disruption, the ideal bead diameter is 0.5 mm for all microorganism cells. After the procedure is complete, the beads can be simply extracted by gravity from the solution, and solvent extraction is then used to recover the lipids. Lipid recovery was increased by more than 40% using a combined enzymatic method of cell disruption that involved bead milling, applying lipase, phospholipase, protease, and cellulase. When enzymatic hydrolysis and bead milling were combined, the extraction of SCO from the microalgae *Chlorella vulgaris* produced a lipid yield of 75% and 88%, respectively. When lipid extraction from the yeast *Yarrowia lipolytica* was compared to microwave and ultrasound methods, bead milling was found to be the most effective and low-energy method that did not alter the lipid profile or degrade the fatty acids.

The processing of dairy products and fruits has made extensive use of the high-pressure homogenization method. This method is also used to disrupt cells during lipid extraction. To do this, high pressure is applied to a suspension of cells and forced through a tiny opening, up to 150–200 MPa. High-pressure homogenization's primary benefit is its simple scalability. All microbial cells can be treated with this technique, with the possible exception of filamentous fungi, which can block the homogenization valve with their mycelium structures. However, this restriction may not occur at low biomass concentrations. When compared to solely solvent maceration, the yield of oil extraction is increased fourfold by including the high-pressure homogenization pretreatment to *Y. lipolytica* cultures. Additionally, some authors have proposed that bead milling and high-pressure homogenization applied together as pretreatments would optimize lipid extraction from *Y. lipolytica* cultures while consuming less energy.

Microwave technology represents a cost-effective, safe, and safe alternative to mechanical pretreatments for cell disruption. Microwaves (between 0 and 300 GHz) cause temperature increases and cell lysis by interacting with free water molecules inside the cells. This causes cell disruption. In SCO extraction, microwave pretreatment shortens extraction times while improving oil yields and quality. After 30 minutes of microwave treatment, the green algae Scenedesmus obliquus produced the maximum amount of oil, accounting for 77% of the total recoverable oil. Moreover, SCO extraction was enhanced by the coupling of microwave pre-treatment with enzymatic hydrolysis, yielding nearly 97% of the total lipids from the *Rhodosporidium toruloides* culture. Ultrasonic-assisted extraction is another non-mechanical pre-treatment that relies on cavitational phenomena. This is produced by a high-frequency sound wave (20 kHz to 10 MHz) that collapses gas bubbles, shears liquid, and disrupts cells. The extraction efficiency and quality (fatty acid composition) of SCO from *Mortierella isabellina* were improved by the combined pretreatments of ultrasound and microwaves, producing nearly 93% of lipids. The drawback of both pretreatment techniques is temperature rise, which in certain situations may change the lipids' chemical profile and have an impact on the quality of the final product. Cell disruption pretreatments in SCO recovery are succeeded by organic solvent extraction (Folch standard extraction methods, primarily developed by Soxhlet, Bligh and Dyer). Solid or dried ground samples are subjected to multiple washing cycles with organic solvents in order to extract the material using the Soxhlet method. In order to prevent damage to the lipids, this method is not appropriate for SCO containing unsaturated fatty acids due to the high temperature of the organic solvents under reflux. The Bligh and Dyer method uses a chloroform-methanol system, where hydrophilic substances are eliminated in the methanol and water phase and lipid fraction occurs in the chloroform phase. While the Folch method is comparable to Bligh and Dyer, it is not the same in terms of solvent volumetric ratios or the salt solution washing step. Research has been done to replace hazardous and toxic solvents with eco-friendly solvents, even though methanol, hexane, and chloroform have high extraction efficiency. The goal is to help the oil industry sector. Isomyl acetate was used in place of hexane at the laboratory scale, increasing the extraction of SCO from Y. lipolytica, followed by high-pressure homogenization and bead milling as a pretreatment. Using supercritical or subcritical fluids, like carbon dioxide, which has been shown to be the most popular solvent for lipid and hydrophobic compound extraction and doesn't cause thermal degradation or contaminate SCO, is another method of replacing toxic solvents. While using organic solvents results in higher yields for lipid extraction when compared to supercritical fluids, some authors have shown that carbon dioxide streamlines the process of refining oil and omits the solvent distillation step. Therefore, in order to maximize the lipid yield in supercritical extraction and broaden the application of this cutting-edge technology in SCO extraction, studies must optimize variables like supercritical fluid, temperature, pressure, and cell disruption pretreatment.

7.3.6. Microbes, Feedstocks and Coproducts for Single-Cell-Oil Biorefineries

Microalgae, yeast, bacteria, and fungi are examples of oleaginous microorganisms that produce single cell oils (SCO), also referred to as glyceride oils. In certain species of these oleaginous bacteria, the intracellular lipid content can reach up to 60–70% of their weight, which is equivalent to about 40% of the dry cell weight.

Oleaginous bacteria aid in the synthesis of lipids. Among the prominent genera of oleaginous bacteria are the Arthrobacter species, the Gordonia species, the Acinetobacter species, and the Rhodococcus species. *Rhodococcus* sp. has drawn the greatest interest out of all of them because it can digest lignin and eventually assimilate lignin-monomeric components into the lipid accumulation pathway. It can also grow on a range of substrates. Wells et al. (2015) looked into the pine organosolv pretreatment effluent and *Rhodococcus opacus* DSM 1069 as the sole source of energy and carbon. After 120 hours at 1.5 w/v percent concentration of solids, the results showed that oleic, palmitic, and stearic fatty acids accounted for a maximum of 26.99 ± 2.88% of its cellular dry weight in oils. Until now, bacteria have not been used for lipid synthesis as much as yeasts, molds, or microalgae.

Oleaginous yeasts are the most promising microbes for the synthesis of oils that are similar to vegetable oils. A few common vegetable oils with lipid profiles similar to those of SCO are sunflower oil, rapeseed oil, and palm oil. Certain oleaginous yeast species, such as those found in the genera *Rhodosporidium, Yarrowia, Cryptococcus, Rhodotorula, Lipomyces, and Trichosporon*, can accumulate lipids up to 80% weight percentage of their dry cell weight. For instance, the oleaginous yeast *Lipomyces starkeyi* can grow in basic media (such as those lacking vitamin supplementation), degrade extracellular polysaccharides, and has a good tolerance to inhibitory substances like alcohols, organic acids, and aldehydes. It can also recycle a small portion of its intracellular lipids and obtain high lipid yields from both pentoses and hexoses.

Probst and Vadlani have demonstrated that SCO can be made from corn and wheat bran by employing the oleaginous yeast *Lipomyces starkeyi* ATCC 56304. The hydrolysates of maize and wheat bran produced the maximum amounts of oil, measuring 126.77 and 124.3 mg oil/g sugar, respectively. Di Fidio et al. (2015) investigated *Lipomyces starkeyi*, an oleaginous yeast. For developing a unique two-step method for turning waste from cellulosic paper mills into SCO. The values for maximum oil productivity, single cell lipid content, lipid yield, and output were 20.2%, 37.5%, 3.7%, and 2.0 g/L/d, respectively.

Utilizing *Lipomyces starkeyi* DSM 70,296 from the giant reed (*Arundo donax* L.] as a carbohydrate source, Di Fidio et al. (2015) investigated a combined hydrolysis cascade method to produce SCO, achieving a lipid content of 30 weight percent and yield values between 15 and 24 weight percent. In order for the hydrolysate of Rundo donax to form SCO, Zuccaro et al. (2021) looked into the green microalga Chloroidium *saccharophilum* and the oleaginous yeast *Lipomyces starkeyi*, both individually and in combination. In every case, the performance of combined cultures was superior to that of individual ones. Zhang et al. provided an overview of recent advancements in yeast-based biorefineries' oleochemical synthesis. The use of alternative renewable feedstocks like xylose and L-arabinose was also covered in (2021).

Research on oleaginous yeast has come a long way. However, most bulk oleochemicals on the market today lack the yields, titers, and rates required for industrial production. It is possible to use the circumstances seen for the modified Y as an illustration of a successful case. *Saccharomyces cerevisiae*, a non-oleaginous yeast, is used in several industrial applications because it is easy to grow and has well-established genetic tools. This lipolytic strain is a good candidate for commercial production since it can manufacture FAMEs at high titers, yields, and rates of 98.9 g/L, 1.3 g/L /h, and 0.27 g/g glucose, respectively. Thus, *Saccharomyces cerevisiae* has also been subjected to metabolic engineering methods and utilized for lipid synthesis. Because of their high oil production, oleaginous fungi are endowed microbes. They have several advantages over other oleaginous microorganisms, such as a high dry biomass content, the capacity to produce large amounts of certain fatty acids, such as γ-linolenic acid, and the capacity to grow on inexpensive feedstocks. Zhao et al. (2020) looked into the fatty acid profiles of 28 taxa belonging to the Mortierellales, Mucorales, and Umbelopsidales. The findings showed that 46% of taxa, 50% of species, and 34% of strains were oleaginous, with Backusella being identified as the species responsible for collecting up to 59.8% and 24.24% of the oils.

Microalgae have a high photosynthetic efficiency and higher oil production productivity because they can store large amounts of lipids (60% of their total lipid content produces more net oils than 50%). A handful of the Chlorella species strains that can accumulate more than 25% oil on glucose are *Chlorella vulgaris*, *Chlorella protothecoides*, and *Chlorella sorokiniana*. However, there aren't many studies looking into how microalgae grow on different carbon substrates, like xylose, which is one of the main components of hemicellulose. Combining the marine alga *Nannochloropsis oceanica* with the oleaginous fungus *Mortierella elongata* allowed Du et al. (2018) to research the production of oil. For N, they generated sizable amounts of total fatty acids and triacylglycerol yielding roughly 15% and 22% of the total dry weight, in that order.

Microbial oil exhibits promising potential as a replacement for integrated fuel and food production processes. However, the high cost of feedstock and competition from less expensive oilseed crops have limited production. An economic analysis of the production of yeast SCO indicates that raw material costs, including feedstock costs, can account for as much as 40% of total costs. Making use of food waste streams could result in manufacturing that is competitively priced for the creation of low-cost SCOs. These include a variety of abundant and visually appealing biomass residues, such as rice straw, wheat straw, maize cobs, corn residues, corn fiber, sugarcane bagasse, sorghum bagasse, and many more. Diwan et al. (2018) assessed a large number of agro-industrial waste materials to SCO so that a workable biorefinery could be started.

Using wastewater from industrial and agricultural processes as a feedstock for cultivation has the potential to mitigate the main environmental risks associated with an imbalance in the ratio of wastewater generation to safe disposal. Chuppa et al. (2018) examined *Aspergillus Niger* growth in residual liquid waste from sugarcane molasses (vinasse), which is produced during the fermentation and distillation processes. The results demonstrated how well the fungus could develop on vinasse and break down this difficult medium. Furthermore, the internal lipids of fungal organisms can be utilized to generate biodiesel from their biomass. Hoarau et al. (2018) have demonstrated the potential of vinasse for the growth of yeast and fungi, the production of lipids, and its suitability for biodiesel requirements. Qin et al. (2017) co-cultivated lignocellulosic biomass hydrolysate, hydrophobic wastes, wastewater, crude glycerol, and microalgae on cheap substrates investigated techniques to enhance lipid synthesis.

It is advised to employ a biorefinery approach, in which SCO is used to produce a variety of co-products, in order to ensure both economic sustainability and optimal environmental outcomes. Oleaginous algae and yeast processes have a great deal of potential to benefit from this strategy because, in addition to lipids, they act as a platform for a variety of different intra- and extracellular products. Among the potential co-products produced with SCO are the co-production of protein, energy co-products, glycerol, organic acids, carotenoids, and other high-value compounds.

Following the treatment of the lipids, the microbe cake can be further processed to yield amino acids or peptides, which can be used to make biomaterials, bioplastic, and biofoam, or it can be fed to animals.

It has been demonstrated that oleaginous yeasts can produce organic acids extracellularly in the succinic, malic, oxalic, and citric varieties. Qian et al. investigated the co-production of gluconic acid and SCO. Using the oleaginous *Cryptococcus podzolicus* DSM 27192 in 2019. The results showed that the economic value of producing biodiesel from microbial lipids has increased.

The results of the proposed co-production approach could also be used as a template for other co-production schemes. Gao et al. (2016) looked into the use of a Y-engineered yeast to synthesize succinic acid from crude glycerol. lipolytica. The results indicated that Y. lipolytica is a potentially helpful microbial factorial cell for a very successful way to address environmental problems related to the production of goods with added value. Certain yeast strains, such as *Rhodotorula glutinis*, *Rhodotorula rubra*, *Phaffia rhodozyma/Xanthophyllomyces dendrorhous*, *Rhodotorula roseus*, and *Sporobolomyces ruberrimus*, are capable of gathering important carotenoids, such as β-carotene, torulene, and astaxanthin.

To make whey protein concentrate, antioxidants, ethanol, tartrate salts, and microbial oil, among other commodity and specialty goods, Kopsahelis et al. (2018) looked into the combined refining of wine lees and cheese whey. Leiva-Candia et al. (2015) used standard sunflower-based biodiesel production facilities and looked into the production of by-products with added value (like protein isolate, extracts high in antioxidants, and microbial oil). Increasing coproduct streams' value and implementing a biorefinery strategy can help SCO become more financially and environmentally viable. Among the various coproduct options, the protein percentage seems to be important in determining the minimum oil selling price and the environmental impact.

7.3.7. Microbial Lipid-Based Products

Microbial lipids derived from bacteria, yeasts, filamentous fungi, and microalgae can be applied in four primary sectors: biofuel, pharmaceutical, cosmetic, and food and feed. The pursuit of fourth-generation biofuels, primarily through the utilization of agricultural waste, has stimulated fundamental and applied research as a substitute for renewable energy sources. The production of crude glycerol, a byproduct of biodiesel, and lignocellulosic sources, which have the potential to spark a related supply chain and soon become commercially and technologically feasible, have not yet received enough attention. Apart from its potential application as a biofuel, it could also be utilized as an additive for automotive lubricants and as a wax ester.

There is a lot of interest in and research on the use of lipids and all the molecules related to them in the pharmaceutical industry. One potential use is as a stabilizing agent for pharmaceutical emulsions. In a similar vein, the specificities of microbial lipids enable a broad range of applications. Certain microbial lipids are known as sophorolipids (SLs), and because of their hydrophobic nature, they can be very useful in drug formulations. In contrast, microbial SLs have been shown to have spermicidal activity in humans. PLs have also been linked to additional pharmacological actions, including immune-response modulators, anti-inflammatory, and antiviral agents. Microbial glycolipids were also implicated in the agent's anti-cancer and antimicrobial properties.

Microbial lipids have two distinct roles in the cosmetics industry: their physiological properties, which include skin cleaner, photoprotector, skin conditioner, antioxidant, and anti-aging effects, and their functional properties, which involve acting as an emulsifier, dispersant, or thickener. Positive outcomes in the synthesis of collagen, the treatment of wrinkles, and the reduction of cellulite were also noted. Permeability is therefore a crucial component of microbial glycolipids, allowing for an effective functional or physiological action, because of their structural resemblance to skin lipids.

Filamentous fungi and yeasts play a major role in the production of microbial lipids for food applications, which can primarily be directed toward the production of dietary supplements like essential fatty acids or oils rich in arachidonic acid. This is true even though several studies have focused on producing polyunsaturated fatty acids from microalgae. In the early 1990s, one of the earliest uses of microbial lipids was to make chocolate instead of cocoa butter.

A significant advancement in the integration of microbial lipids as a dietary supplement was made in 2001. "Oil of Javanicus" has been approved by the Food and Drug Administration (FDA) as safe (GRAS) to use in baby formulas. Accordingly, it's estimated that by 2010, over 24 million infants were fed food that contained microbial lipids. In addition, combining microbial lipids with nanotechnology is a crucial step in enhancing their stability during storage, making their incorporation into food matrices easier, or boosting their bioavailability post-ingestion.

Microbial lipids are produced or used by a number of companies, including DSM (Switzerland), Cargill Alking Bioengineering (China), Nestle S.A, Fermentalg (France), Algomed (Germany), Phytolutions (Germany), Parry Nutraceuticals (India), Oilgae (India), Algatechnologies (Israel), Sunchlorella (Japan), Algae Technology Solutions (Mexico), Fitoplancton Marino (Spain), Simris (Sweden), Vedan (Taiwan), Aurora Algae (United States), Algaeon (United States), Altech Algae (United States), and Algae to Omega Holdings (United States). At the moment, DSM is the business that produces the greatest quantity of microbial products for commercial use. Some commercial products with a lipid base of microorganisms can be seen in Table 7.4 (Silva et al., 2023).

Table 7.4. *Microbiological Lipid-Based Commercial Products*

Product	Source	Company/Country	Applications
Javanicus oil	*Mucor javanicus*	J & E Sturge, North Yorkshire, UK	Oil rich in linolenic fatty acid, used as a substitute for evening primrose oil
DHASCO-B	*microalgae Crypthecodinium cohnii*	DSM, Grenzach, Switzerland	Benefits for the optimal development of the child's brain, eyes and nervous system. Support a healthy pregnancy
SUNT-GA40S in Japan and Cabio oil in China	Produced by *Crypthecodinium cohnii and various strains and species of thrautochyrids*	Produced by Suntory, Osaka, Japan and sold in China by Cargill	Enrichment of infant formulas
life's OMEGA	*Crypthecodinium cohnii and Schizochytrium* sp.	DSM, Grenzach Switzerland	Prevention of cardiovascular diseases
Eicooil	*microalgae Y. lipolytica genetically modified*	Qualitas Health Ltd., Jerusalem, Israel	Benefits for blood pressure, platelet aggregation and various inflammatory process
Simris Algae ômega-3	*Chlorella vulgaris*	Simris, Simrishamn, Sweden	Prevention of cardiovascular diseases
BioAstin SCE5 and BioAstin SCE10	*Haematococcus pluvialis*	Parry Nutraceuticals, Chennai, India	antioxidant effects

Source" Silva JdMe, Martins LHdS, Moreira DKT, Silva LdP, Barbosa PdPM, Komesu A, Ferreira NR, Oliveira JARd. (2023). Microbial lipid based biorefinery concepts: A review of status and prospects. Foods, 12(10):2074. https://doi.org/10.3390/foods12102074.

7.3.8. Perspectives for Microbial Lipid-Based Products

The main determinants of microbial lipid perspectives must be technical and financial viability. The primary factor is the decrease in production costs. The impact of production costs is related to the product's intended use. For instance, in the pharmaceutical and cosmetics industries, products with higher added values are typically less influenced. Except for a few unresolved technical issues, like the production of inhibitory products during treatment, several studies suggest that using agricultural residues as a growth medium for oleaginous microorganisms can be a viable and sustainable technology, particularly for abundant raw materials like lignocellulosic residues. As can be observed, these microbial highly adaptive metabolism allows for the production of a variety of interesting products. The potential of this market is thus indicated by the number of companies that produce microbial lipids in a consolidated manner (Silva et al., 2023).

REFERENCES

1. Abomohra, A. E. F., El-Sheekh, M., & Hanelt, D. (2017). Screening of marine microalgae isolated from the hypersaline Bardawil lagoon for biodiesel feedstock. *Renewable Energy, 101*, 1266–1272.
2. Abou-Zeid, A. M., Abd El-Zaher, E. H. F., Saad-Allah, K. M., & Ahmed, R. U. (2019). Screening and optimization of some Egyptian soil-borne fungi for lipids production as a possible source for biofuel. *Egyptian Journal of Experimental Biology, 15*, 243–250.
3. Ageitos, J. M., Vallejo, J. A., Veiga-Crespo, P., & Villa, T. G. (2011). Oily yeasts as oleaginous cell factories. *Applied Microbiology and Biotechnology, 90*, 1219–1227.
4. Ahmad, F. B., Zhang, Z., Doherty, W. O. S., & O'Hara, I. M. (2019). The prospect of microbial oil production and applications from oil palm biomass. *Biochemical Engineering Journal, 143*, 9–23.
5. Alavijeh, R. S., Karimi, K., Wijffels, R. H., van den Berg, C., & Eppink, M. (2020). Combined bead milling and enzymatic hydrolysis for efficient fractionation of lipids, proteins, and carbohydrates of *Chlorella vulgaris* microalgae. *Bioresource Technology, 309*, 123321.
6. Alvarez, H. M. (2003). Relationship between β-oxidation pathway and the hydrocarbon-degrading profile in *Actinomycetes* bacteria. *International Biodeterioration and Biodegradation, 52*, 35–42.
7. Alvarez, H. M. (2010). Biotechnological production and significance of triacylglycerols and wax esters. In K. N. Timmis (Ed.), *Handbook of Hydrocarbon and Lipid Microbiology* (pp. 2995–3002). Springer Berlin Heidelberg.
8. Alvarez, H. M., & Steinbüchel, A. (2002). Triacylglycerols in prokaryotic microorganisms. *Applied Microbiology and Biotechnology, 60*, 367–376.
9. Alvarez, H. M., Hernández, M. A., Lanfranconi, M. P., Silva, R. A., & Villalba, M. S. (2021). *Rhodococcus* as biofactories for microbial oil production. *Molecules, 26*, 4871.
10. Alvarez, H. M., Kalscheuer, R., & Steinbüchel, A. (1997). Accumulation of storage lipids in species of *Rhodococcus* and *Nocardia* and effect of inhibitors and polyethylene glycol. *Fett/Lipid, 99*, 239–246.
11. Alvarez, H. M., Mayer, F., Fabritius, D., & Steinbüchel, A. (1996). Formation of intracytoplasmic lipid inclusions by *Rhodococcus opacus* strain PD630. *Archives of Microbiology, 165*, 377–386.
12. Alvarez, H. M., Pucci, O. H., & Steinbüchel, A. (1997). Lipid storage compounds in marine bacteria. *Applied Microbiology and Biotechnology, 47*, 132–139.

13. Alzagameem, A., Bergs, M., Do, X. T., Klein, S. E., Rumpf, J., Larkins, M., Monakhova, Y., Pude, R., & Schulze, M. (2019). Low-input crops as lignocellulosic feedstock for second-generation biorefineries and the potential of chemometrics in biomass quality control. *Applied Sciences, 9*, 2252.
14. Ananthi, V., Siva Prakash, G., Chang, S. W., Ravindran, B., Nguyen, D. D., Vo, D. V. N., La, D. D., Bach, Q. V., Wong, J. W. C., Kumar Gupta, S., et al. (2019). Enhanced microbial biodiesel production from lignocellulosic hydrolysates using yeast isolates. *Fuel, 256*, 115932.
15. Arora, N., Yen, H. W., & Philippidis, G. P. (2020). Harnessing the power of mutagenesis and adaptive laboratory evolution for high lipid production by oleaginous microalgae and yeasts. *Sustainability, 12*, 5125.
16. Babu, M. V., Murthy, K. M., & Rao, G. A. P. (2020). Production process optimization of Sterculia Foetida methyl esters (biodiesel) using response surface methodology. *International Journal of Ambient Energy, 43*, 1837–1846.
17. Balasubramanian, S., Allen, J. D., Kanitkar, A., & Boldor, D. (2011). Oil extraction from *Scenedesmus obliquus* using a continuous microwave system—Design, optimization, and quality characterization. *Bioresource Technology, 102*, 3396–3403.
18. Bandhu, S., Bansal, N., Dasgupta, D., Junghare, V., Sidana, A., Kalyan, G., Hazra, S., & Ghosh, D. (2019). Overproduction of single cell oil from xylose-rich sugarcane bagasse hydrolysate by an engineered oleaginous yeast *Rhodotorula mucilaginosa* IIPL32. *Fuel, 254*, 115653.
19. Barney, B. M., Wahlen, B. D., Garner, E., Wei, J., & Seefeldt, L. C. (2012). Differences in substrate specificities of five bacterial wax ester synthases. *Applied and Environmental Microbiology, 78*, 5734–5745.
20. Béligon, V., Christophe, G., Fontanille, P., & Larroche, C. (2016). Microbial lipids as potential source to food supplements. *Current Opinion in Food Science, 7*, 35–42.
21. Bellou, S., Triantaphyllidou, I. E., Aggeli, D., Elazzazy, A. M., Baeshen, M. N., & Aggelis, G. (2016). Microbial oils as food additives: Recent approaches for improving microbial oil production and its polyunsaturated fatty acid content. *Current Opinion in Biotechnology, 37*, 24–35.
22. Bettencourt, S., Miranda, C., Pozdniakova, T. A., Sampaio, P., Franco-Duarte, R., Pais, C. (2020). Single cell oil production by oleaginous yeasts grown in synthetic and waste-derived volatile fatty acids. *Microorganisms, 8*, 1809.

23. Bhadoriya, S. S., & Madoriya, N. (2013). Biosurfactants: A new pharmaceutical additive for solubility enhancement and pharmaceutical development. *Biochemical Pharmacology, 2*, 5.
24. Bharathiraja, B., Sridharan, S., Sowmya, V., Yuvaraj, D., & Praveenkumar, R. (2017). Microbial oil—A plausible alternate resource for food and fuel application. *Bioresource Technology, 233*, 423–432.
25. Bharti, R. K., Srivastava, S., & Thakur, I. S. (2014). Extraction of extracellular lipids from chemoautotrophic bacteria *Serratia* sp. ISTD04 for production of biodiesel. *Bioresource Technology, 165*, 201–204. https://doi.org/10.1016/j.biortech.2014.02.075.
26. Bharti, R. K., Srivastava, S., & Thakur, I. S. (2014). Production and characterization of biodiesel from carbon dioxide concentrating chemolithotrophic bacteria, *Serratia* sp. ISTD04. *Bioresource Technology, 153*, 189–197.
27. Biermann, U., Bornscheuer, U., Meier, M. A. R., Metzger, J. O., & Schäfer, H. J. (2011). Oils and fats as renewable raw materials in chemistry. *Angewandte Chemie International Edition, 50*, 3854–3871.
28. Blazeck, J., Hill, A., Liu, L., Knight, R., Miller, J., Pan, A., Otoupal, P., & Alper, H. S. (2014). Harnessing *Yarrowia lipolytica* lipogenesis to create a platform for lipid and biofuel production. *Nature Communications, 5*, 3131.
29. Bokinsky, G., Peralta-Yahya, P. P., George, A., Holmes, B. M., Steen, E. J., Dietrich, J., Lee, T. S., Tullman-Ercek, D., Voigt, C. A., Simmons, B. A., & Keasling, J. D. (2011). Synthesis of three advanced biofuels from ionic liquid-pretreated switchgrass using engineered *Escherichia coli*. *Proceedings of the National Academy of Sciences, 108*, 19949–19954.
30. Bonturi, N., Matsakas, L., Nilsson, R., Christakopoulos, P., Miranda, E. A., Berglund, K. A., & Rova, U. (2015). Single cell oil producing yeasts *Lipomyces starkeyi* and *Rhodosporidium toruloides*: Selection of extraction strategies and biodiesel property prediction. *Energies, 8*, 5040–5052.
31. Borand, M. N., & Karaosmanoğlu, F. (2018). Effects of organosolv pretreatment conditions for lignocellulosic biomass in biorefinery applications: A review. *Journal of Renewable and Sustainable Energy, 10*, 033104.
32. Bredemeier, R., Hulsch, R., Metzger, J. O., & Berthe-Corti, L. (2003). Submersed culture production of extracellular wax esters by the marine bacterium *Fundibacter jadensis*. *Marine Biotechnology, 5*, 579–583.
33. Bruder, S., Hackenschmidt, S., Moldenhauer, E. J., & Kabisch, J.

(2018). Chapter 12—Conventional and oleaginous yeasts as platforms for lipid modification and production. In U. T. Bornscheuer (Ed.), *Lipid Modification by Enzymes and Engineered Microbes* (pp. 257–292). AOCS Press.

34. Bruno, L., Di Pippo, F., Antonaroli, S., Gismondi, A., Valentini, C., & Albertano, P. (2012). Characterization of biofilm-forming cyanobacteria for biomass and lipid production. *Journal of Applied Microbiology, 113*, 1052–1064.
35. Bryn, K., Jantzen, E., & Bøvre, K. (1977). Occurrence and patterns of waxes in *Neisseriaceae*. *Journal of General Microbiology, 102*, 33–43.
36. Cameotra, S. S., & Makkar, R. S. (2004). Recent applications of biosurfactants as biological and immunological molecules. *Current Opinion in Microbiology, 7*, 262–266.
37. Cao, Y., Liu, W., Xu, X., Zhang, H., Wang, J., & Xian, M. (2014). Production of free monounsaturated fatty acids by metabolically engineered *Escherichia coli*. *Biotechnology for Biofuels, 7*, 59–70.
38. Caporusso, A., Capece, A., & De Bari, I. (2021). Oleaginous yeasts as cell factories for the sustainable production of microbial lipids by the valorization of agri-food wastes. *Fermentation, 7*, 50.
39. Caporusso, A., De Bari, I., Giuliano, A., Liuzzi, F., Albergo, R., Pietrafesa, R., Siesto, G., Romanelli, A., Braccio, G., & Capece, A. (2023). Optimization of wheat straw conversion into microbial lipids by *Lipomyces tetrasporus* DSM 70314 from bench to pilot scale. *Fermentation, 9*, 180.
40. Carsanba, E., Papanikolaou, S., & Erten, H. (2018). Production of oils and fats by oleaginous microorganisms with an emphasis given to the potential of the nonconventional yeast *Yarrowia lipolytica*. *Critical Reviews in Biotechnology, 38*, 1230–1243.
41. Chang, L., Lu, H., Chen, H., Tang, X., Zhao, J., Zhang, H., Chen, Y. Q., & Chen, W. (2022). Lipid metabolism research in oleaginous fungus *Mortierella alpina*: Current progress and future prospects. *Biotechnology Advances, 54*, 107794.
42. Chebbi, H., Leiva-Candia, D., Carmona-Cabello, M., Jaouani, A., & Dorado, M. P. (2019). Biodiesel production from microbial oil provided by oleaginous yeasts from olive oil mill wastewater growing on industrial glycerol. *Industrial Crops and Products, 139*, 111535.
43. Chen, J., Song, X., Zhang, H., & Qu, Y. (2006). Production, structure elucidation and anticancer properties of sophorolipid from *Wickerhamiella domercqiae*. *Enzyme and Microbial Technology, 39*, 501–506.

44. Cheng, P., Ji, B., Gao, L., Zhang, W., Wang, J., & Liu, T. (2013). The growth, lipid and hydrocarbon production of *Botryococcus braunii* with attached cultivation. *Bioresource Technology, 138*, 95–100.
45. Cheng, P., Okada, S., Zhou, C., Chen, P., Huo, S., Li, K., Addy, M., Yan, X., & Ruan, R. R. (2019). High-value chemicals from *Botryococcus braunii* and their current applications: A review. *Bioresource Technology, 291*, 121–911.
46. Cheng, Y. S., Mutrakulcharoen, P., Chuetor, S., Cheenkachorn, K., Tantayotai, P., Panakkal, E. J., & Sriariyanun, M. (2020). Recent situation and progress in biorefining process of lignocellulosic biomass: Toward green economy. *Applied Sciences and Engineering Progress, 13*, 299–311.
47. Chisti, Y. (2007). Biodiesel from microalgae. *Biotechnology Advances, 25*, 294–306.
48. Cho, H. U., & Park, J. M. (2018). Biodiesel production by various oleaginous microorganisms from organic wastes. *Bioresource Technology, 256*, 502–508.
49. Chopra, I., & Roberts, M. (2001). Tetracycline antibiotics: Mode of action, applications, molecular biology, and epidemiology of bacterial resistance. *Microbiology and Molecular Biology Reviews, 65*, 232–260.
50. Chu, Q., Tong, W., Chen, J., Wu, S., Jin, Y., Hu, J., & Song, K. (2021). Organosolv pretreatment assisted by carbocation scavenger to mitigate surface barrier effect of lignin for improving biomass saccharification and utilization. *Biotechnology for Biofuels, 14*, 136.
51. Chuppa-Tostain, G., Hoarau, J., Watson, M., Adelard, L., Shum Cheong Sing, A., Caro, Y., Grondin, I., Bourven, I., François, J. M., Girbal-Neuhauser, E., et al. (2018). Production of *Aspergillus niger* biomass on sugarcane distillery wastewater: Physiological aspects and potential for biodiesel production. *Fungal Biology and Biotechnology, 5*, 1.
52. Comba, S., Sabatini, M., Menendez-Bravo, S., Arabolaza, A., & Gramajo, H. (2014). Engineering a *Streptomyces coelicolor* biosynthesis pathway into *Escherichia coli* for high yield triglyceride production. *Biotechnology for Biofuels, 7*, 172–183.
53. Cordeiro, R. S., Vaz, I. C. D., Magalhães, S. M. S., & Barbosa, F. A. R. (2017). Effects of nutritional conditions on lipid production by cyanobacteria. *Anais da Academia Brasileira de Ciências, 89*, 2021–2031.
54. De Andrés, C., Espuny, M. J., Robert, M., Mercadé, M. E., Manresa, A., & Guinea, J. (1991). Cellular lipid accumulation by *Pseudomonas aeruginosa* 44T1. *Applied Microbiology and Biotechnology, 35*, 813–816.

55. Deeba, F., Pruthi, V., & Negi, Y. S. (2016). Converting paper mill sludge into neutral lipids by oleaginous yeast *Cryptococcus vishniaccii* for biodiesel production. *Bioresource Technology, 213*, 96–102.

56. Di Fidio, N., Dragoni, F., Antonetti, C., De Bari, I., Raspolli Galletti, A. M., & Ragaglini, G. (2020). From paper mill waste to single cell oil: Enzymatic hydrolysis to sugars and their fermentation into microbial oil by the yeast *Lipomyces starkeyi*. *Bioresource Technology, 315*, 123790.

57. Di Fidio, N., Ragaglini, G., Dragoni, F., Antonetti, C., & Raspolli Galletti, A. M. (2021). Integrated cascade biorefinery processes for the production of single cell oil by *Lipomyces starkeyi* from *Arundo donax* L. hydrolysates. *Bioresource Technology, 325*, 124635.

58. Dias, C., Santos, J., Reis, A., & Lopes da Silva, T. (2019). Yeast and microalgal symbiotic cultures using low-cost substrates for lipid production. *Bioresource Technology Reports, 7*, 100261.

59. Diwan, B., Parkhey, P., & Gupta, P. (2018). From agro-industrial wastes to single cell oils: A step towards prospective biorefinery. *Folia Microbiologica, 63*, 547–568.

60. Dourou, M., Aggeli, D., Papanikolaou, S., & Aggelis, G. (2018). Critical steps in carbon metabolism affecting lipid accumulation and their regulation in oleaginous microorganisms. *Applied Microbiology and Biotechnology, 102*, 2509–2523.

61. Drévillon, L., Koubaa, M., Nicaud, J. M., & Vorobiev, E. (2019). Cell disruption pre-treatments towards an effective recovery of oil from *Yarrowia lipolytica* oleaginous yeast. *Biomass and Bioenergy, 128*, 105320.

62. Du, Z. Y., Alvaro, J., Hyden, B., Zienkiewicz, K., Benning, N., Zienkiewicz, A., Bonito, G., & Benning, C. (2018). Enhancing oil production and harvest by combining the marine alga *Nannochloropsis oceanica* and the oleaginous fungus *Mortierella elongata*. *Biotechnology for Biofuels, 11*, 174.

63. Duan, Y., Zhi, Z., Ke, C., Xiaoming, T., & Xuefeng, L. (2011). De novo biosynthesis of biodiesel by *Escherichia coli* in optimized fed-batch cultivation. *PLoS ONE, 6*, 1–7.

64. Eberly, J. O., Ringelberg, D. B., & Indest, K. J. (2013). Physiological characterization of lipid accumulation and *in vivo* ester formation in *Gordonia* sp. KTR9. *Journal of Industrial Microbiology and Biotechnology, 40*, 201–208.

65. Elbahloul, Y., & Steinbüchel, A. (2010). Pilot-scale production of fatty acid ethyl esters by an engineered *Escherichia coli* strain harboring the

p(Microdiesel) plasmid. *Applied and Environmental Microbiology, 76*, 4560–4565.

66. Elrayies, G. M. (2018). Microalgae: Prospects for greener future buildings. *Renewable and Sustainable Energy Reviews, 81*, 1175–1191.
67. Fan, L., Liu, J., Nie, K., Liu, L., Wang, F., Tan, T., & Deng, L. (2013). Synthesis of medium chain length fatty acid ethyl esters in engineered *Escherichia coli* using endogenously produced medium chain fatty acids. *Enzyme and Microbial Technology, 53*, 128–133.
68. Feng, P., Xu, Z., Qin, L., Alam, M. A., Wang, Z., & Zhu, S. (2020). Effects of different nitrogen sources and light paths of flat plate photobioreactors on the growth and lipid accumulation of *Chlorella* sp. GN1 outdoors. *Bioresource Technology, 301*, 122–762.
69. Fixter, L. M., Nagi, M. N., McCormack, J. G., & Fewson, C. A. (1986). Structure, distribution and function of wax esters in *Acinetobacter calcoaceticus*. *Journal of General Microbiology, 132*, 3147–3157.
70. Francisco, E. C., Franco, T. T., Wagner, R., & Jacob-Lopes, E. (2014). Assessment of different carbohydrates as exogenous carbon source in cultivation of cyanobacteria. *Bioprocess and Biosystems Engineering, 37*, 1497–1505.
71. Gálvez-López, D., Chávez-Meléndez, B., Vázquez-Ovando, A., & Rosas-Quijano, R. (2019). The metabolism and genetic regulation of lipids in the oleaginous yeast *Yarrowia lipolytica*. *Brazilian Journal of Microbiology, 50*, 23–31.
72. Gao, C., Yang, X., Wang, H., Rivero, C. P., Li, C., Cui, Z., Qi, Q., & Lin, C. S. K. (2016). Robust succinic acid production from crude glycerol using engineered *Yarrowia lipolytica*. *Biotechnology for Biofuels, 9*, 179.
73. Gao, R., Li, Z., Zhou, X., Bao, W., Cheng, S., & Zheng, L. (2020). Enhanced lipid production by *Yarrowia lipolytica* cultured with synthetic and waste-derived high-content volatile fatty acids under alkaline conditions. *Biotechnology for Biofuels, 13*, 3–16.
74. Gaykawad, S. S., Ramanand, S. S., Blomqvist, J., Zimmermann, B., Shapaval, V., Kohler, A., Oostindjer, M., & Boccadoro, C. (2021). Submerged fermentation of animal fat by-products by oleaginous filamentous fungi for the production of unsaturated single cell oil. *Fermentation, 7*, 300.
75. Gonçalves, F. A. G., Colen, G., & Takahashi, J. A. (2014). *Yarrowia lipolytica* and its multiple applications in the biotechnological industry. *Science World Journal, 2014*, e476207.

76. Görner, C., Redai, V., Bracharz, F., Schrepfer, P., Garbe, D., & Brück, T. (2016). Genetic engineering and production of modified fatty acids by the non-conventional oleaginous yeast *Trichosporon oleaginosus* ATCC 20509. *Green Chemistry, 18*, 2037–2046.
77. Goswami, L., Tejas Namboodiri, M. M., Vinoth Kumar, R., Pakshirajan, K., & Pugazhenthi, G. (2017). Biodiesel production potential of oleaginous *Rhodococcus opacus* grown on biomass gasification wastewater. *Renewable Energy, 105*, 400–406.
78. Gouda, M. K., Omar, S. H., & Aouad, L. M. (2008). Single cell oil production by *Gordonia* sp. DG using agro-industrial wastes. *World Journal of Microbiology and Biotechnology, 24*, 1703–1711.
79. Guerfali, M., Ayadi, I., Belhassen, A., Gargouri, A., & Belghith, H. (2018). Single cell oil production by *Trichosporon cutaneum* and lignocellulosic residues bioconversion for biodiesel synthesis. *Process Safety and Environmental Protection, 113*, 292–304.
80. Guerrand, D. (2017). Lipases industrial applications: Focus on food and agroindustries. *OCL, 24*, D403.
81. Guldhe, A., Ansari, F. A., Singh, P., & Bux, F. (2017). Heterotrophic cultivation of microalgae using aquaculture wastewater: A biorefinery concept for biomass production and nutrient remediation. *Ecological Engineering, 99*, 47–53.
82. Guo, D., Zhu, J., Deng, Z., & Liu, T. (2014). Metabolic engineering of *Escherichia coli* for production of fatty acid short-chain esters through combination of the fatty acid and 2-keto acid pathways. *Metabolic Engineering, 22*, 69–75.
83. Han, S., Kim, G. Y., & Han, J. I. (2019). Biodiesel production from oleaginous yeast, *Cryptococcus* sp. by using banana peel as carbon source. *Energy Reports, 5*, 1077–1081.
84. Hayat, S., & Christias, C. (2010). Isolation and fusion of protoplasts from the phytopathogenic fungus *Sclerotium rolfsii* (Sacc.). *Brazilian Journal of Microbiology, 41*, 253–263.
85. He, J., Zhang, W., Liu, X., Xu, N., & Xiong, P. (2016). Optimization of prehydrolysis time and substrate feeding to improve ethanol production by simultaneous *Saccharomyces cerevisiae* clarification and fermentation of furfural process residue. *Journal of Bioscience and Bioengineering, 122*, 563–569.
86. Hernández, M. A., Comba, S., Arabolaza, A., Gramajo, H., & Alvarez, H. M. (2015). Overexpression of a phosphatidic acid phosphatase type 2 leads to an increase in triacylglycerol production in oleaginous *Rhodococcus*

strains. *Applied Microbiology and Biotechnology, 99*, 2191–2207.

87. Hernández, M. A., Mohn, W. W., Martínez, E., Rost, E., Alvarez, A. F., & Alvarez, H. M. (2008). Biosynthesis of storage compounds by *Rhodococcus jostii* RHA1 and global identification of genes involved in their metabolism. *BMC Genomics, 9*, 600–614.
88. Hernández-Almanza, A., Cesar Montanez, J., Aguilar-González, M. A., Martínez-Ávila, C., Rodríguez-Herrera, R., & Aguilar, C. N. (2014). *Rhodotorula glutinis* as source of pigments and metabolites for food industry. *Food Bioscience, 5*, 64–72.
89. Hetzler, S., & Steinbüchel, A. (2013). Establishment of cellobiose utilization for lipid production in *Rhodococcus opacus* PD630. *Applied and Environmental Microbiology, 79*, 3122–3125.
90. Himmel, M. E., Ding, S. Y., Johnson, D. K., Adney, W. S., Nimlos, M. R., Brady, J. W., & Foust, T. D. (2007). Biomass recalcitrance: Engineering plants and enzymes for biofuels production. *Science, 315*, 804–880.
91. Hoarau, J., Grondin, I., Caro, Y., & Petit, T. (2018). Sugarcane distillery spent wash, a new resource for third-generation biodiesel production. *Water, 10*, 1623.
92. Huang, C., Chen, X., Xiong, L., Chen, X., Ma, L., & Chen, Y. (2013). Single cell oil production from low-cost substrates: The possibility and potential of its industrialization. *Biotechnology Advances, 31*, 129–139.
93. Huang, X., Wang, Y., Liu, W., & Bao, J. (2011). Biological removal of inhibitors leads to the improved lipid production in the lipid fermentation of corn stover hydrolysate by *Trichosporon cutaneum*. *Bioresource Technology, 102*, 9705–9709.
94. Ilic-Tomic, T., Genčić, Ž., Vasiljevic, B., Djokic, L., Nikodinovic-Runic, J., & Radulović. (2015). Structural diversity and possible functional roles of free fatty acids of the novel soil isolate *Streptomyces* sp. NP10. *Applied Microbiology and Biotechnology, 99*, 4815–4833.
95. Imatoukene, N., Koubaa, M., Perdrix, E., Benali, M., & Vorobiev, E. (2020). Combination of cell disruption technologies for lipid recovery from dry and wet biomass of *Yarrowia lipolytica* and using green solvents. *Process Biochemistry, 90*, 139–147.
96. Ishak, M. Z., Kuan, M. L., & Ismail, K. S. K. (2021). Optimization of upstream processing for the production of bioethanol from *Leucaena leucocephala* seeds using *Kluyveromyces marxianus* UniMAP 1-1. *IOP Conference Series: Earth and Environmental Science, 765*, 012053.
97. Ishige, T., Tani, A., & Sakai, Y. (2003). Wax ester production by bacteria. *Current Opinion in Microbiology, 6*, 244–250.

98. Ishige, T., Tani, A., Takabe, K., Kawasaki, K., Sakai, Y., & Kato, N. (2002). Wax ester production from n-alkanes by *Acinetobacter* sp. strain M-1: Ultrastructure of cellular inclusions and role of acyl Coenzyme A reductase. *Applied and Environmental Microbiology, 68*, 1192–1195.

99. Janßen, H. J., & Steinbüchel, A. (2014). Production of triacylglycerols in *Escherichia coli* by deletion of the diacylglycerol kinase gene and heterologous overexpression of atfA from *Acinetobacter baylyi* ADP1. *Applied Microbiology and Biotechnology, 98*, 1913–1924.

100. Je, S., & Yamaoka, Y. (2022). Biotechnological approaches for biomass lipid production using microalgae *Chlorella*: Its future perspectives. *Journal of Microbiology and Biotechnology, 32*, 1357–1372.

101. Jin, G., Yang, F., Hu, C., Shen, H., & Zhao, Z. K. (2012). Enzyme-assisted extraction of lipids directly from the culture of the oleaginous yeast *Rhodosporidium toruloides*. *Bioresource Technology, 111*, 378–382.

102. Jin, M., Gunawan, C., Balan, V., Lau, M. W., & Dale, B. E. (2012). Simultaneous *Saccharomyces cerevisiae* clarification and co-fermentation (SSCF) of AFEXTM pretreated corn stover for ethanol production using commercial enzymes and *Saccharomyces cerevisiae* 424A(LNH-ST). *Bioresource Technology, 110*, 587–594.

103. Jin, M., Slininger, P. J., Dien, B. S., Waghmode, S., Moser, B. R., Orjuela, A., da Costa Sousa, L., & Balan, V. (2015). Microbial lipid-based lignocellulosic biorefinery: Feasibility and challenges. *Trends in Biotechnology, 33*, 43–54.

104. Kaiwan-arporn, P., Dong Hai, P., Thi Thu, N., & Annachhatre, A. P. (2012). Cultivation of cyanobacteria for extraction of lipids. *Biomass and Bioenergy, 44*, 142–149.

105. Kalscheuer, R., & Steinbüchel, A. (2003). A novel bifunctional wax ester synthase/acyl-CoA acyltransferase mediates wax ester and triacylglycerol biosynthesis in *Acinetobacter calcoaceticus* ADP1. *Journal of Biological Chemistry, 278*, 8075–8082.

106. Kalscheuer, R., Stölting, T., & Steinbüchel, A. (2006). Microdiesel: *Escherichia coli* engineered for fuel production. *Microbiology (SGM), 152*, 2529–2536.

107. Kalscheuer, R., Stöveken, T., Malkus, U., Reichelt, R., Golyshin, P. N., Sabirova, J. S., Ferrer, M., Timmis, K. N., & Steinbüchel, A. (2007). Analysis of storage lipid accumulation in *Alcanivorax borkumensis*: Evidence for alternative triacylglycerol biosynthesis routes in bacteria. *Journal of Bacteriology, 189*, 918–928.

108. Kanno, M., Katayama, T., Morita, N., Tamaki, H., Hanada, S., & Kamagata,

Y. (2015). *Catenisphaera adipataccumulans* gen. nov., sp. nov., a member of the family *Erysipelotrichaceae* isolated from an anaerobic digester. *International Journal of Systematic and Evolutionary Microbiology, 65*, 805–810.

109. Karatay, S. E., & Dönmez, G. (2011). Microbial oil production from thermophile cyanobacteria for biodiesel production. *Applied Energy, 88*, 3632–3635.
110. Katayama, T., Kanno, M., Morita, N., Hori, T., Narihiro, T., Mitani, Y., & Kamagata, Y. (2014). An oleaginous bacterium that intrinsically accumulates long-chain free fatty acids in its cytoplasm. *Applied and Environmental Microbiology, 80*, 1126–1131.
111. Katiyar, R., & Arora, A. (2020). Health promoting functional lipids from microalgae pool: A review. *Algal Research, 46*, 101800.
112. Khan, M. I., Shin, J. H., & Kim, J. D. (2018). The promising future of microalgae: Current status, challenges, and optimization of a sustainable and renewable industry for biofuels, feed, and other products. *Microbial Cell Factories, 17*, 36.
113. Kikukawa, H., Sakuradani, E., Ando, A., Shimizu, S., & Ogawa, J. (2018). Arachidonic acid production by the oleaginous fungus *Mortierella alpina* 1S-4: A review. *Journal of Advanced Research, 11*, 15–22.
114. Kim, J., & Yun, S. (2006). Discovery of cellulose as a smart material. *Macromolecules, 39*, 4202–4206.
115. Kolouchová, I., Maťátková, O., Sigler, K., Masák, J., & Řezanka, T. (2016). Lipid accumulation by oleaginous and non-oleaginous yeast strains in nitrogen and phosphate limitation. *Folia Microbiologica, 61*, 431–438.
116. Kommoji, S., Gopinath, M., Satya Sagar, P., Yuvaraj, D., Iyyappan, J., Jaya Varsha, A., & Sunil, V. (2021). Lipid bioproduction from delignified native grass (*Cyperus distans*) hydrolysate by *Yarrowia lipolytica*. *Bioresource Technology, 324*, 124659.
117. Kopsahelis, N., Dimou, C., Papadaki, A., Xenopoulos, E., Kyraleou, M., Kallithraka, S., Kotseridis, Y., Papanikolaou, S., & Koutinas, A. A. (2018). Refining of wine lees and cheese whey for the production of microbial oil, polyphenol-rich extracts, and value-added co-products. *Journal of Chemical Technology and Biotechnology, 93*, 257–268.
118. Kosa, M., & Ragauskas, A. J. (2011). Lipids from heterotrophic microbes: Advances in metabolism research. *Trends in Biotechnology, 29*, 53–60.
119. Kosa, M., & Ragauskas, A. J. (2012). Bioconversion of lignin model

compounds with oleaginous *Rhodococci. Applied Microbiology and Biotechnology, 93*, 891–900.

120. Kosa, M., & Ragauskas, A. J. (2013). Lignin to lipid bioconversion by oleaginous *Rhodococci. Green Chemistry, 15*, 2070–2074.

121. Koubaa, M., Imatoukene, N., Drévillon, L., & Vorobiev, E. (2020). Current insights in yeast cell disruption technologies for oil recovery: A review. *Chemical Engineering and Processing: Process Intensification, 150*, 107868.

122. Koutinas, A. A., Chatzifragkou, A., Kopsahelis, N., Papanikolaou, S., & Kookos, I. K. (2014). Design and techno-economic evaluation of microbial oil production as a renewable resource for biodiesel and oleochemical production. *Fuel, 116*, 566–577.

123. Kumar, D., Singh, B., & Korstad, J. (2017). Utilization of lignocellulosic biomass by oleaginous yeast and bacteria for production of biodiesel and renewable diesel. *Renewable and Sustainable Energy Reviews, 73*, 654–671.

124. Kumar, S., Gupta, N., & Pakshirajan, K. (2015). Simultaneous lipid production and dairy wastewater treatment using *Rhodococcus opacus* in a batch bioreactor for potential biodiesel application. *Journal of Environmental Chemical Engineering, 3*, 1630–1636.

125. Kurosawa, K., Laser, J., & Sinskey, A. J. (2015c). Tolerance and adaptive evolution of triacylglycerol-producing *Rhodococcus opacus* to lignocellulose-derived inhibitors. *Biotechnology for Biofuels, 8*, 76–90.

126. Kurosawa, K., Plassmeier, J., Kalinowski, J., Rückert, C., & Sinskey, A. J. (2015a). Engineering L-arabinose metabolism in triacylglycerol-producing *Rhodococcus opacus* for lignocellulosic fuel production. *Metabolic Engineering, 30*, 89–95.

127. Kurosawa, K., Radek, A., Plassmeier, J. K., & Sinskey, A. J. (2015b). Improved glycerol utilization by a triacylglycerol producing *Rhodococcus opacus* strain for renewable fuels. *Biotechnology for Biofuels, 8*, 31–42.

128. Kurosawa, K., Wewetzer, S. J., & Sinskey, A. J. (2013). Engineering xylose metabolism in triacylglycerol-producing *Rhodococcus opacus* for lignocellulosic fuel production. *Biotechnology for Biofuels, 6*, 134–147.

129. Kurosawa, K., Wewetzer, S. J., & Sinskey, A. J. (2014). Triacylglycerol production from corn stover using a xylose-fermenting *Rhodococcus opacus* strain for lignocellulosic biofuels. *Journal of Microbial and Biochemical Technology, 6*, 254–259.

130. Lamers, D., van Biezen, N., Martens, D., Peters, L., van de Zilver, E.,

Jacobs-van Dreumel, N., Wijffels, R. H., & Lokman, C. (2016). Selection of oleaginous yeasts for fatty acid production. *BMC Biotechnology, 16*, 45.

131. Le, R. K., Das, P., Mahan, K. M., Anderson, S. A., Wells, T., Yuan, J. S., & Ragauskas, A. J. (2017). Utilization of simultaneous *Saccharomyces cerevisiae* clarification and fermentation residues as feedstock for lipid accumulation in *Rhodococcus opacus*. *AMB Express, 7*, 185.
132. Lee, A. K., Lewis, D. M., & Ashman, P. J. (2012). Disruption of microalgal cells for the extraction of lipids for biofuels: Processes and specific energy requirements. *Biomass and Bioenergy, 46*, 89–101.
133. Lee, D. S., Lee, Y. G., Cho, E. J., Song, Y., & Bae, H. J. (2021). Hydrolysis pattern analysis of xylem tissues of woody plants pretreated with hydrogen peroxide and acetic acid: Rapid *Saccharomyces cerevisiae* clarification of softwood for economical bioconversion. *Biotechnology for Biofuels, 14*, 37.
134. Lee, J. W., Kim, J. Y., Jang, H. M., Lee, M. W., & Park, J. M. (2015). Sequential dilute acid and alkali pretreatment of corn stover: Sugar recovery efficiency and structural characterization. *Bioresource Technology, 182*, 296–301.
135. Lee, J. Y., Li, P., Lee, J., Ryu, H. J., & Oh, K. K. (2013). Ethanol production from *Saccharomyces cerevisiae Arina Japonica* using an optimized extremely low acid pretreatment followed by simultaneous *Saccharomyces cerevisiae* clarification and fermentation. *Bioresource Technology, 127*, 119–125.
136. Leiva-Candia, D. E., Tsakona, S., Kopsahelis, N., García, I. L., Papanikolaou, S., Dorado, M. P., & Koutinas, A. A. (2015). Biorefining of by-product streams from sunflower-based biodiesel production plants for integrated synthesis of microbial oil and value-added co-products. *Bioresource Technology, 190*, 57–65.
137. Lenneman, E. M., Ohlert, J. M., Palani, N. P., & Barney, B. M. (2013). Fatty alcohols for wax esters in *Marinobacter aquaeolei* VT8: Two optional routes in the wax biosynthesis pathway. *Applied and Environmental Microbiology, 79*, 7055–7062.
138. Lennen, R. M., & Pfleger, B. F. (2012). Engineering *Escherichia coli* to synthesize free fatty acids. *Trends in Biotechnology, 30*, 659–667.
139. Li, J., Wei, X., Wang, Q., Chen, J., Chang, G., Kong, L., Su, J., & Liu, Y. (2012). Homogeneous isolation of nanocellulose from sugarcane bagasse by high pressure homogenization. *Carbohydrate Polymers, 90*, 1609–1613.
140. Li, S., Yu, H., Liu, Y., Zhang, X., & Ma, F. (2018). The lipid strategies

in *Cunninghamella echinulata* for an allostatic response to temperature changes. *Process Biochemistry, 76*, 85–94.

141. Liang, Y., Tang, T., Umagiliyage, A. L., Siddaramu, T., McCarroll, M., & Choudhary, R. (2012). Utilization of sorghum bagasse hydrolysates for producing microbial lipids. *Applied Energy, 91*, 451–458.
142. Liang, Z., Zhi, H., Fang, Z., & Zhang, P. (2021). Genetic engineering of yeast, filamentous fungi, and bacteria for terpene production and applications in the food industry. *Food Research International, 147*, 110487.
143. Lin, F., Chen, Y., Levine, R., Lee, K., Yuan, Y., & Lin, X. N. (2013). Improving fatty acid availability for bio-hydrocarbon production in *Escherichia coli* by metabolic engineering. *PLOS ONE, 8*, e78595.
144. Liu, H., Yu, C., Feng, D., Cheng, T., Meng, X., Liu, W., Zou, H., & Xian, M. (2012). Production of extracellular fatty acid using engineered *Escherichia coli*. *Microbial Cell Factories, 11*, 41–54.
145. Liu, T., Vora, H., & Khosla, C. (2010). Quantitative analysis and engineering of fatty acid biosynthesis in *E. coli*. *Metabolic Engineering, 12*, 378–386.
146. Liu, W., Wang, Y., Yu, Z., & Bao, J. (2012). Simultaneous *Saccharomyces cerevisiae* clarification and microbial lipid fermentation of corn stover by oleaginous yeast *Trichosporon cutaneum*. *Bioresource Technology, 118*, 13–18.
147. Liu, X., Sheng, J., & Curtis, R. (2011). Fatty acid production in genetically modified cyanobacteria. *Proceedings of the National Academy of Sciences, 108*, 6899–6904.
148. Liu, Z., Fels, M., Dragone, G., & Mussatto, S. I. (2021). Effects of inhibitory compounds derived from lignocellulosic biomass on the growth of the wild-type and evolved oleaginous yeast *Rhodosporidium toruloides*. *Industrial Crops and Products, 170*, 113799.
149. Lu, X., Vora, H., & Khosla, C. (2008). Overproduction of free fatty acids in *E. coli*: Implications for biodiesel production. *Metabolic Engineering, 10*, 333–339.
150. Ma, Y., Gao, Z., Wang, Q., & Liu, Y. (2018). Biodiesels from microbial oils: Opportunity and challenges. *Bioresource Technology, 263*, 631–641.
151. Matsakas, L., Giannakou, M., & Vörös, D. (2017). Effect of synthetic and natural media on lipid production from *Fusarium oxysporum*. *Electronic Journal of Biotechnology, 30*, 95–102.
152. Matsakas, L., Sterioti, A. A., Rova, U., & Christakopoulos, P. (2014). Use

of dried sweet sorghum for the efficient production of lipids by the yeast *Lipomyces starkeyi* CBS 1807. *Industrial Crops and Products, 62*, 367–372.

153. Maza, D. D., Viñarta, S. C., Su, Y., Guillamón, J. M., & Aybar, M. J. (2020). Growth and lipid production of *Rhodotorula glutinis* R4, in comparison to other oleaginous yeasts. *Journal of Biotechnology, 310*, 21–31.
154. McCarthy, C. (1971). Utilization of palmitic acid by *Mycobacterium avium*. *Infection and Immunity, 4*, 199–204.
155. McClements, D. J., & Gumus, C. E. (2016). Natural emulsifiers—biosurfactants, phospholipids, biopolymers, and colloidal particles: Molecular and physicochemical basis of functional performance. *Advances in Colloid and Interface Science, 234*, 3–26.
156. McClements, D. J., & Öztürk, B. (2022). Utilization of nanotechnology to improve the application and bioavailability of phytochemicals derived from waste streams. *Journal of Agricultural and Food Chemistry, 70*, 6884–6900.
157. Meullemiestre, A., Breil, C., Abert-Vian, M., & Chemat, F. (2016). Microwave, ultrasound, thermal treatments, and bead milling as intensification techniques for extraction of lipids from oleaginous *Yarrowia lipolytica* yeast for a biojetfuel application. *Bioresource Technology, 211*, 190–199.
158. Miao, Z., Tian, X., Liang, W., He, Y., & Wang, G. (2020). Bioconversion of corncob hydrolysate into microbial lipid by an oleaginous yeast *Rhodotorula taiwanensis* AM2352 for biodiesel production. *Renewable Energy, 161*, 91–97.
159. Miazek, K., Remacle, C., Richel, A., & Goffin, D. (2014). Effect of lignocellulose related compounds on microalgae growth and product biosynthesis: A review. *Energies, 7*, 4446–4481.
160. Michou, S., Tsouko, E., Vastaroucha, E. S., Diamantopoulou, P., & Papanikolaou, S. (2022). Growth potential of selected yeast strains cultivated on xylose-based media mimicking lignocellulosic wastewater streams: High production of microbial lipids by *Rhodosporidium toruloides*. *Fermentation, 8*, 713.
161. Modiri, S., Sharafi, H., Alidoust, L., Hajfarajollah, H., Haghighi, O., Azarivand, A., Zamanzadeh, Z., Zahiri, H. S., Vali, H., & Noghabi, K. A. (2015). Lipid production and mixotrophic growth features of cyanobacterial strains isolated from various aquatic sites. *Microbiology, 161*, 662–673.
162. Mu, J., Li, S., Chen, D., Xu, H., Han, F., Feng, B., & Li, Y. (2015).

Enhanced biomass and oil production from sugarcane bagasse hydrolysate (SBH) by heterotrophic oleaginous microalga *Chlorella protothecoides*. *Bioresource Technology, 185*, 99–105.

163. Muller, E. E. L., Sheik, A. R., & Wilmes, P. (2014). Lipid-based biofuel production from wastewater. *Current Opinion in Biotechnology, 30*, 9–16.
164. Němcová, A., Gonová, D., Samek, O., Sipiczki, M., Breierová, E., Márová, I. (2021). The use of Raman spectroscopy to monitor metabolic changes in stressed *Metschnikowia* sp. yeasts. *Microorganisms, 9*, 277.
165. Ngamsirisomsakul, M., Reungsang, A., & Kongkeitkajorn, M. B. (2021). Assessing oleaginous yeasts for their potentials on microbial lipid production from sugarcane bagasse and the effects of physical changes on lipid production. *Bioresource Technology Reports, 14*, 100650.
166. Ninomiya, K., Abe, M., Tsukegi, T., Kuroda, K., Omichi, M., Takada, K., Noguchi, M., Tsuge, Y., Ogino, C., Taki, K., et al. (2017). Ionic liquid pretreatment of bagasse improves mechanical property of bagasse/ polypropylene composites. *Industrial Crops and Products, 109*, 158–162.
167. Nisha, A., Udaya Sankar, K., & Venkateswaran, G. (2012). Supercritical CO2 extraction of *Mortierella alpina* single cell oil: Comparison with organic solvent extraction. *Food Chemistry, 133*, 220–226.
168. Ntimbani, R. N., Farzad, S., & Görgens, J. F. (2021). Furfural production from sugarcane bagasse along with co-production of ethanol from furfural residues. *Biomass Conversion and Biorefinery, 12*, 5257–5267.
169. Ochsenreither, K., Glück, C., Stressler, T., Fischer, L., & Syldatk, C. (2016). Production strategies and applications of microbial single cell oils. *Frontiers in Microbiology, 7*, 1539.
170. Okamura, Y., Nakai, S., Ohkawachi, M., Suemitsu, M., Takahashi, H., Aki, T., Matsumura, Y., Tajima, T., Nakashimada, Y., & Matsumoto, M. (2016). Isolation and characterization of bacterium producing lipid from short-chain fatty acids. *Bioresource Technology, 201*, 215–221.
171. Olukoshi, E. R., & Packter, N. M. (1994). Importance of stored triacylglycerols in *Streptomyces*: Possible carbon source for antibiotics. *Microbiology (SGM), 140*, 931–943.
172. Osman, M. E., Abdel-Razik, A. B., Zaki, K. I., Mamdouh, N., & El-Sayed, H. (2022). Isolation, molecular identification of lipid-producing *Rhodotorula diobovata*: Optimization of lipid accumulation for biodiesel production. *Journal of Genetic Engineering and Biotechnology, 20*, 32.
173. Pádrová, K., Lukavský, J., Nedbalová, L., Čejková, A., Cajthaml, T., Sigler, K., Vítová, M., Řezanka, T. (2015). Trace concentrations of

iron nanoparticles cause overproduction of biomass and lipids during cultivation of cyanobacteria and microalgae. *Journal of Applied Phycology, 27*, 1443–1451.

174. Palazzolo, M. A., & Garcia-Perez, M. (2022). Microbial lipid biosynthesis from lignocellulosic biomass pyrolysis products. *Biotechnology Advances, 54*, 107791.

175. Palmqvist, E., & Hahn-Hägerdal, B. (2000). Fermentation of lignocellulosic hydrolysates. II: Inhibitors and mechanisms of inhibition. *Bioresource Technology, 74*, 25–33.

176. Papadaki, A., Mallouchos, A., Efthymiou, M. N., Gardeli, C., Kopsahelis, N., Aguieiras, E. C. G., Freire, D. M. G., Papanikolaou, S., & Koutinas, A. A. (2017). Production of wax esters via microbial oil synthesis from food industry waste and by-product streams. *Bioresource Technology, 245*, 274–282.

177. Park, B. G., Kim, M., Kim, J., Yoo, H., & Kim, B. G. (2017). Systems biology for understanding and engineering of heterotrophic oleaginous microorganisms. *Biotechnology Journal, 12*, 1600104.

178. Parsons, S., Allen, M. J., & Chuck, C. J. (2020). Coproducts of algae and yeast-derived single cell oils: A critical review of their role in improving biorefinery sustainability. *Bioresource Technology, 303*, 122862.

179. Patel, A., & Matsakas, L. (2018). A comparative study on de novo and ex novo lipid fermentation by oleaginous yeast using glucose and sonicated waste cooking oil. *Ultrasonics Sonochemistry, 52*, 364–374.

180. Patel, A., Arora, N., Sartaj, K., Pruthi, V., & Pruthi, P. A. (2016). Sustainable biodiesel production from oleaginous yeasts utilizing hydrolysates of various non-edible lignocellulosic biomasses. *Renewable and Sustainable Energy Reviews, 62*, 836–855.

181. Patel, A., Hrůzová, K., Rova, U., Christakopoulos, P., & Matsakas, L. (2019). Sustainable biorefinery concept for biofuel production through holistic valorization of food waste. *Bioresource Technology, 294*, 122–247.

182. Patel, A., Karageorgou, D., Rova, E., Katapodis, P., Rova, U., Christakopoulos, P., & Matsakas, L. (2020). An overview of potential oleaginous microorganisms and their role in biodiesel and omega-3 fatty acid-based industries. *Microorganisms, 8*, 434.

183. Patel, A., Matsakas, L., Rova, U., & Christakopoulos, P. (2018). Heterotrophic cultivation of *Auxenochlorella protothecoides* using forest biomass as a feedstock for sustainable biodiesel production. *Biotechnology for Biofuels, 11*, 169.

184. Patel, A., Mikes, F., Bühler, S., & Matsakas, L. (2018). Valorization of brewers' spent grain for the production of lipids by oleaginous yeast. *Molecules, 23*, 3052.

185. Patel, A., Sarkar, O., Rova, U., Christakopoulos, P., & Matsakas, L. (2021). Valorization of volatile fatty acids derived from low-cost organic waste for lipogenesis in oleaginous microorganisms—A review. *Bioresource Technology, 321*, 124457.

186. Patel, V. K., Maji, D., Singh, A. K., Suseela, M. R., Sundaram, S., & Kalra, A. (2014). A natural plant growth promoter, calliterpenone, enhances growth and biomass, carbohydrate, and lipid production in cyanobacterium *Synechocystis PCC 6803*. *Journal of Applied Phycology, 26*, 279–286.

187. Patnayak, S., & Sree, A. (2005). Screening of bacterial associates of marine sponges for single cell oil and PUFA. *Letters in Applied Microbiology, 40*, 358–363.

188. Perez-Garcia, O., Escalante, F. M. E., de-Bashan, L. E., & Bashan, Y. (2011). Heterotrophic cultures of microalgae: Metabolism and potential products. *Water Research, 45*, 11–36.

189. Probst, K. V., & Vadlani, P. V. (2015). Production of single cell oil from *Lipomyces starkeyi* ATCC 56304 using biorefinery by-products. *Bioresource Technology, 198*, 268–275.

190. Qian, X., Gorte, O., Chen, L., Zhang, W., Dong, W., Ma, J., Jiang, M., Xin, F., & Ochsenreither, K. (2019). Co-production of single cell oil and gluconic acid using oleaginous *Cryptococcus podzolicus* DSM 27192. *Biotechnology for Biofuels, 12*, 127.

191. Qin, L., Liu, L., Zeng, A. P., & Wei, D. (2017). From low-cost substrates to single cell oils synthesized by oleaginous yeasts. *Bioresource Technology, 245*, 1507–1519.

192. Rakicka, M., Lazar, Z., Dulermo, T., Fickers, P., & Nicaud, J. M. (2015). Lipid production by the oleaginous yeast *Yarrowia lipolytica* using industrial by-products under different culture conditions. *Biotechnology for Biofuels, 8*, 104.

193. Ranganathan, S., Wei Tee, T., Chowdhury, A., Zomorrodi, A. R., Moon Yoon, J., Fu, Y., Shanks, J. V., & Maranas, C. D. (2012). An integrated computational and experimental study for overproducing fatty acids in *Escherichia coli*. *Metabolic Engineering, 14*, 687–704.

194. Ratledge, C., & Cohen, Z. (2008). Microbial and algal oils: Do they have a future for biodiesel or as commodity oils? *Lipid Technology, 20*, 155–160.

195. Ren, Y., Deng, J., Huang, J., Wu, Z., Yi, L., Bi, Y., & Chen, F. (2021).

Using green alga *Haematococcus pluvialis* for astaxanthin and lipid co-production: Advances and outlook. *Bioresource Technology, 340*, 125736.

196. Robles-Iglesias, R., Naveira-Pazos, C., Fernández-Blanco, C., Veiga, M. C., & Kennes, C. (2023). Factors affecting the optimisation and scale-up of lipid accumulation in oleaginous yeasts for sustainable biofuels production. *Renewable and Sustainable Energy Reviews, 171*, 113043.
197. Röttig, A., & Steinbüchel, A. (2013). Acyltransferases in bacteria. *Microbiology and Molecular Biology Reviews, 77*, 277–321.
198. Röttig, A., Hauschild, P., Madkour, M. H., Al-Ansari, A. M., Almakishah, N. H., & Steinbüchel, A. (2016). Isolation of lipid-accumulating bacteria from desert soil. *Submitted for publication.*
199. Röttig, A., Wenning, L., Bröker, D., & Steinbüchel, A. (2010). Fatty acid alkyl esters: Perspectives for production of alternative biofuels. *Applied Microbiology and Biotechnology, 85*, 1713–1733.
200. Röttig, A., Zurek, P. J., & Steinbüchel, A. (2015). Assessment of bacterial acyltransferases for an efficient lipid production in metabolically engineered strains of *Escherichia coli. Metabolic Engineering, 32*, 195–206.
201. Rucker, J., Paul, J., Pfeifer, B. A., & Lee, K. (2013). Engineering *E. coli* for triglyceride accumulation through native and heterologous metabolic reactions. *Applied Microbiology and Biotechnology, 97*, 2753–2759.
202. Rude, M. A., & Schirmer, A. (2009). New microbial fuels: A biotech perspective. *Current Opinion in Microbiology, 12*, 274–281.
203. Saini, R., Osorio-Gonzalez, C. S., Hegde, K., Kaur Brar, S., & Vezina, P. (2022). A co-fermentation strategy with wood hydrolysate and crude glycerol to enhance the lipid accumulation in *Rhodosporidium toruloides*-1588. *Bioresource Technology, 364*, 127821.
204. Salcedo-Vite, K., Sigala, J. C., Segura, D., Gosset, G., & Martinez, A. (2019). *Acinetobacter baylyi* ADP1 growth performance and lipid accumulation on different carbon sources. *Applied Microbiology and Biotechnology, 103*, 6217–6229.
205. Santala, S., Efimova, E., Kivinen, V., Larjo, A., Aho, T., Karp, M., & Santala, V. (2011). Improved triacylglycerol production in *Acinetobacter baylyi* ADP1 by metabolic engineering. *Microbial Cell Factories, 10*, 36–46.
206. Schneiter, R., & Daum, G. (2006). Extraction of yeast lipids. In W. Xiao (Ed.), *Yeast Protocol* (pp. 41–45). Methods in Molecular Biology. Humana Press.

207. Senanayake, S. P. J. N., & Fichtali, J. (2006). Single-cell oils as sources of nutraceutical and specialty lipids: Processing technologies and applications. In *Nutraceutical and Specialty Lipids and Their Co-Products*. CRC Press.

208. Serero, A., Bedrat, A., Baulande, S., Bastianelli, G., Colavizza, D., Desfougères, T., Pignède, G., Quipourt-Isnard, A. D., & Nicolas, A. (2021). Recombination in a sterile polyploid hybrid yeast upon meiotic return-to-growth. *Microbial Research, 250*, 126789.

209. Shah, A. M., Mohamed, H., Zhang, Z., & Song, Y. (2021). Isolation, characterization and fatty acid analysis of *Gilbertella persicaria* DSR1: A potential new source of high-value single-cell oil. *Biomass and Bioenergy, 151*, 106156.

210. Shah, V., Doncel, G. F., Seyoum, T., Eaton, K. M., Zalenskaya, I., Hagver, R., Azim, A., Gross, R. (2005). Sophorolipids, microbial glycolipids with anti-human immunodeficiency virus and sperm-immobilizing activities. *Antimicrobial Agents and Chemotherapy, 49*, 4093–4100.

211. Shen, H., Zhang, X., Gong, Z., Wang, Y., Yu, X., Yang, X., & Zhao, Z. K. (2017). Compositional profiles of *Rhodosporidium toruloides* cells under nutrient limitation. *Applied Microbiology and Biotechnology, 101*, 3801–3809.

212. Shields-Menard, S. A., Amirsadeghi, M., Sukhbaatar, B., Revellame, E., Hernandez, R., Donaldson, J. R., & French, W. T. (2015). Lipid accumulation by *Rhodococcus rhodochrous* grown on glucose. *Journal of Industrial Microbiology and Biotechnology, 42*, 693–699.

213. Silva, J. D. M., Martins, L. H. D. S., Moreira, D. K. T., Silva, L. D. P., Barbosa, P. D. P. M., Komesu, A., Ferreira, N. R., & Oliveira, J. A. R. D. (2023). Microbial lipid-based biorefinery concepts: A review of status and prospects. *Foods, 12*(10), 2074. https://doi.org/10.3390/foods12102074.

214. Singh, A., & Bishnoi, N. R. (2013). Ethanol production from pretreated wheat straw hydrolyzate by *Saccharomyces cerevisiae* via sequential statistical optimization. *Industrial Crops and Products, 41*, 221–226.

215. Singh, J., Suhag, M., & Dhaka, A. (2015). Augmented digestion of lignocellulose by steam explosion, acid and alkaline pretreatment methods: A review. *Carbohydrate Polymers, 117*, 624–631.

216. Singh, N., Batghare, A. H., Choudhury, B. J., Goyal, A., & Moholkar, V. S. (2020). Microalgae based biorefinery: Assessment of wild freshwater microalgal isolate for simultaneous biodiesel and β-carotene production. *Bioresource Technology Reports, 11*, 100440.

217. Singh, P., Kumari, S., Guldhe, A., Misra, R., Rawat, I., & Bux, F. (2016). Trends and novel strategies for enhancing lipid accumulation and quality

in microalgae. *Renewable and Sustainable Energy Reviews, 55*, 1–16.

218. Steen, E. J., Kang, Y., Bokinsky, G., Hu, Z., Schirmer, A., McClure, A., del Cardayre, S. B., & Keasling, J. D. (2010). Microbial production of fatty-acid-derived fuels and chemicals from plant biomass. *Nature, 463*, 559–562.
219. Stephanopoulos, G. (2007). Challenges in engineering microbes for biofuels production. *Science, 315*, 801–804.
220. Su, H., Feng, J., Lv, J., Liu, Q., Nan, F., Liu, X., & Xie, S. (2021). Molecular mechanism of lipid accumulation and metabolism of oleaginous *Chlorococcum sphacosum* GD from soil under salt stress. *International Journal of Molecular Sciences, 22*, 1304.
221. Subhash, G. V., & Mohan, S. V. (2011). Biodiesel production from isolated oleaginous fungi *Aspergillus sp.* using corncob waste liquor as a substrate. *Bioresource Technology, 102*, 9286–9290.
222. Subramaniam, R., Dufreche, S., Zappi, M., & Bajpai, R. (2010). Microbial lipids from renewable resources: Production and characterization. *Journal of Industrial Microbiology and Biotechnology, 37*, 1271–1287.
223. Sukhang, S., Choojit, S., Reungpeerakul, T., & Sangwichien, C. (2019). Bioethanol production from oil palm empty fruit bunch with SSF and SHF processes using *Kluyveromyces marxianus* yeast. *Cellulose, 27*, 301–314.
224. Sun, Z., Ramsay, J. A., Guay, M., & Ramsay, B. A. (2007). Fermentation process development for the production of medium-chain-length poly-3-hydroxyalkanoates. *Applied Microbiology and Biotechnology, 75*, 475–485.
225. Tang, S., Dong, Q., Fang, Z., Cong, W., & Zhang, H. (2020). Microbial lipid production from rice straw hydrolysates and recycled pretreated glycerol. *Bioresource Technology, 312*, 123580.
226. Tao, H., Guo, D., Zhang, Y., Deng, Z., & Liu, T. (2015). Metabolic engineering of microbes for branched-chain biodiesel production with low-temperature property. *Biotechnology for Biofuels, 8*, 92–103.
227. Tomás-Pejó, E., Morales-Palomo, S., & González-Fernández, C. (2021). Microbial lipids from organic wastes: Outlook and challenges. *Bioresource Technology, 323*, 124–612.
228. Tomás-Pejó, E., Oliva, J. M., González, A., Ballesteros, I., & Ballesteros, M. (2009). Bioethanol production from wheat straw by the thermotolerant yeast *Kluyveromyces marxianus* CECT 10875 in a simultaneous saccharification and fermentation fed-batch process. *Fuel, 88*, 2142–2147.
229. Traversier, M., Gaslondes, T., Milesi, S., Michel, S., & Delannay, E.

(2018). Polar lipids in cosmetics: Recent trends in extraction, separation, analysis, and main applications. *Phytochemistry Reviews, 17*, 1179–1210.

230. Trivedi, J., Aila, M., Bangwal, D. P., Kaul, S., & Garg, M. O. (2015). Algae based biorefinery—How to make sense? *Renewable and Sustainable Energy Reviews, 47*, 295–307.
231. Valdés, G., Mendonça, R. T., & Aggelis, G. (2020). Lignocellulosic biomass as a substrate for oleaginous microorganisms: A review. *Applied Sciences, 10*, 7698.
232. van Smeden, J., Janssens, M., Gooris, G. S., & Bouwstra, J. A. (2014). The important role of stratum corneum lipids for the cutaneous barrier function. *Biochimica et Biophysica Acta (BBA)–Molecular and Cell Biology of Lipids, 1841*, 295–313.
233. Varga, E., Szengyel, Z., & Réczey, K. (2002). Chemical pretreatments of corn stover for enhancing enzymatic digestibility. In M. Finkelstein, J. D. McMillan, & B. H. Davison (Eds.), *Biotechnology for Fuels and Chemicals: The Twenty-Third Symposium* (pp. 73–87). *Applied Biochemistry and Biotechnology*. Humana Press.
234. Voss, I., & Steinbüchel, A. (2001). High cell density cultivation of *Rhodococcus opacus* for lipid production at a pilot-plant scale. *Applied Microbiology and Biotechnology, 55*, 547–555.
235. Wältermann, M., Hinz, A., Robenek, H., Troyer, D., Reichelt, R., Malkus, U., Galla, H. J., Kalscheuer, R., Stöveken, T., von Landenberg, P., & Steinbüchel, A. (2005). Mechanism of lipid-body formation in prokaryotes: How bacteria fatten up. *Molecular Microbiology, 55*, 750–763.
236. Wang, C., Chen, L., Rakesh, B., Qin, Y., & Lv, R. (2012). Technologies for extracting lipids from oleaginous microorganisms for biodiesel production. *Frontiers in Energy, 6*, 266–274.
237. Wang, F., Gao, B., Dai, C., Su, M., & Zhang, C. (2020). Comprehensive utilization of the filamentous oleaginous microalga *Tribonema utriculosum* for the production of lipids and chrysolaminarin in a biorefinery concept. *Algal Research, 50*, 101973.
238. Wang, J., Ledesma-Amaro, R., Wei, Y., Ji, B., & Ji, X. J. (2020). Metabolic engineering for increased lipid accumulation in *Yarrowia lipolytica*—A review. *Bioresource Technology, 313*, 123707.
239. Wayman, M., Jenkins, A. D., & Kormendy, A. G. (1984). Bacterial production of fats and oils. In C. Ratledge, P. Dawson, & J. Rattray (Eds.), *Biotechnology for the Oils and Fats Industry*. American Oil Chemists' Society (AOCS).

240. Wei, Z., Zeng, G., Huang, F., Kosa, M., Huang, D., & Ragauskas, A. J. (2015b). Bioconversion of oxygen-pretreated kraft lignin to microbial lipid with oleaginous *Rhodococcus opacus* DSM 1069. *Green Chemistry, 17*, 2784–2789.

241. Wei, Z., Zeng, G., Huang, F., Kosa, M., Sun, Q., Meng, X., Huang, D., & Ragauskas, A. J. (2015a). Microbial lipid production by oleaginous *Rhodococci* cultured in lignocellulosic autohydrolysates. *Applied Microbiology and Biotechnology, 99*, 7369–7377.

242. Wells, T. Jr., Wei, Z., & Ragauskas, A. J. (2015). Bioconversion of lignocellulosic pretreatment effluent via oleaginous *Rhodococcus opacus* DSM 1069. *Biomass and Bioenergy, 72*, 200–205.

243. Woiciechowski, A. L., Neto, C. J. D., de Souza Vandenberghe, L. P., de Carvalho Neto, D. P., Sydney, A. C. N., Letti, L. A. J., Karp, S. G., Torres, L. A. Z., & Soccol, C. R. (2020). Lignocellulosic biomass: Acid and alkaline pretreatments and their effects on biomass recalcitrance—Conventional processing and recent advances. *Bioresource Technology, 304*, 122848.

244. Woong Cho, K., & Jun Mo, S. (1999). Screening and characterization of eicosapentaenoic acid-producing marine bacteria. *Biotechnology Letters, 21*, 215–218.

245. Wu, H., Karanjikar, M., & San, K. Y. (2014a). Metabolic engineering of *Escherichia coli* for efficient free fatty acid production from glycerol. *Metabolic Engineering, 25*, 82–91.

246. Wu, H., Lee, J., Karanjikar, M., & San, K. Y. (2014b). Efficient free fatty acid production from woody biomass hydrolysate using metabolically engineered *Escherichia coli*. *Bioresource Technology, 169*, 119–125.

247. Wynn, J. P., & Ratledge, C. (2005). Oils from microorganisms. In F. Shahidi (Ed.), *Bailey's Industrial Oil and Fat Products*. John Wiley & Sons.

248. Wynn, J., Behrens, P., Sundararajan, A., Hansen, J., & Apt, K. (2010). 6—Production of single cell oils by dinoflagellates. In Z. Cohen & C. Ratledge (Eds.), *Single Cell Oils* (2nd ed., pp. 115–129). AOCS Press.

249. Xue, Z., Sharpe, P. L., Hong, S. P., Yadav, N. S., Xie, D., Short, D. R., & Zhu, Q. (2013). Production of omega-3 eicosapentaenoic acid by metabolic engineering of *Yarrowia lipolytica*. *Nature Biotechnology, 31*, 734–740.

250. Yamada, R., Kashihara, T., & Ogino, H. (2017). Improvement of lipid production by the oleaginous yeast *Rhodosporidium toruloides* through UV mutagenesis. *World Journal of Microbiology and Biotechnology, 33*, 99.

251. Yeiser, M., Harris, C. L., Kirchoff, A. L., Patterson, A. C., Wampler, J. L., Zissman, E. N., & Berseth, C. L. (2016). Growth and tolerance of infants fed formula with a new algal source of docosahexaenoic acid: Double-blind, randomized, controlled trial. *Prostaglandins, Leukotrienes and Essential Fatty Acids (PLEFA), 115*, 89–96.

252. Yu, Y., Xu, Z., Chen, S., & Jin, M. (2020). Microbial lipid production from dilute acid and dilute alkali pretreated corn stover via *Trichosporon dermatis*. *Bioresource Technology, 295*, 122253.

253. Zhang, F., Carothers, J. M., & Keasling, J. D. (2012). Design of a dynamic sensor-regulator system for production of chemicals and fuels derived from fatty acids. *Nature Biotechnology, 30*, 354–359.

254. Zhang, X., Li, M., Agrawal, A., & San, K. Y. (2011). Efficient free fatty acid production in *Escherichia coli* using plant acyl-ACP thioesterases. *Metabolic Engineering, 13*, 713–722.

255. Zhang, Y., Nielsen, J., & Liu, Z. (2021). Yeast-based biorefineries for oleochemical production. *Current Opinion in Biotechnology, 67*, 26–34.

256. Zhao, H., Lv, M., Liu, Z., Zhang, M., Wang, Y., Ju, X., Song, Z., Ren, L., Jia, B., Qiao, M., et al. (2020). High-yield oleaginous fungi and high-value microbial lipid resources from *Mucoromycota*. *BioEnergy Research, 14*, 1196–1206.

257. Zuccaro, G. (2022). Single cell oil production: A new approach in biorefinery. Retrieved from: http://www.fedoa.unina.it/11804/ (accessed on 8 November 2024).

258. Zuccaro, G., del Mondo, A., Pinto, G., Pollio, A., & De Natale, A. (2021). Biorefinery-based approach to exploit mixed cultures of *Lipomyces starkeyi* and *Chloroidium Saccharomyces cerevisiae arophilum* for single cell oil production. *Energies, 14*, 1340.

Chapter 8

Vitamins

Contents

INTRODUCTION

It was only a "short step" to consider using microorganisms for vitamin production, as the value of yeasts and other microorganisms as sources of the B-group vitamins has been known for more than 50 years. Only vitamin B_{12} and riboflavin have been produced on a commercial scale by microbial processes, despite a plethora of reports in the scientific and patent literature suggesting that micro-organisms can be used to produce large amounts of B-group vitamins, including thiamin, biotin, folic acid, pantothenic acid, and pyridoxine. Since chemical synthesis is not cost-effective and isolation from liver and other sources is more expensive than using fermentations, all of the vitamin B_{12} that is commercially available is actually made by fermentation processes. Riboflavin fermentation processes are a major source of feed-grade material used to supplement animal and poultry feeds, and they compete very well with those that use chemical synthesis.

Due to the intricacy of its chemical synthesis, vitamin B_{12} is a compound that is widely used in the feed and food, healthcare, and medical industries. It can only be produced through fermentation. For this reason, industrial producers have spent the last few decades concentrating on developing better producer strains and streamlining their bioprocesses. In this chapter, we first give a brief history of vitamin B_{12} research as well as an overview of the key biosynthetic properties of the two families of microorganisms that are commonly used to produce it industrially: strains of Propionibacterium freudenreichii and strains associated with *Pseudomonas denitrificans*. Afterwards, a thorough overview of the current situation of industrial vitamin B_{12} production is provided, along with an analysis of the key developments and obstacles to improvement. A particular emphasis is placed on bioprocess optimization, which aims to boost production while maintaining sustainability.

Humans require riboflavin, an essential micronutrient, which can only be acquired externally through diet or supplementation. Riboflavin has been implicated in the prevention and treatment of a number of diseases, according to numerous studies. Riboflavin synthesis can be achieved through three main strategies: total chemical synthesis, chemical semi-synthesis, and microbial fermentation. Of these, the latter is currently the most promising. Flavinogenic microbes have garnered more and more attention in the past few years. The best cell factories for riboflavin overproduction are fungi, such as *Eremothecium ashbyii* and *Ashbya gossypii*, and bacteria, such as Bacillus subtilis, Escherichia coli, and lactic acid bacteria. As a result, they make excellent choices for raising the riboflavin content of fermented foods or creating new, riboflavin-bio-enriched foods that have better nutritional value and/or health benefits. The function of riboflavin in human health and the history of its industrial production are briefly covered in this chapter. After that, it discusses the biosynthesis of riboflavin in bacteria and fungi and concludes with a summary of methods for riboflavin overproduction that include both fermentation condition optimization and the creation of strains that produce more riboflavin through metabolic engineering and chemical mutagenesis. All things considered, this chapter offers a current understanding of riboflavin biosynthesis and can encourage the investigation and creation of riboflavin-rich fermented food products.

8.1. MICROBIAL PRODUCTION OF VITAMIN B12

Cobalamin, another name for vitamin B_{12}, is a water-soluble molecule that is necessary for many organisms' metabolism. With more than 30 biotransformation steps, its biosynthesis is intricate and has a complex structure. Although the phyla that possess this biosynthetic pathway are not necessarily related, mammals, including humans, are not able to synthesize vitamin B_{12} because it is only present in a small number of bacteria and archaea.

The 1920s saw the start of research into vitamin B_{12} in relation to pernicious anemia, a disease that was initially identified in 1824. Fatigue, headaches, weight loss, and, in more severe cases, dementia, memory loss, muscle weakness, and peripheral neuropathy—all of which can be fatal in the absence of treatment—were the primary symptoms of this illness. Patients with pernicious anemia could effectively recover from the illness by following a special diet that included large amounts of lightly cooked liver and muscle meat, as Minot and Murphy demonstrated in 1926. They postulated that an unidentified "extrinsic factor" found in animal livers was the reason the treatment worked. Though it took more than 20 years for the so-called "extrinsic factor" to be discovered and isolated, they were granted the 1934 Nobel Prize in Physiology or Medicine for this discovery. This happened in 1948 when researchers from two pharmaceutical companies—Smith at Glaxo and Folkers at Merck, Sharp and Dohme—isolated a cobalt compound from animal livers that had the ability to treat pernicious anemia on its own, essentially at the same time. One year later, additional sources of the same compound, including milk, beef, and multiple bacterial cultures, were also found to contain it. Vitamin B_{12} was the name given to this red, crystalline, octahedral cobalt compound. Years later, it was interestingly found that this compound was actually one of the many isoforms of the cobalamin (Cbl) family, specifically cyanocobalamin (CNCbl), an artificially produced physiologically inactive form of cobalamin produced during the industrial process of liver extraction and isolation. Furthermore, the structure of CNCbl, the first cobalamin isoform, was solved by Dorothy Hodgkin in 1955.

The same group also determined the structure of adenosylcobalamin (AdoCbl) in 1957, which is one of the two active forms of cobalamin. It was because of these findings that Hodgkin was granted the 1960 Chemistry Nobel Prize. Another physiologically active form of vitamin B_{12}, methylcobalamin (MetCbl), was found two years later. In the years that followed, a large number of MetCbl- and AdoCbl-dependent enzymes were identified and described. It was discovered that both AdoCbl and MetCbl functioned as cofactors in a number of enzymes. A number of them, including the majority of the molecules in charge of vitamin B_{12} transport in mammals, as well as methionine synthase from *Escherichia coli* and L-methylmalonyl-CoA mutase from *Propionibacterium shermanii*, were crystallized.

Following more than 10 years of research, Woodward et al. described the full chemical synthesis of vitamin B_{12} in 1973. With more than 60 steps and protection and deprotection reactions, the process was intricate and produced very low yields—less than 1%.

8.1.1. Structure of Cobalamin Derivatives and Functions as Enzyme Cofactors

The term "vitamin B_{12}" refers to a group of substances known as cobalamins, or Cbl, that have the same general structure: a tetrapyrrolic corrinic ring with four nitrogen atoms coordinated to a central cobalt atom. This scaffold bears similarities to other prosthetic groups, including cytochrome P450 and heme in hemoglobin. This structure enables the molecule to perform its various tasks by utilizing the redox state of the central metallic atom, cobalt.

Four pyrrole units (C_4H_5N), joined on opposite sides by a C-CH_3 methylene link, a C-H link on one side, and two pyrrole units directly joined form the corrin ring, as illustrated in Figure 8.1. The methine bridge between the A and D subunits, which is present in other known porphyrins like hemoglobin, is absent from this corrin ring. The oxidation state, the central metallic atom, the quantity and kind of side chains, and other structural characteristics set this molecule apart from others (Calvillo et al., 2022).

The central Co ion is connected to two additional ligands in addition to the four N atoms of the pyrrole units. The lower ligand is the base 5,6-dimethylbenzimidazole (DMBI), which is connected to the central Co ion in an α-axial conformation by the N7-atom. Additionally, DMBI is connected to a side chain of the central corrinic structure. This side chain is the propionic acid side chain of the pyrrole unit D of the corrinic ring, which is connected to the phosphate group of DMBI.

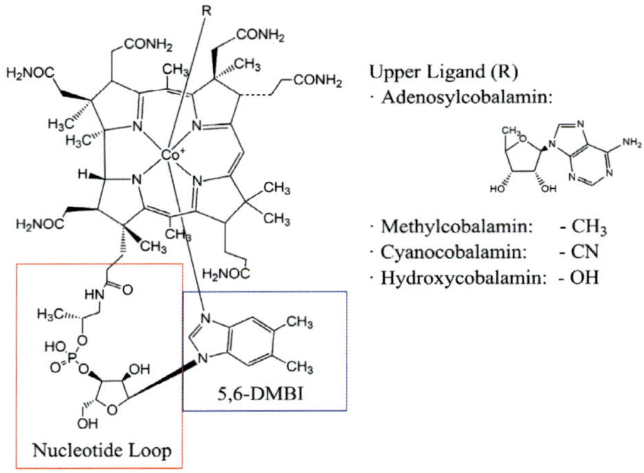

Figure 8.1. *Schematical representation of the structure shared by all cobalamin isoforms. The main groups that usually act as the upper ligand are also shown. 5,6-DMBI: 5,6-dimethylbenzimidazole.*

Source: Álvaro Calvillo,Teresa Pellicer, Marc Carnicer, and Antoni Planas. Bioprocess strategies for vitamin b_{12} *production by microbial fermentation and its market applications. Bioengineering, https://doi.org/10.3390/bioengineering9080365.*

Lastly, Co is connected to the sixth ligand in the β-axial orientation. This chemical group exhibits variable physiological and catalytic functions due to its variable nature. For instance, a methyl group in this position forms the methylcobalamin (MetCbl) isoform, whereas a 5-deoxyadenosyl group forms adenosylcobalamin (AdoCbl). The C-Co present in AdoCbl was the first bond of its kind to be reported in a biological molecule. While cyanocobalamin (CNCbl) and hydroxocobalamin (OHCbl) are the two isoforms that exhibit physiological activity as cofactors in humans, there are other forms as well, including those with less common upper ligands like nitritocobalamin (NitCbl), sulfitocobalamin (SulCbl), and glutathionylcobalamin (GlutCbl).

Methionine synthase, an essential cellular housekeeping enzyme involved in two major metabolic pathways—the tetrahydrofolate-dependent one-carbon cycle and the last stage in the conversion of methionine from homocysteine—is one of the methyltransferases that MetCbl cofactors with. On the other hand, L-methylmalonyl-CoA mutase (MMCM), a crucial enzyme for propionate catabolism and degradation of odd-chain fatty acids, several amino acids (valine, isoleucine, methionine, threonine), and cholesterol, uses AdoCbl as a cofactor by several enzymes, mostly mutases. However, only one AdoCbl dependent enzyme is found in mammals. The active forms of cobalamin are AdoCbl and MetCbl, as was previously mentioned, however they are also known to be light sensitive. Because of this, the most widely used commercial form of vitamin B_{12} is CNCbl, which the body can convert into an active coenzyme form more easily and steadily.

8.1.2. Biosynthesis of Vitamin B_{12}: The Aerobic and Anaerobic Pathways

The biosynthetic pathways of organisms that produce Cbl have drawn the attention of numerous researchers since the 1970s, when the structure and biological functions of the various vitamin B_{12} compounds were discovered. It was eventually determined that the structural complexity resulted from a complex and extensive biosynthesis involving over 30 genes and numerous enzymatic steps for the "de novo" synthesis of the molecule. Since no eukaryotic organism has been shown to possess the ability to produce any isoform of Cbl through genetic evidence, this pathway is assumed to be exclusive to certain bacteria and archaea. The entire *Pseudomonas denitrificans* biosynthesis pathway was not described until the 1990s, despite the discovery and isolation of a few intermediates earlier. The prefix cob and a letter designating each gene position in the operon were assigned to the genes involved in Cbl synthesis. The French company Rhône-Poulenc Santé, which is now Sanofi, was responsible for characterizing and isolating the cob enzymes and cobalamin intermediates of *P. denitrificans* in the subsequent years.

Genes implicated in the biosynthesis of Cbl were later identified in other organisms, including Salmonella enterica, *Bacillus megaterium*, and the previously investigated *Propionibacterium freudenreichii*. It was evident right away that the pathway in the later organisms shared similarities with *P. denitrificans*, despite having distinct genetic makeup. The absence of a monooxygenase and the distinct cobaltochelatase were two important distinctions. In light of this, two distinct pathways for the biosynthesis of Cbl were identified: (i) an aerobic pathway for the late insertion of cobalt, carried out primarily by *P. denitrificans* and, as was later discovered, by other microorganisms like *Ensifer casida* and *Sinorhizobium meliloti*; and (ii) an anaerobic pathway for the early insertion of cobalt, carried out primarily by *P. freudenreichii*, *B. megaterium*, and *S. enterica*.

Tetrapyrrole synthesis starts with the synthesis of 5-aminolaevulinic acid (ALA), regardless of the biosynthetic pathway. Three distinct enzymatic reactions are then used to convert ALA into a tetrapyrrolic macrocycle structure. Porphobilinogen (PBG) is first formed by the condensation reaction between two ALA molecules, which is catalyzed by the enzyme ALA dehydratase (EC 4.2.1.24), which is dependent on Zn^{2+} and Mg^{2+}. A linear tetrapyrrole is then created by the polymerization of four PBG molecules by a PBG deamynase

(EC 4.3.1.8). Uroporphyrinogen III is an unsymmetrical hexahydro porphyrin isomer that is finally formed when a uroporphyrinogen III synthase (EC 4.2.1.75) can invert the last pyrrole unit and link it to the first pyrrole unit of the linear tetrapyrrole. This molecule is the final intermediate that is shared by the chlorophyll and heme groups and other prosthetic groups.

Uroporphyrinogen III methyltransferase (EC 2.1.1.107) catalyzes the conversion of uroporphyrinogen III to precorrin-2, the first molecule in Figure 8.2. The methyl donor for this reaction is S-adenosyl-L-methione (SAM). To be more precise, the enzyme methylates the uroporphyrinogen III at position C-2 to form precorrin I. Subsequently, it can methylate at position C-7 to form precorrin-2, which is the final common intermediate for coenzyme siroheme, P450, and vitamin B_{12}. This process occurs after a phototrophic tautomerization.

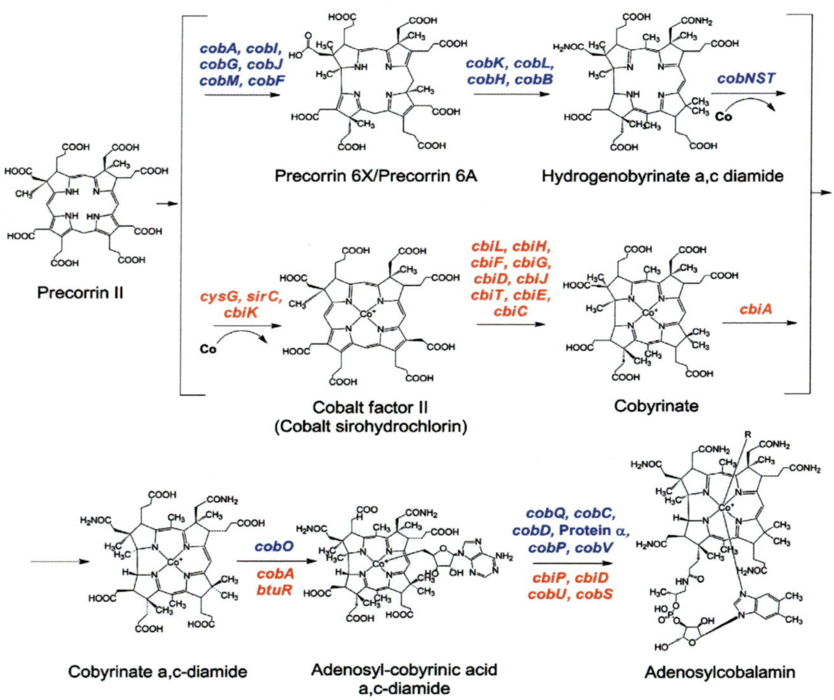

Figure 8.2. *Summary of anaerobic and aerobic adenosylcobalamin biosynthesis. Genes encoding the proteins from the aerobic and anaerobic pathways are shown in blue and red, respectively, except for protein* ***α whose*** *coding gene is not known.*

Source*: Álvaro Calvillo, Teresa Pellicer, Marc Carnicer, and Antoni Planas. Bioprocess Strategies for vitamin* b_{12} *production by microbial fermentation and its market applications. Bioengineering, https://doi.org/10.3390/bioengineering9080365.*

The ring contraction and cobalt chelation phases are where the aerobic and anaerobic pathways diverge most (*see* Figure 8.2). A molecule of oxygen and a monooxygenase (CobG) are needed for the aerobic pathway's ring contraction to form precorrin-3B, an intermediate hydroxylated γ-lactone that goes through a masked pinacol rearrangement during the contraction, extruding the methylated C20 position. After that, the ring completely contracts, releasing an acetic acid molecule in the process. However, after the cobalt has already been incorporated into the molecule, the ring contraction occurs later in the anaerobic pathway. The cbiH gene-encoded enzyme catalyzes this step without the need for molecular oxygen. After that, methylation at position C17 that is SAM-dependent occurs, encouraging the extrusion of the methylated C20 position and creating an δ-lactone ring.

In both pathways, cobalt chelation differs significantly as well. The cobalt atom enters the aerobic pathway "late," after the ring has completely contracted. In the presence of magnesium, an ATP-dependent multienzyme complex called cobNST catalyzes this insertion. This stage occurs earlier in the anaerobic pathway when a cobalt chelatase, which is catalyzed by the cbiX or cbiK genes, catalyzes the cobalt insertion while the ring is still not contracted.

Furthermore, DMBI is synthesized independently of the corrin ring's biosynthesis pathway and is subsequently attached in an α-axial conformation. According to the requirements for oxygen, lower ligand synthesis can be achieved through two distinct pathways, the anaerobic and the aerobic pathway, which have been recently described.

On the one hand, the 5,6-dimethylbenzimidazole synthase BluB (EC 1.13.11.79) catalyzes the aerobic biosynthesis of DMBI by performing the cleavage of the ribityl tail and the contraction and fragmentation of the bound flavin mononucleotide cofactor in the presence of molecular oxygen, resulting in the formation of DMBI and D-erythrose 4-phosphate. Afterwards, the DMBI is introduced by the phosphoribosyltransferase CobU/T (EC 2.4.2.21) through a nucleophilic substitution reaction. This pathway was initially identified for *S. meliloti* and subsequently discovered in most bacteria that produce Cbl, including *P. freudenreichii* and *P. denitrificans*, the two most significant industrial producers. This observation demonstrated that *P. freudenreichii* cannot produce Cbl anaerobically in its entirety without the addition of DMBI from the outside. However, the *bzaA-bzaB-cobT-bzaC-bzaD-bzaE* operon's gene products catalyze the anaerobic biosynthesis of DMBI, which leads to the formation of DMBI with the intermediates 5-hydroxybenzimidazole, 5-methoxybenzimidazole, and 5-methoxy-6-methylbenzimidazole. The obligate anaerobic bacteria *Eubacterium limosum* and *Acetobacterium woodii* were found to use this pathway (Mathur et al., 2020; Calvillo et al., 2022).

8.1.3. Microbial Production of Vitamin B_{12}: Bioprocess Optimization for Cyanocobalamin Production

The food, beverage, dietary, and nutraceutical industries have seen a sharp rise in demand for cobalamin in recent years, owing to both the general public's growing health consciousness and the growing acceptance of alternative diets like vegan and vegetarian diets. This is why over the years; a lot of work has gone into optimizing the strain and process for the production of cyanocobalamin.

In addition to the three cobalamin forms—AdoCbl, MetCbl, and CNCbl—there are numerous other cobamides with various lower axial ligands that function as essential cofactors for corrinoid-dependent enzymes that are crucial for various processes, including the gut microbiota. This section only addresses cyanocobalamin, the form currently produced industrially and absorbed and utilized by humans, despite their significance.

In the past, high naturally productive strains of *P. freudenreichii* and *P. denitrificans*, along with related strains like *Pseudomonas nitroreducens* and *E. casida*, were employed for the industrial production of cobalamin.

For many years, using random mutagenesis techniques to boost vitamin B_{12} productivity or resistance to toxic intermediates present in media has been a common strategy to improve these strains. However, downregulating many genes, heterologous expression of foreign genes, and overexpression of cobalamin biosynthesis-related genes have also produced improved producer strains. It's also important to take note of the emergence of fresh, fruitful strains that show promise, like *Acetobacter pasteurianus* and *B. megaterium*, as well as the heterologous expression of the biosynthetic pathway in other well-known cell platforms, like *E. coli*. To increase cobalamin production in various cell factories, Balabanova and colleagues have recently conducted a thorough review of the genetic and biosynthetic regulation as well as the genetic tools that have been employed (Balabanova et al., 2021).

On the other hand, there are also numerous instances of improvements in microbial production of vitamin B12 through bioprocess optimization. The most pertinent bioprocess innovations for higher Cbl production as well as fresh approaches for Cbl production on new platforms or media are compiled in Table 8.1. This section aims to give an update on the main bioprocess innovations for biotechnological cobalamin production, so studies focusing on increasing production via genetic engineering of known strains are not presented. Included is a synopsis of media requirements, volumetric productions and productivities, and cultural conditions.

Table 8.1. Summary of Reported Productions with the Industrial Cobalamin-Producing Strains

Microorganism/Strain	**Main Media Components**	**Scale**	**Summary/Innovation**	**Volumetric Production**	**Volumetric Productivity (mg/L/h)**
B. megaterium DSM 319	Terrific broth media	250 mL shake flask	Precursor supplementation and pO_2 control	0.21 mg/L	0.006 mg/L/h
Lactobacillus reuteri ZJ03	Soymilk	250 mL shake flask	Different carbon source supplementation	0.204 mg/L	0.003 mg/L/h
P. freudenreichii subsp. *Shermanii* NRRL-B-4327, 3523 and NRRL-B-3524	Sodium lactate broth	250 mL shake flask	Vitamin B_{12} analogue addition	31 mg/L	0.51 mg/L
P. freudenreichii CICC 10019	Glucose, CSL	7 L stirred tank bioreactor	Expanded-bed bioreactor (EBAB) with crop stark hydrolysates	47.6 mg/L	0.18 mg/L/h
P. freudenreichii CICC 10019	Glucose, CSL	7 L stirred tank bioreactor	EBAB bioreactor	43.4 mg/L	0.27 mg/L/h
P. freudenreichii CICC 10019	Glucose, CSL	1.5 L stirred tank bioreactor	EBAB bioreactor and DMBI addition	58.8 mg/L	0.59 mg/L/h

P. freudenreichii CICC 10019	Glucose/ glycerol, CSL	5 L stirred tank bioreactor	EBAB bioreactor, glycerol as carbon source and crop stalk hydrolysate as nitrogen source	43 mg/L	0.36 mg/L/h
P. freudenreichii DF13	Supplemented whey permeate	1 L stirred tank bioreactor	Co-culture with *Lactobacillus plantarum* SM39 for simultaneous folate and Cbl production	0.75 mg/L	0.004 mg/L/h
P. freudenreichii DSM 20271//*Lactobacillus brevis* ATCC 14869	Wheat bran dough	n.d.	Co-fermentation in wheat bran dough for in situ production of Vitamin B_{12}	332 ng/g	n.d.
P. freudenreichii IFO 12424// *Ralstonia eutropha* H16 (ATCC17699)	Polypeptone, casein, yeast extract	5 L stirred tank bioreactor	Cell recycling system and co-culture with *Ralstonia eutropha* for decreasing propionic acid inhibition	8 mg/L	0.14 mg/L/h
P. freudenreichii PTCC 1674.	Tryptone, yeast extract, different carbon sources	100 cm^3	Waste frying sun oil as a carbon source for vitamin B_{12} production	2.74 mg/L	0.02 mg/L/h
P. freudenreichii subsp. *shermanii* ATCC 13673	Glucose, yeast extract	2 L stirred tank bioreactor	Inoculum volume, pH control and substrate concentration optimization	0.087 mg/L	0.002 mg/L/h

P. freudenreichii subsp. *shermanii* CICC 10019	Glucose, CSL	100 L fermenter	Addition of DMBI precisely with Ado-Cbl control strategy	39.15 mg/L	0.32 mg/L/h
P. freudenreichii subsp. *shermanii*	Glycerol, tryptone, casein, DMBI	200 mL shake flask	Media optimization by design of experiments with crude glycerol as the main carbon source	4.01 mg/L	0.024 mg/L/h
P. freudenreichii subsp. *shermanii*	Whey based media	20 mL tubes	DMBI, Nicotinamide and Riboflavin supplementation	5.3 mg/L	0.03 mg/L/h
P. freudenreichii subsp. *shermanii*	Food-like media (cereal matrices)	n.d.	Precursor supplementation in different cereal-like matrices	1.5 mg/Kg	0.009 mg/Kg/h
P. freudenreichii subsp. *shermanii* 2067	Cheese-based propionic media/ whey-based liquid media	50 mL shake flask	Production in food-like conditions without DMBI addition	0.124 mg/L	0.0013 mg/L/h
P. freudenreichii CICC10019	Glucose, yeast extract, CSL	100 mL flasks	Media optimization by statistical analysis	8.32 mg/L	0.068 mg/L/h
P. freudenreichii CICC10019	Glucose, CSL	7 L fermenter	Membrane separation-coupled fed-batch fermentation	21.6 mg/L	0.16 mg/L/h

P. denitrificans	Maltose, peptone, betaine	250 mL shake flask	Addition of rotenone as a respiration inhibitor for enhanced production	54.7 mg/L	0.57 mg/L/h
P. denitrificans	Beet molasses, sucrose, betaine	120 m^3 fermenter	Glucose-betaine feeding, pH control strategy	214.13 mg/L	1.27 mg/L/h
P. denitrificans	Glucose, CSL, betaine	120 m^3 fermenter	Stepwise oxygen uptake rate control strategy	188 mg/L	1.12 mg/L/h
P. denitrificans	Glucose, CSL, betaine	50 L fermenter	Effects of specific oxygen consumption rate on cell morphology and production	213.1 mg/L	1.88 mg/L/h
P. denitrificans	Maltose, peptone, betaine	250 mL shake flask	Betaine supplementation	58.61 mg/L	0.48 mg/L/h
P. denitrificans	Maltose syrup, CSL, betaine	120 m^3 fermenter	Maltose syrup and CSL as the main substrates	198.27 mg/L	1.10 mg/L/h
P. denitrificans	Glucose, CSL, betaine	120 m^3 fermenter	pO_2 stepwise control	198.80 mg/L	1.18 mg/L/h

Source: Álvaro Calvillo, Teresa Pellicer, Marc Carnicer, and Antoni Planas. Bioprocess strategies for VITAMIN B_{12} production by microbial fermentation and its market applications. Bioengineering, https://doi.org/10.3390/bioengineering9080365.

Overall, there are notable variations in cobalamin production between the various microorganisms that produce it. This means that while *P. freudenreichii* productions vary greatly between strains and culture conditions, volumetric productions and productivities obtained with *P. denitrificans* are unquestionably superior to those obtained with other producers. Strategies aimed at lessening the inhibitory effect of propionic acid appear to be the most successful in the latter case.

8.1.3.1. Microbial Production in Pseudomonas denitrificans

A Gram-negative bacterium called *P. denitrificans* makes vitamin B_{12} through the aerobic biosynthesis pathway. *P. denitrificans* is currently the primary vitamin B_{12} producer used by industrial manufacturers, such as Sanofi in Europe and the Huarong Pharmacy Corporation in China, despite not having a Generally Recognized as Safe (GRAS) status.

On the one hand, the Sanofi strain was created initially through a combination of molecular biology and random mutagenesis techniques. While its volumetric production is unknown, considering other aerobic strains, one is tempted to assume that it could produce 200–300 mg/L. The natural, high-producing strain known as MB-580, which was initially described and granted a patent in 1962 (US3018225A), served as the source of the optimized strain. The cob genes involved in the biosynthesis of vitamin B_{12} were amplified over a number of years by the French company Rhône-Poulenc, leading to the creation of SBL27 and, eventually, SC510, two high-producing strains. Today, Sanofi, formerly known as Rhône-Poulenc, is the primary producer of vitamin B_{12} in Europe.

However, the leading global producers of vitamin B_{12} on the market are based in China and include the Chinese CSPC Huarong Pharmaceutical Company, the North China Pharmaceutical Company, the Henan Luyuan Pharmaceutical Company, and the Hebei Yuxing Bio-Engineering Company. Their combined reported production of vitamin B_{12} in 2020 was estimated to be about 31.41 tons, with an estimated value of USD 339.8 million. The exact origin of the strains used in their industrial productions is unknown, but it is believed to be an aerobic strain based on various publications from Huarong Pharmaceutical Company-affiliated research groups and the most recent patents on P-based bioprocess optimization. Volumetric productions of up to 281 mg/L are claimed by denitrificans that were presented in China.

In addition to genetic modifications, optimization of culture media and modifications to bioprocessing conditions have also been successful in improving vitamin B_{12} productivity in *P. denitrificans*. For instance, tests have been conducted on the effects of various supplements, pH, dissolved oxygen control, and trace element levels in media. Accordingly, it has been observed that supplementing with Zn^{2+} significantly increases the synthesis of ALA and PBG, two of the primary precursors of cobalamin, whereas DMBI, the base incorporated into the nucleotide loop, and Co^{2+} also positively affects production. Production of cobalamin increased by 13% when the starting concentrations of these three compounds were optimized through experimental design. Vitamin production was significantly impacted by the media composition, which also had an impact on the pH stability of cultures. A feeding strategy with betaine as a methyl donor and glucose as a carbon source was developed to better control pH. When applied to 120 m^3 bioreactor cultures, this feeding strategy was found to be beneficial for vitamin production. Furthermore, betaine can also inhibit cell growth at high concentrations, although it is well known that it improves the formation of several important intermediates, including ALA, glutamate, glycine, and methionine, and acts as a methyl donor for vitamin B_{12} biosynthesis. To counteract the detrimental effects on cell growth and the beneficial effects on cobalamin production, an appropriate betaine feeding strategy was subsequently developed and successfully applied at the industrial scale (Li et al., 2008).

In *P. denitrificans*, the oxygen transfer rate (OTR) has also been a key topic of bioprocess optimization. Lower OTRs in later stages were found to be essential for higher productivity, whereas higher OTRs during the early stages of culture promote cell growth. Subsequent research found that the higher production seen under lower oxygenation conditions may be linked to changes in cell morphology, which promote the transition of cells from the growth phase to an elongation state, which exhibits higher production of vitamin B_{12}. This was taken into consideration when developing many multi-step dissolved oxygen control strategies. These strategies involved gradually reducing agitation and aeration until dissolved oxygen values dropped below 2 percent, which resulted in a 17% increase in production. In addition, even though respiratory chain inhibitors like rotenone have a negative effect on cell growth, they may also improve vitamin production (Cheng et al., 2014).

Lastly, less expensive substitutes for more costly refined sucrose and glucose have been tested, including glucose, maltose syrup, beet molasses, and corn steep liquor. Certain compounds have the potential to adversely impact pH stability and, consequently, the ultimate production of vitamins. However, it has been noted that a blend of betaine, corn steep liquor, and maltose syrup works well and is less expensive than the conventional media compositions (Xia et al., 2015).

8.1.3.2. Microbial Production in Propionibacterium freudenreichii

The Gram-positive rod-shaped bacteria known as *P. freudenreichii* strains get their name from their ability to produce large amounts of propionic acid through the Wood–Werkman pathway. Compared to producers of aerobic vitamin B_{12}, *P.*

freudenreichii enjoys the benefit of both FDA-granted GRAS status and EFSA-granted Qualified Presumption of Safety (QPS) status.

In *P. freudenreichii*, several genetic engineering techniques were tested to increase vitamin B_{12} levels. For instance, it has been reported that improving cobalamin production involves overexpressing a few of the key genes involved in cobalamin synthesis as well as using a genome-shuffling strategy. To obtain better cobalamin producers, however, the main industrial strains were typically obtained by random mutagenesis using various mutagenic agents, such as UV light or chemical compounds. These high-yield strains of *P. freudenreichii* typically exhibit increased resistance and tolerance to propionic acid.

P. freudenreichii are facultative anaerobic strains that produce cobalamin by anaerobic biosynthesis. The synthesis of DMBI and its attachment to the corrinic ring require oxygen, even though they only produce high cobalamin yields under extremely low oxygen conditions. Because of this, the culture is typically split into two phases: a first phase during which the cells are cultivated entirely anaerobically, and a second phase that occurs 72–96 hours later and involves gentle agitation-induced aeration to produce the microaeration required for the synthesis of DMBI and cobalamin.

These vitamin B_{12} producers were able to broaden their market reach due to their GRAS status, which permits them to be used directly in food product production. In this regard, food-like media, such as cheese-like propionic medium or whey-based liquid medium, cereal matrices, and in situ fortification of tempeh, have been successfully used to test *P. freudenreichii* for edible fortification. In the context of food fortification, it permits an increase in cobalamin content using non-traditional sources and the achievement of the recommended daily consumption levels of vitamin B_{12} with only small amounts of fermented products, despite final cell densities and reported production levels being low compared to other, traditional media.

As previously noted, these bacteria are capable of producing large amounts of propionic acid, which eventually turns toxic and inhibits the growth of new cells. Consequently, a number of bioprocess optimization techniques have been investigated to reduce propionic accumulation. Specifically, the simultaneous production of vitamin B_{12} and propionic acid has demonstrated encouraging results when using in situ product removal (ISPR) techniques. The utilization of high biocompatibility resins, like ZGA330, in expanded-bed adsorption bioreactors (EBABs) has been reported to facilitate vitamin B_{12} volumetric production levels ranging from 40 mg/L to 60 mg/L among ISPR techniques. Propionic acid is kept in the resin of EBABs while adsorption takes place when the column is expanded, enabling the culture to flow through the chromatographic column without clogging. It has been investigated whether it is possible to increase the production of propionic acid and vitamin B_{12} at the same time by varying the culture conditions, the sources of carbon and nitrogen, and the addition of media supplements like DMBI. The combination of glucose, glycerol, and corn stalk hydrosylates has been shown to be an effective carbon source in an EBAB system; volumetric CNCbl production levels of 43.2 mg/L and 47.6 mg/L, respectively, have been reported.

Propionic acid concentration can also be lowered by co-fermenting *P. freudenreichii* with other microorganisms that can metabolize the acid. This is an intriguing method. For instance, cobalamin production increased from 6.73 mg/L to nearly 19 mg/L in the coculture of *P. freudenreichii* and *Ralstonia eutropha*. Furthermore, co-fermentation has been effectively used to either simultaneously produce multiple products or strengthen other cell cultures in addition to reducing propionic acid. The co-cultivation of *P. freudenreichii* and Lactobacillus plantarum (now called *Lactiplantibacillus plantarum*) resulted in the simultaneous production of folate and vitamin B_{12}. More recently, a patent for the co-fermentation of a Basidiomycota strain plus *P. freudenreichii* produced vitamin D and B_{12} simultaneously. The in-situ production of vitamin B_{12} in bread dough in a whey-based medium, along with the co-cultivation of *P. freudenreichii* and *Lactobacillus brevis* (now known as *Levilactobacillus brevis*) to ensure microbial safety and stability, is an example of food fortification.

Another popular method for raising productivity is supplementing with cobalamin precursors. It is frequently stated that adding common precursors and necessary compounds, like ALA and Co^{2+}, is advantageous for the synthesis of vitamins. While all strains of *P. freudenreichii* are able to synthesize DMBI on their own, there is little biosynthesis of this base. Furthermore, because oxygen is necessary for the synthesis of DMBI, its formation cannot occur under strictly anaerobic conditions. The cells start to accumulate incomplete forms of vitamin B_{12}, such as cobinamide or pseudovitamin B_{12}, if the availability of DMBI is restricted, preventing the formation of the active form. Therefore, it has been repeatedly reported that the addition of DMBI, or even DMBI precursors like riboflavin or nicotiamide, is beneficial for cobalamin production. Furthermore, it has been observed by other groups that supplementing with vitamin B_{12} analogs can reduce feedback inhibition and boost the production of cobalamin.

Lastly, the ability of *P. freudenreichii* cultures to grow in a variety of complex carbon and nitrogen sources, as well as waste and spent media, such as molasses, crude glycerol, leftover frying sunflower oil, tomato pomace, liquid acid protein residue of soybean, and vegetable juice spent media, makes them interesting in industrial settings (Calvillo et al., 2022).

8.1.4. Vitamin B_{12} Downstream Processing and Post-Modification Strategies

To the best of the authors' knowledge, vitamin B12 recovery is a well-documented procedure that hasn't altered at the industrial level over the previous few decades (*see* Figure 8.3 for a classical bioprocess scheme). In summary, culture broth undergoes multiple stages of separation and purification, such as filtration, adsorption, and extraction, all of which have an effect on the overall yield and viability of the process. The first step in traditional downstream processing is a biomass concentration, which is often achieved by centrifugation and greatly reduces the volume. However, Cbl may also be detected extracellularly, depending on the bioprocess, so purification may begin with the entire broth.

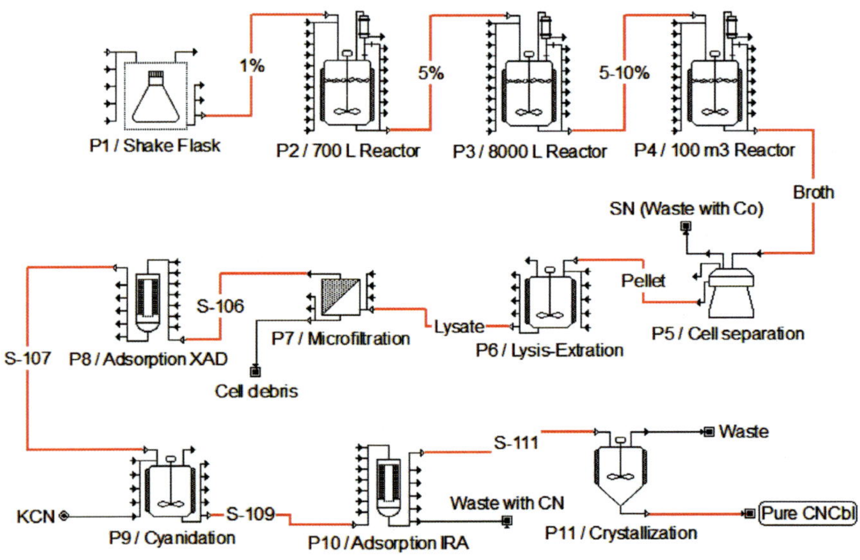

Figure 8.3. *Classical bioprocess to obtain highly pure cyanocobalamin. The main CNCbl stream is highlighted in red. Intermediate storage vessels are omitted to simplify the figure.*

Note: P1 to P11 represent Process 1 to Process 11, respectively. Bioprocess represented using SuperPro Designer® V9 Academic Site Edition, Intelligen, Inc. (Scotch Plains, NJ, USA).

Source: Álvaro Calvillo, Teresa Pellicer, Marc Carnicer, and Antoni Planas. Bioprocess Strategies for Vitamin B_{12} Production by Microbial Fermentation and Its Market Applications. Bioengineering, https://doi.org/10.3390/bioengineering9080365.

In either case, heating for 10–30 minutes at 80–120°C and a pH of 6.5–8.5 is used to extract all species of corrinoids. Cyanidation can be carried out either after the first stages of filtration and adsorption or during the extraction process. In both situations, the addition of potassium cyanide or thiocyanate converts the various corrinoids into CNCbl. Heat and sodium nitrite are typically present during this process.

One or more filtration (microfiltration and/or nanofiltration) and adsorption (XAD resin) procedures are then used to clarify the CNCbl solution. The vitamin solution is frequently treated with zinc chloride and precipitated using organic solvents, such as acetone, if the generated Cbl is intended for animal feed. If higher purity is needed, say, for medicinal applications, additional adsorption processes using various resins (e.g., IRA, and Alumina) are frequently required to produce a pure end product. A traditional bioprocess for obtaining extremely pure CNCbl is shown in Figure 8.3.

To increase its bioavailability, vitamin B_{12} may go through various post-modifications before being used as an oral medication or food supplement. In situations where there is a non-functional intrinsic factor, which results in extremely low Cbl bioavailability, protecting these compounds may be particularly interesting. In this regard, a number of methods have been established to safeguard oral supplements against particular ailments present in the digestive system. Microencapsulation is one of these methods that has been shown to increase vitamin B_{12} stability. It is currently widely used in the pharmaceutical and cosmetic industries and can be applied to food-grade $W_1/O/W_2$ emulsions, liposomes, or a variety of food-grade encapsulating agents, including chitosan, Arabic gum, sodium alginate, carrageenan, maltodextrin, modified starch, cyanobacterial extracellular polymeric, xanthan, and pectin. Furthermore, Fidaleo and colleagues have examined the use of nanocarriers as a promising nanotechnology that could lead to the improvement of vitamin B_{12} therapies, lowering costs and side effects while also improving patient outcomes (Calvillo et al., 2022).

8.2. BIOTECHNOLOGICAL STRATEGIES OF RIBOFLAVIN BIOSYNTHESIS IN MICROBES

Vitamin B_2, or riboflavin, is a water-soluble B-group vitamin with high thermostability that melts and breaks down at 280°C (Figure 8.4). While riboflavin is very stable in hot neutral or acidic solutions for a brief period of time, it is readily destroyed by heating in alkaline solutions despite its thermostability. In addition, riboflavin is a photosensitive molecule that degrades when exposed to short-wave radiation (< 400 nm). This molecule can be activated by light. As one of the micronutrients that the human body needs for a variety of physiological processes, riboflavin has been identified by the World Health Organization (WHO) as one of the six primary indicators for evaluating the nutritional status, growth, and development of humans. Adults who consume less than 0.2–0.3 mg of riboflavin per day are considered to be riboflavin deficient, with the average amount needed for good health being 0.3–1.8 $mg{\cdot}d^{-1}$. Pregnancy, the adolescent growth period, mental stress, and intense physical labor are among the circumstances that raise the need for riboflavin. Concerns are currently raised by riboflavin deficiency in both developed and developing nations.

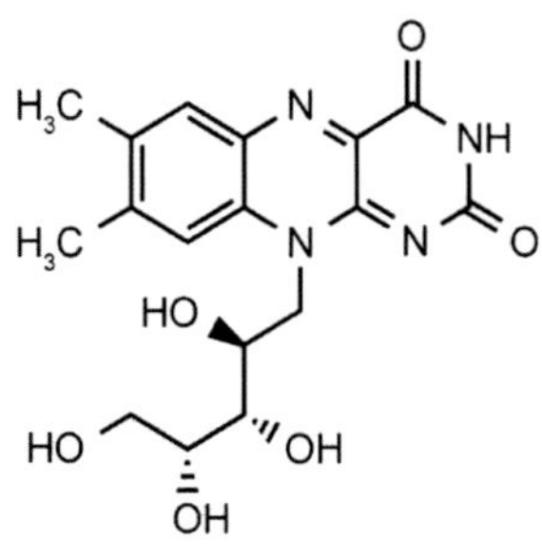

Figure 8.4. *The chemical structure of riboflavin.*

Source: By Calvero. – Selfmade with ChemDraw., Public Domain, https://commons.wikimedia.org/w/index.php?curid=1690365.

A necessary nutrient for proper cellular operation is riboflavin. Following uptake, flavokinase and FAD synthetase intracellularly convert it to flavin mononucleotide (FMN) and flavin adenine dinucleotide (FAD), respectively. These two main biologically active riboflavin derivatives play important roles in human metabolism by participating in a variety of redox reactions. In addition to serving as vital cofactors for several enzymes that catalyze redox reactions, such as glucose, aldehyde, and xanthine oxidases, succinic and fumaric dehydrogenases, and cytochrome c reductase, these riboflavin derivatives also function as electron carriers. Riboflavin has been shown in animal studies to influence the metabolism and absorption of iron; iron absorption can be improved by supplementing with riboflavin along with zinc. Given the critical roles that iron and zinc play in the proliferation of cells, riboflavin indirectly promotes growth. On the other hand, riboflavin deficiency can cause a decrease in the amount of iron that is absorbed, stored, and used, which can stunt human growth. Furthermore, it is well recognized that riboflavin excess or poor transport can result in cataract development, neurological conditions, abnormalities of the cardiovascular system, and even cancer. Thus, to avoid disorders linked to riboflavin deficiencies, it is imperative to ensure a regular dietary intake of riboflavin.

Riboflavin, like most vitamins, is not biosynthesized by living things. It comes mostly from food and possibly from the microbiota in the intestines. Naturally occurring in many different foods at varying concentrations are riboflavin-containing foods, including animal sources like internal organs, milk, and eggs. When it comes to riboflavin, these are superior to natural plant foods. Furthermore, certain green veggies and beans have riboflavin in them in certain amounts. Due to its water solubility, riboflavin is easily absorbed through food but is not stored in significant amounts in the human body. If ingested in excess, riboflavin may also be eliminated through the urine and stool.

8.2.1. Current Strategies of Riboflavin Production

In 2012, riboflavin was produced and sold at a global annual rate of over 9,000 t, of which about 1,800 t were used in beverages and pharmaceuticals and about 7,200 t in animal feed. The demand for riboflavin rises along with the growth of the functional food and animal feed industries. There are primarily three methods for producing riboflavin: microbial fermentation, chemical semi-synthesis, and total chemical synthesis. Moreover, there are two primary categories of microbial fermentation: fermentation using genetically modified microbes and conventional flavinogenic microbial fermentation.

8.2.1.1. Total Chemical Synthesis of Riboflavin

Using D-ribose or glucose as the starting material, riboflavin is synthesized through a series of six to nine chemical reaction steps, which include oxidation, displacement, transposition, acidification, lactonization, reduction, condensation, coupling, and cyclization. It is clear that this is an expensive and environmentally hazardous procedure that takes a long time and is laborious (Figure 8.5). Additionally, it is typically difficult to remove contaminants with a certain level of toxicity from the final products. Thus, the strategy of microbial fermentation has gradually supplanted total chemical synthesis.

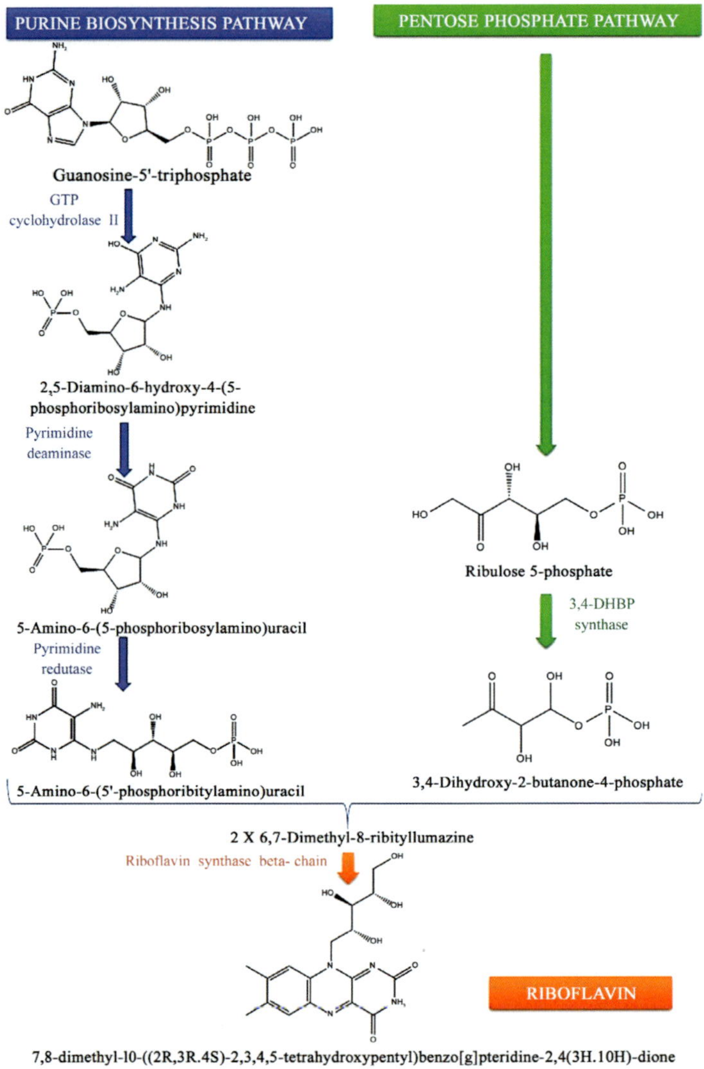

Figure 8.5. *The biochemical synthesis of riboflavin.*

***Source**: Aragão MÂ, Pires L, Santos-Buelga C, Barros L, Calhelha RC. (2024). Revitalizing riboflavin: unveiling its timeless significance in human physiology and health. Foods, 13(14), 2255. https://doi.org/10.3390/foods13142255.*

8.2.1.2. Chemical Semi-Synthesis of Riboflavin

Microbial fermentation and total chemical synthesis are used in tandem to produce riboflavin through chemical semi-synthesis. This strategy uses D-ribose as the primary raw material for the chemical synthesis of riboflavin, with D-glucose first fermented by microorganisms. The main distinction between chemical semi-synthesis and total chemical synthesis is that the fermentation of D-ribose from D-glucose can simplify and lower the cost of riboflavin production. However, because it is difficult to remove chemical additives, which leave high levels of residues in the finished products, chemical semi-synthesis is not appropriate for producing riboflavin on a large scale.

8.2.1.3. Production of Riboflavin by Microbial Fermentation

During the mid-1900s, three riboflavin-producing microbes were first used in the commercial production of riboflavin through microbial fermentation: one bacterium (*Clostridium acetobutylicum*; *C. acetobutylicum*) and two fungi (*Ashbya gossypii*; *A. gossypii* and *Eremothecium ashbyii*; *E. ashbyii*). The lengthy cycle and poor yield of the early microbial fermentation processes, however, prevented them from competing with chemical synthesis. As a result, mass marketing proved to be challenging. *Bacillus subtilis* (*B. subtilis*) and Corynebacterium *ammoniagenes* are two examples of genetically modified bacteria that have been successfully created for the production of riboflavin through research and development of genetic engineering technology. These bacteria are able to convert D-glucose into riboflavin efficiently, which can shorten the production cycle and increase the yield of riboflavin. As a result, their application may ultimately help bring riboflavin production through microbial fermentation to the market.

Overall, the current method of producing riboflavin through microbial fermentation has a number of benefits over total chemical synthesis and chemical semi-synthesis, including a shorter production cycle, the use of simple raw materials, a higher yield, and a significantly lower production cost.

8.2.2. Microbial Biosynthesis of Riboflavin

For the biosynthesis of one riboflavin molecule in flavinogenic organisms like fungi, yeasts, and eubacteria, two molecules of ribulose-5-phosphate (Ru5P) and one guanosine triphosphate (GTP) are needed. As seen in Figure 8.5, GTP cyclohydrolase II (enzyme 1) catalyzes the riboflavin biosynthesis pathway, which begins with GTP (1) as the substrate. In order to produce 2,5-diamino-6-ribosylamino-4(3H)-pyrimidinone 5′-phosphate (DARPP, 2), this step entails opening the imidazole ring and hydrolytically releasing inorganic pyrophosphate from the side-chain of the ribose moiety of GTP. Next, through two consecutive reactions involving reduction and deamination, DARPP (2) is converted into 5-amino-6-ribitylamino-2,4(1H, 3H)-pyrimidinedione 5′-phosphate (ArPP, 5). Different microbes have different reactions in this sequence. In fungi, the intermediate 2,5-diamino-5-ribitylamino-4(3H)-pyrimidinone 5′-phosphate (DArPP, 4) is deaminated to ArPP (5) after the ribosyl side-chain of DARPP (2) is first reduced to the ribityl side-chain. The ribofuranose moiety of the intermediate ARPP (3) is then reductively ring-opened to yield ArPP (5). In contrast, DARPP (2) is first deaminated to 5-amino-6-ribosylamino-2,4(1H,3H)-pyrimidinedione 5′-phosphate (ARPP, 3).

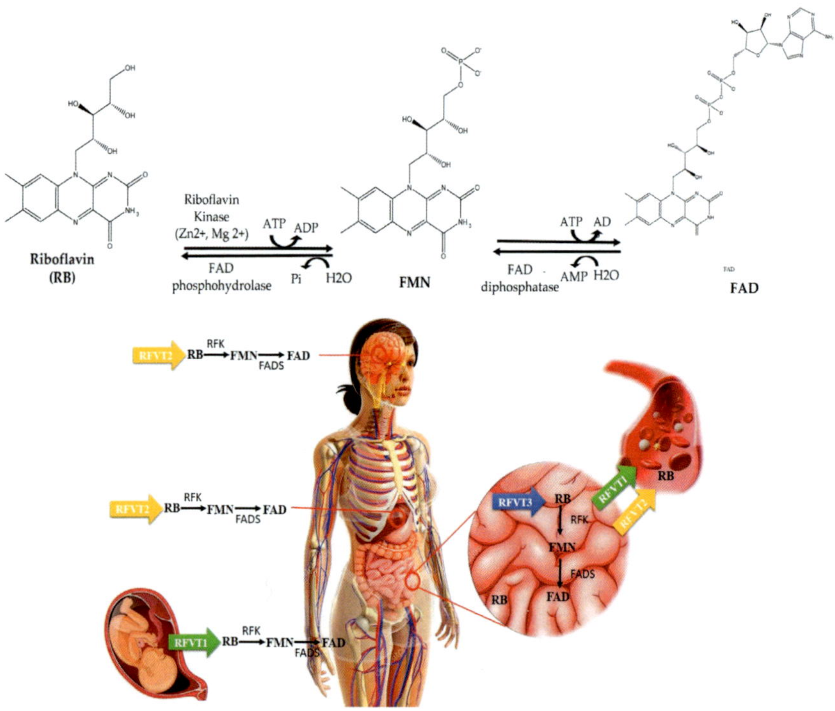

Figure 8.6. *Metabolism of riboflavin.*

Source: Aragão MÂ, Pires L, Santos-Buelga C, Barros L, Calhelha RC. (2024). Revitalizing riboflavin: Unveiling its timeless significance in human physiology and health. Foods, 13(14), 2255. https://doi.org/10.3390/foods13142255.

The reduction and deamination reactions in fungi and bacteria happen in the opposite order. Through the catalytic action of a non-specific phosphatase (enzyme 4), which is still unknown, ArPP (5) is further dephosphorylated into 5-amino-6-ribitylamino-2,4(1H,3H)-pyrimidinedione (ArP, 6). It was discovered recently that members of the haloacid dehalogenase (HAD) superfamily catalyze the dephosphorylation of ArPP (5) in certain species, including *B. subtilis, Escherichia coli*, and *E. bacteria*, and *Bacteroides thetaiotaomicron*, which acts as enzyme 4. However, through a skeletal rearrangement catalyzed by the DHBP synthase (enzyme 5), another initial substrate, Ru5P (7), can be converted into 3,4-dihydroxy-2-butanone-4-phosphate (DHBP, 8). Further lumazine synthase (or riboflavin synthase β-chain, enzyme 6) catalyzes the condensation of Arp (6) with DHBP (8) to give 6,7-dimethyl-8-ribityllymazine (DMRL, 9), which can be ultimately transformed into riboflavin (10) and ArPP (5) via a highly unusual dismutation that is catalyzed by the riboflavin synthase (or riboflavin synthase α-chain, enzyme 7), and the product ArPP (5) may be recycled for the biosynthesis of another riboflavin molecule.

Microbial fermentation is a common method for synthesizing riboflavin using bacteria and fungi. Present research primarily centers on the application of three bacteria—B. subtilis, *E. coli*, and lactic acid bacteria (LAB)—as well as two yeast-like fungi, A. gossypii and *E. ashbyii*. The control of riboflavin biosynthesis, which can be accomplished by directed evolution of the fungi via chemical mutagenesis and genetic engineering, may improve the riboflavin yield for fungi like *E. ashbyii* and A. gossypii. Still, the resultant fungi should be grown in a multicomponent growth medium due to their high viscosity and lengthy fermentation process. Due to all of these unfavorable aspects, riboflavin separation and purification will become more challenging in the later stages, and production costs will rise accordingly. Additionally, in order to improve the fungi's capacity for biosynthesis, exogenous matter (unsaturated fatty acids) should be continuously added to the growth medium. In contrast, there are several benefits to using bacteria, including a quick fermentation process, an easy-to-use culture medium, and the availability of genetic engineering technologies that can be used with well-established bacteria.

8.2.2.1. Fungi

8.2.2.1.1. E. ashbyii

One of the primary strains presently employed in the industrial fermentation process for riboflavin synthesis is *E. ashbyii*. Kalingan and Krishnan showed how various carbon and nitrogen sources, as well as the initial concentrations of those sources, affected the production of riboflavin by *E. ashbyii* NRRL 1363. An ideal industrial riboflavin production process has been created based on their findings. Kolonne et al. examined the impact of pH on the exocellular riboflavin production of *E. ashbyii*. Their findings indicated that at pH 4.5 and 5.5, a high yield of riboflavin production was achievable, while at pH 3.5 and 8.5, essentially no riboflavin was produced. Pujari and Chandra presented a high-yield mutant of *E. ashbyii* DT1 generated by UV radiation-induced mutation. However, this strain's low yield and unstable genetic makeup have prevented further development for strain production. The metabolic regulation mechanism in microbial states that using mutants resistant to analogues of riboflavin and its metabolites can lessen the feedback inhibition of some important enzymes involved in the riboflavin biosynthesis pathway. This could lead to a higher accumulation of the finished goods. However, *E. ashbyii* cannot grow in inorganic salt culture medium, and finding an appropriate synthetic culture medium for its growth is also a challenge. Finding an appropriate synthetic culture, therefore, offers a good chance of screening *E. ashbyii* mutants with a high and stable riboflavin production capacity, which further utilization can improve the riboflavin production yield.

8.2.2.1.2. Candida famata

Formerly employed for industrial riboflavin production, yeast *C. flareri* is a naturally occurring overproducer of riboflavin. Even though this yeast hasn't been utilized on an industrial scale recently, a number of genetic engineering research advancements point to the possibility that it could become a new platform for riboflavin overproduction.

Yatsyshyn et al. created the plasmid pTFMN1–FAD1, which was then transformed into *Candida famata* (*C. famata*) strain T-OP 13–76, a recombinant strain with increased FMN1 gene expression. The plasmid contained the FAD1 gene, which encodes FAD synthetase, and an additional copy of the FMN1 gene, which encodes riboflavin kinase, from *Debaryomyces hansenii* (*D. hansenii*) under the control of the TEF1 promoter. One of the resulting strains, *C. famata* T-FD-FM 27, accumulated 451 $mg{\cdot}L^{-1}$ of FAD in an optimized 40-hour batch fermenter cultivation. The first report on obtaining yeast strains that overproduce FAD was this one.

In a different work, Dmytruk et al. created a strain that overproduced riboflavin by co-overexpressing the genes SEF1, RIB1, and RIB7 in a strain of *C. famata* AF-4 that produced riboflavin without reverting. The created strain, which represents one of the most well-known active riboflavin producers, accumulated up to 16.4 $g{\cdot}L^{-1}$ of riboflavin in optimized medium under fed-batch fermentation in a 7 L laboratory bioreactor. Moreover, riboflavin production doubles in this created strain upon introduction and co-expression of the two modified genes from *D. hansenii*, PRS3 (encoding Phosphoribosyl pyrophosphate (PRPP) synthetase) and ADE4 (encoding PRPP amidotransferase).

8.2.2.1.3. A. gossypii

Originally isolated from sick cotton plants (*Gossypium* sp.), the filamentous hemiascomycete *A. gossypii* was later discovered to be a naturally occurring overproducer of riboflavin. The mycelium produced by this fungus under nutritional stress differentiates into sporangia, which produce riboflavin concurrently with sporangia spore generation. The sporangia's function is likely to shield the hyaline and UV-sensitive spores. The previous multi-step chemical riboflavin production process was replaced in 1990 when high specific productivity was attained through the establishment of biotechnical riboflavin production using *A. gossypii*, following decades of strain improvements. For a number of reasons, a gossypii has already been used for more than 20 years in the industrial riboflavin production process. It has the capacity to naturally produce riboflavin, and it is highly regarded in science due to its extensive use as a model organism in studies on the biology of fungal development and evolution, as well as its potential for biotechnology. The development of molecular and in silico tools for *A. gossypii* high-level genetic engineering has been aided by the productive outcomes of basic and industrial research conducted over the past few years. This has allowed for the logical exploitation of the plant's full biotechnological potential.

In recent years, there has been a notable progress in the utilization of *A. gossypii* for biotechnological purposes other than riboflavin synthesis. Despite this, many researchers continue to focus on developing high riboflavin-producing *A. gossypii* strains and optimizing the riboflavin production process using this strain. Most of their current research is done at the molecular level. Schwechheimer et al. calculated carbon fluxes in the riboflavin-producing fungus *A. gossypii* B_2 using ^{13}C isotope experiments. Vegetable oil was utilized as the raw material in this study, and the production process was carried out under intricate industrial cultivation conditions. While glycine was only utilized as a two-carbon donor of the pyrimidine ring of riboflavin, formate and serine were found to be one-carbon donors and to have beneficial effects on riboflavin biosynthesis, based on the ^{13}C labeling data. Furthermore, during the first stage of riboflavin synthesis, a significant bottleneck of the one-carbon metabolism within the cells was seen. This bottleneck could be overcome by the accumulation of intrinsic formate in the subsequent stages of production. Furthermore, a 45% increase in riboflavin titer could result from the time-resolved supplementation of minute amounts of formate and serine to increase their intracellular availability.

In later work, Schwechheimer et al. have successfully achieved the first quantitative calculation of carbon fluxes in the riboflavin biosynthetic phase in an industrial riboflavin production setup, as well as in the growth phase of *A. gossypii* B_2 on rapeseed oil and complex medium. The ensuing data demonstrated that the primary carbon source for strain growth and one that significantly influenced riboflavin biosynthesis was yeast extract, a component commonly found in industrial media. Nonetheless, the primary carbon source for riboflavin synthesis was found to be rapeseed oil, which consequently has the biggest influence on riboflavin synthesis. These findings demonstrated the significance of a carefully chosen carbon source for the synthesis of riboflavin. New insights into the metabolism of *A. gossypii* B_2 under demanding industrial cultivation conditions were provided by the study, and these insights were helpful for strain improvement and process optimization in the future. *A. gossypii* wild strain (ATCC 10895) and an *A. gossypii* mutant strain (W122032, a riboflavin overproducer obtained from *A. gossypii* ATCC 10895 by disparity mutagenesis) were subjected to ^{13}C-metabolic flux analysis (MFA) in order to characterize the riboflavin biosynthetic pathway. Differences in their central carbon pathway were analyzed. It was demonstrated that the mutant strain's metabolic flux into pentose-5-phosphate through the pentose phosphate pathway (PPP) was 9% higher than that of the wild strain, and that the mutant strain's metabolic flux into riboflavin through the purine synthetic pathway (PSP) was 1.6%, 16 times higher than that of the wild strain (0.1 percent). These findings showed that the increase of GTP flux through the PPP and PSP enhanced the production of riboflavin in the mutant strain of *A. gossypii*. They also suggested that one way to enhance riboflavin production is through expression reinforcement of the genes associated with the PPP and PSP. Silva et al. studied the impact of blocking the de novo pyrimidine biosynthesis pathway in *A. gossypii* Agura, a recently created uridine/uracil auxotrophic strain. It was discovered that this strain produced more riboflavin on standard solid complex medium and that additional uridine/uracil supplementation inhibited riboflavin production. Additionally, it was discovered that high uracil concentration inhibited the growth of both this strain and its parent strain (*A. gossypii* ATCC 10895), whereas excess uridine accelerated growth. This was the first time that the effects of genetic alterations in the pyrimidine biosynthesis pathway on *A. gossypii* ability to produce riboflavin were described using experimental data.

As a riboflavin overproducer by nature, *A. gossypii* needs to maintain a robust flow through the guanine nucleotide pathway to boost the bioavailability of GTP, a limiting precursor of riboflavin. To further enhance riboflavin production in this fungal species, a thorough characterization of the essential enzymes in the GTP biosynthesis pathway is required. The enzyme inosine-5′-monophosphate dehydrogenase (IMPDH) catalyzes the oxidative transformation of inosine-5′-monophosphate (IMP) into xanthosine-5′-monophosphate (XMP), a reaction step that is rate-limiting in the purine biosynthetic pathway. IMP is the common precursor of adenine and guanine nucleotides. Finally, the resulting XMP is transformed into GTP through a series of steps. As a result, IMPDH is a crucial purine pathway metabolic bottleneck. Furthermore, this enzyme is technically amenable to additional genetic modification through metabolic engineering. The IMPDH enzyme from the fungus *A. gossypii* underwent a thorough functional and structural characterization by Buey et al. The results demonstrated that overexpression of the IMPDH gene enhanced the metabolic flux through the guanine biosynthetic pathway in *A. gossypii* ATCC 10895, ultimately leading to a 40% increase in riboflavin production. The creation of a metabolic engineering toolbox intended to boost riboflavin production in *A. gossypii* was greatly aided by this study. Ledesma-Amaro et al. determined that the low transcription levels of the rib genes and the bioavailability of GTP substrate were the main causes of the inhibition of riboflavin production in *A. gossypii* by examining the relative contributions of each rib gene to riboflavin production in that organism. Moreover, simultaneous overexpression of the rib genes (up to five) resulted in a notable increase in the yield of riboflavin production. Additionally, it was discovered that riboflavin production was positively impacted by the inactivation and underexpression of the ADE12 gene, which encodes adenylosuccinate synthase and regulates the first step of the adenosine-5′-monophosphate (AMP) branch. Additionally, through metabolic engineering, an Ashbya strain (A330) was created that combined the overexpression of all rib genes with the underexpression of the ADE12 gene. This strain produced a total yield of 523 $mg \cdot L^{-1}$ of riboflavin, which was 5.4 times higher than that of the strain *A. gossypii* ATCC 10895. This study offered a scalable and controlled method to boost riboflavin production in the industrial setting.

Disparity mutagenesis is a novel genetic mutation technique that was utilized by Park et al. to obtain a mutant strain (W122032) of *A. gossypii* that had increased productivity of riboflavin. This strain produced 13.7 $g \cdot L^{-1}$ of riboflavin in a 3 L. fermenter in an optimized medium; this yield is nine times higher than that of the wild strain (A gossypii ATTC 10895). Genes involved in the purine and riboflavin biosynthesis pathways were found to be upregulated, while pathways linked to the assimilation of carbon sources, energy production, and glycolysis were found to be downregulated, as revealed by proteomic and DNA microarray analysis. These findings suggest that a change in carbon flux from β-oxidation to the riboflavin biosynthesis pathway is responsible for the enhanced riboflavin production in the produced mutant.

8.2.2.2. Bacteria

Numerous studies have been conducted on the biosynthesis of riboflavin in both Gram-positive and Gram-negative bacteria, with a focus on *E. coli* and *B. subtilis*. The so-called riboflavin biosynthesis operon (rib operon), which is made up of five genes (ribGBAHT) that encode the riboflavin biosynthesis enzymes, regulates the production of riboflavin in riboflavin-producing bacteria (Figure 8.7). The gene order of the rib operon is different from the enzymatic reactions in the riboflavin biosynthetic pathway (Figure 8.6). Three rib genes, ribA and ribB, respectively, encode two different enzymes: GTP cyclohydrolase II and 3,4-dihydroxyphosphobutanone synthase, which functions as a rate-limiting enzyme. The first enzymatic reaction is catalyzed by the enzyme's GTP cyclohydrolase II activity. The first rib gene that codes for a second bifunctional enzyme is called ribG. The second and third enzymatic reactions are catalyzed successively by the riboflavin-specific deaminase/reductase. The penultimate enzymatic reaction is catalyzed by lumazine synthase, also known as riboflavin synthase β subunit, which is encoded by the operon's final gene, ribH. The operon's second gene, ribB, codes for the α subunit of riboflavin synthase, which catalyzes the final enzymatic reaction. Additionally, the operon includes the gene ribT, whose purpose is still unknown. It was shown that inactivating this gene would result in a marked reduction in the riboflavin yield rather than riboflavin auxotrophy, despite the fact that it was annotated to encode a putative N-acetyltransferase. The gene ribT in *B. subtilis* belongs to the general control nonderepressible-5 (GCN5)-related N-acetyltransferase (GNAT) superfamily, whose members catalyze the transfer of an acetyl group from acetyl coenzyme A (acetyl-CoA) to a range of substrates, according to a recent structural study.

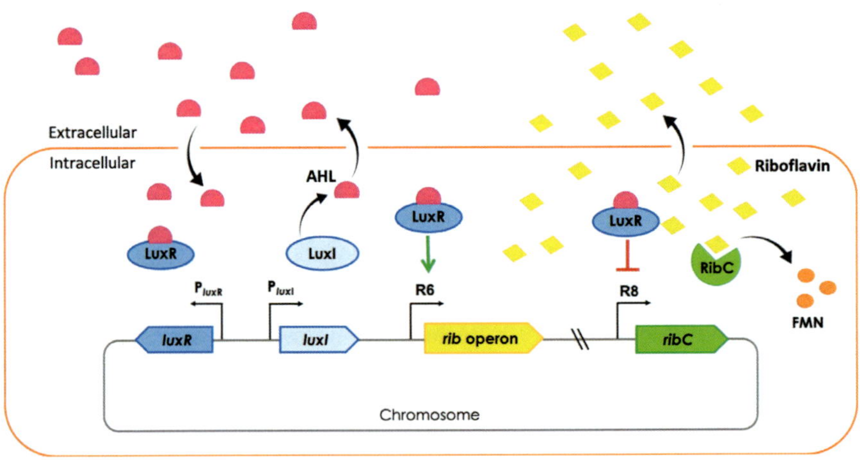

Figure 8.7. *Engineered B. subtilis carrying an autoinduction device.*

Source: Silva BF, Corrêa GG, Zocca VFB, Picheli FP, Lins MRCR, Pedrolli DB. (2023). Expanding the functionality of an autoinduction device for repression of gene expression in Bacillus subtilis. International Journal of Molecular Sciences, 24(1), 84. https://doi.org/10.3390/ijms24010084.

In *B. subtilis*, ribC, ribR, and ribO are the three regulatory regions that are located upstream of the rib operon. These three regions are thought to be presumed repressors of riboflavin biosynthesis because mutation in any one of them causes an excess of riboflavin to be produced. Among these, ribC and ribR encode a monofunctional flavokinase and a bifunctional flavokinase/FAD-synthetase, respectively. These two enzymes play a part in the sequential formation of FMN and FAD from riboflavin. The noncoding region known as ribO is situated between the ribP1 promoter and the gene ribG. The transcriptional product of ribG can be folded into an FMN riboswitch, a conserved RNA secondary structure consisting of a base stem and four hairpins. The regulation of the rib operon is largely dependent on the interaction between flavons and the FMN riboswitch. While under repression conditions, FMN binding with the FMN riboswitch results in the formation of a transcription terminator, which ends transcription, in the relieving state, FMN concentration is too low and prevents it from binding to the FMN riboswitch, forming an anti-terminator to guarantee the ready read-through of the subsequent sequence.

The rib operon additionally has three promoters: two additional internal promoters, P2 and P3 (ribP2 and ribP3), and a major promoter, P1 (ribP1). The first structure gene, ribG, has the ribP1 promoter upstream, while the ribP2 promoter is situated in the distal part of the ribB gene. Furthermore, the ribH and ribT genes are separated by the ribP3 promoter, which is located in between them. The ribP1 and regulatory region situated at the 5′ termini of the rib operon primarily regulates the transcription of the four rib genes (ribGBAH), while the ribP2 promoter and the RFN regulatory region are responsible for the transcription of the final two genes, ribA, and ribH.

8.2.2.2.1. C. acetobutylicum

For over a century, the completely anaerobic, Gram-positive bacterium *C. acetobutylicum* has been widely employed in the fermentation process to produce biobutanol. The average mass ratio at which this bacterium synthesizes ethanol, acetone, and butanol is 6:3:1. Since the 1940s, *C. acetobutylicum* has been known for its primary function as a butanol overproducer as well as for being a natural source of riboflavin, as evidenced by the discovery of some riboflavin during the butanol–acetone–ethanol (ABE) fermentation. The economic value of the ABE fermentation process would be greatly enhanced if the production yield of riboflavin, a valuable by-product or second product, could be achieved on an industrial scale at a rate of 0.5–1 $g \cdot L^{-1}$. The amount of riboflavin produced by *C. acetobutylicum* during the ABE fermentation process is still insufficient, despite decades of research. Recent work by Zhao et al. demonstrated that adding additional sodium acetate (NaAc) to the bacterial growth medium containing xylose as a carbon source significantly increased the production of riboflavin during the ABE fermentation by *C. acetobutylicum* ATCC 824. The riboflavin biosynthesis rate (ribA flux rate) increased to nearly

10 times the control level after supplementing with 60 mmol·L^{-1} NaAc. Initially undetectable, the final concentration of riboflavin increased to 0.2 g·L^{-1} (0.53 mmol·L^{-1}). This work offered a novel method for using *C. acetobutylicum* to co-produce riboflavin and biobutanol, increasing the commercial value of the ABE fermentation process.

8.2.2.2.2. E. coli

The genome of the Gram-negative bacterium *E. coli* is well-characterized, and a wide range of sophisticated molecular tools suitable for its genetic manipulation are established. It is a common workhorse in basic biological research. Furthermore, because of its many benefits, this bacterium has long been employed as a common host for the effective synthesis of a wide range of materials, including bulk chemicals, biofuels, and amino acids. For instance, *E. coli* can easily grow in culture medium made up of inexpensive components and grows quickly under ideal culture conditions, doubling in 20 minutes. Despite the fact that wild-type *E. coli* cannot accumulate riboflavin in the wild, these appealing features made *E. coli* a viable and efficient host for the production of riboflavin. Numerous metabolic engineering techniques were used in a series of studies to transform *E. coli* into a strain that can produce riboflavin.

Presently, K-12 or B. strains that are not pathogenic are the source of nearly all laboratory strains of *E. coli*. Comparative genomic analysis revealed a high degree of similarity and close relationship between strains B. and K-12, despite the discovery of numerous cellular metabolic and physiological differences between them. B. strains grow more quickly in minimal media and express more recombinant proteins than K-12 strains do, but they produce less acetate. The B. strain that has been engineered specifically, *E. coli* BL21 (DE3), is the most commonly used strain in research and industrial applications. Improved utilization of K-12 and B. strains in metabolic engineering will result from a clearer understanding of the distinctions in their regulative mechanism and metabolism.

For the first time, riboflavin accumulation in *E. coli* BL21 (DE3) was demonstrated by Wang et al. in standard culture conditions. After comparing the sequences of the enzymes involved in riboflavin biosynthesis between BL21 (DE3) and K-12 MG1655, a single site mutation on the 115 residues of the ribF gene was found in BL21 (DE3). This His115Leu mutation caused an insufficient activity of the bifunctional riboflavin kinase/FMN adenylyltransferase encoded by ribF gene, which may have contributed to the riboflavin accumulation in this strain. Nevertheless, additional quantitative polymerase chain reaction (PCR) analysis revealed upregulation of all riboflavin biosynthesis genes, indicating that this explanation was more likely to be the cause of the excessive riboflavin accumulation. This mechanism may upregulate all riboflavin biosynthesis genes. According to this research, *E. coli* BL21 (DE3) could be employed to produce riboflavin as well.

It has been previously reported that altering a few important genes in *B. subtilis*' purine biosynthesis pathway may increase the availability of riboflavin precursors, thereby promoting riboflavin biosynthesis. Similar metabolic engineering studies have also been carried out with *E. coli* in recent years. In a riboflavin-producing *E. coli* RF01S engineered from wild-type E coli MG1655, Xu et al. modified the ribB gene and five important purine pathway genes (ndk, gmk, purA, purF, and prs) in six steps. The final strain, RF18S, which combined overexpression of all the six genes, produced 387.6 $mg{\cdot}L^{-1}$ riboflavin in shake-flask fermentation, with a riboflavin production yield of 44.8 $mg{\cdot}g^{-1}$ glucose. The strain RF18S yielded a 55.6% higher yield and a 72.2% higher riboflavin titer than the strain RF01S. This study showed that an effective method to greatly increase riboflavin production in *E. coli* is to concurrently modify the DHBP synthase and GTP biosynthetic pathway through rational metabolic engineering.

The strain RF01S was able to accumulate 229.1 $mg{\cdot}L^{-1}$ riboflavin after Lin et al. used the metabolic engineering method to create a plasmid p20C-EC10 that contained a riboflavin operon controlled by the inducible trc promoter (Ptrc) and was introduced into wild-type *E. coli* MG1655. Then, strain RF01S was subjected to a stepwise deletion of the genes pgi (encoding glucose-6-phosphate isomerase), Edd (encoding phosphogluconate dehydratase), and Eda (encoding multifunctional 2-keto-3-deoxygluconate-6-phosphate aldolase/2-keto-4-hydroxyglutarate aldolase/oxaloacetate decarboxylase), followed by the insertion of the Acs (encoding acetyl-CoA synthetase) promoter in strain RF01S. The resultant strain, RF05S, displayed a riboflavin titer of 585 $mg{\cdot}L^{-1}$. To decrease the transformation of riboflavin into FMN, the expression of the ribF gene, which codes for a bifunctional riboflavin kinase/FMN adenylyltransferase, was finally modulated in strain RF05S. As a result, the final engineered strain, RF05S-M40, was able to produce 1036 $mg{\cdot}L^{-1}$ of riboflavin in Luria–Bertani (LB) medium at 37°C. After that, the fermentation conditions were optimized, and strain RF05S-M40 produced 2703 $mg{\cdot}L^{-1}$ of riboflavin in shake-flask fermentation at the ideal conditions. This was almost 12 times more than strain RF01S, yielding 136 mg of riboflavin per gram of glucose. *E. coli* strain RF05S produced the most riboflavin during shake-flask fermentation of any strain that has been reported to be able to produce the substance. Later work from the same lab successfully created a novel strain of *E. coli* that produced riboflavin and was engineered from the wild-type MG1655. To increase the carbon flux through PPP, the genes pfka (encoding 6-phosphogluconolactonase I), Edd, and Eda were deleted in this strain. This newly constructed riboflavin operon expression plasmid, pLS01, was found to be more stable than p20C-EC10 and was present in the engineered strain LS02T. This strain produced 667 mg·L^{-1} of riboflavin in maximum sustainable yield (MSY) medium with 10 g glucose per liter in shake-flask fermentation, and 56.8 mg·g^{-1} glucose was produced in fed-batch fermentation, yielding a riboflavin titer of 10.4 g·L^{-1}. For the first time, an engineered strain of *E. coli* that could accumulate more than 10 g of riboflavin per liter of culture medium through fed-batch processing was reported in this work. The outcomes of the two studies mentioned above showed that *E. coli* is a potentially effective species for producing riboflavin.

8.2.2.2.3. B. subtilis

Among all the Gram-positive bacteria, *B. subtilis* is a common and naturally occurring bacterium that is one of the best characterized species. This bacterium has been granted qualified presumption of safety (QPS) status by European Food Safety Authority (EFSA) for certain applications in animal feeds and in the production of foods and supplements consumed by humans, such as vitamin K2, and is therefore regarded as a safe and stable producer. Furthermore, *B. subtilis* was effectively engineered as a riboflavin-producing cell factory based on the thorough characterization of the enzymes involved in the riboflavin biosynthetic pathway at the biochemical, physiological, and genetic levels. As a result, *B. subtilis* has long been the preferred species for the industrial bacterial production of riboflavin and is currently the most significant one.

Metabolic engineering has been used in many studies over the past 20 years to increase *B. subtilis*'s capacity to produce riboflavin. The majority of these studies concentrated on the supply of precursors and the control of key bottleneck enzyme expression, which were thought to be the two main factors limiting the production of riboflavin.

In the oxidative branch of PPP, glucose-6-phosphate dehydrogenase (G6PD) and 6-phosphogluconate dehydrogenase (6GPD), two consecutive oxidized forms of nicotinamide adenine dinucleotide phosphate (NADP$^+$)-dependent enzymes, catalyze reactions that yield Ru5P, the first precursor of riboflavin. Ru5P, however, and a few other intracellular metabolites, including fructose-1,6-biphosphate (FBP) and the reduced form of nicotinamide adenine dinucleotide phosphate (NADPH), inhibit the activities of these two enzymes. Intracellular metabolites' allosteric inhibition may result in an inadequate supply of precursors for the synthesis of riboflavin. The genes zwf (encodes G6PD) and gnd (encodes 6GPD) from *Corynebacterium glutamicum* ATCC 13032 were successfully cloned by Wang et al. in order to eliminate their feedback inhibition by site-directed mutagenesis, thereby increasing the availability of Ru5P through genetic modifications. To measure the impact of the two mutant genes' expressions on riboflavin production, *B. subtilis* RH33 was used to express the genes separately or in combination. The resulting data demonstrated that metabolites from the riboflavin biosynthetic pathway were strongly increased in the engineered strains Ru5P (up 46%), DMRL, aminoimidazole, and intracellular riboflavin, in comparison to the parent strain RH33. Individual expression of the genes zwf (an engineered strain of *B. subtilis* SVZ) and gnd (an engineered strain of *B. subtilis* SVG) in shake-flask fermentation led to an approximate increase in riboflavin production of 18% and 22%, respectively. Furthermore, the co-expression of the two genes (in the engineered strain known as *B. subtilis* VGZ) increased the production of riboflavin by 31%. Additionally, riboflavin production improved by an average of 39% when *B. subtilis* VGZ was used in subsequent fed-batch fermentation. This study showed that the primary factor limiting riboflavin production in *B. subtilis* was precursor supply, and metabolic engineering was a useful technique for rerouting carbon flux toward PPP.

Shi et al. genetically modified the rib operon in the riboflavin-producing *B. subtilis* strain BS77 to increase the amount of carbon flux through the riboflavin biosynthetic pathway. This involved overexpressing the ribA gene, replacing the native promoter ribP1 with the strong promoter P43 derived from *B. subtilis* 168, and deleting the RFN regulatory region ribO. The strain BS89, which overexpressed the ribA gene, produced the highest yield of riboflavin (506 $mg{\cdot}L^{-1}$), 1.4-fold more than that of BS77 (210 $mg{\cdot}L^{-1}$). The purine biosynthesis pathway in the rib operon of strain BS89 was then deregulated using a sequential optimization strategy. This included: 1) removing transcription repression by deleting the pur operon repressor PurR (encoded by the purR gene) and the 5′-untranslated region (UTR) of the pur operon using a guanine-sensing riboswitch; 2) removing the product-feedback inhibition of PRPP aminotransferase (encoded by the purF gene) utilizing site-directed mutagenesis. The purine biosynthetic pathway's metabolic flux was successfully achieved with the help of these genetic modifications, leading to the highest riboflavin production yield of 827 $mg{\cdot}L^{-1}$ in the engineered stain with purF-VQW mutation (BS110) in shake-flask fermentation. This study showed that a workable method for increasing metabolite yield is the logical deregulation of the purine biosynthesis pathway in *B. subtilis* for the removal of both transcription repression and product-feedback inhibition.

To improve the production of metabolites of interest, sensible metabolic engineering strategies necessitate a deeper understanding of the central carbon metabolism. One of the most significant routes in central carbon metabolism is gluconeogenesis (GNG), a biological process that produces glucose from non-sugar carbon substrates. In *Corynebacterium glutamicum*, its alteration has been effectively used to redirect the carbon flux toward the PPP. Furthermore, earlier research has shown that it may be possible to further increase riboflavin production by downregulating the important enzymes involved in the gluconeogenic pathway.

Using overexpression of the three essential gluconeogenic genes in parental strain *B. subtilis* RH33—gapB (encodes NADPH-dependent glyceraldehyde 3-phosphate dehydrogenase), pckA (encodes phosphoenolpyruvate carboxykinase), and fbp (encodes fructose-1,6-biphosphatase—Wang et al. conducted a systematic study to examine the effects of GNG deregulation on the improvement of riboflavin production. The findings demonstrated that co-overexpression of the genes gapB and fbp resulted in increases of 21.9% and 27.8%, respectively, in riboflavin production in shake-flask and fed-batch fermentation, as well as a significant increase in riboflavin titers up to 4.89 $g{\cdot}L^{-1}$ in shake-flash fermentation compared to the strain RH33. According to this study, using glucose as a carbon source, deregulating gluconeogenesis is a useful tactic to increase the production of riboflavin and other metabolites directly from PPP and NADPH-dependent compounds.

The impact of ribA and ribH gene modifications on *B. subtilis* riboflavin production was investigated by Zhang et al. The outcome of a 60-hour shake-flask fermentation demonstrated that individual overexpression of the gene ribA in the engineered strain LXZ-2 increased riboflavin production by 99% when compared to the parent strain LXZ-1, with a riboflavin titer of 0.47 $g{\cdot}L^{-1}$. However, mass spectrometry analysis revealed a 30% reduction in biomass and cell autolysis due to the accumulation of 5-amino-6-(1-D-ribitylamino) uracil. When the genes ribA and ribH were co-overexpressed in the modified strain LXZ-3, riboflavin production rose by 280% and 91%, respectively, in comparison to strains LXZ-1 and LXZ-2. Additionally, there was no discernible reduction in biomass or cell autolysis. Then, strain LXZ-3 was created by transforming a low copy-number plasmid containing the entire rib operon (pMX45). The resulting strain LXZ-3/pMX45 was then utilized to investigate the effects of various carbon sources, such as glucose, sucrose, xylose, and a mixture of xylose and sucrose in varying ratios, on the production of riboflavin. The highest riboflavin titer (1.6 $g{\cdot}L^{-1}$) was found in a 60-hour shake-flask fermentation when a mixture of sucrose and xylose in a ratio of 1.5/6.5 (total addition of sugar, 8 percent, w/v) was used as carbon source. During large-scale fermentation carried out in a 5 L. fermenter, the highest riboflavin titer of 3.6 $g{\cdot}L^{-1}$ was obtained from the use of the same mixed carbon source after 70 h incubation. This work showed that co-metabolism with xylose and sucrose can improve the supply of precursors, which in turn improves the synthesis of riboflavin.

One streptothricin antibiotic that can prevent protein synthesis is norseothricin (NTC). As a result, it is frequently employed for screening plant, fungus, bacterial, and yeast cell resistance. By inserting the gene sat, which codes for streptothricin acetyltransferase, into the expression plasmid pMA5, Cheng et al. created a novel NTC-resistant plasmid known as pMA5-sat. The zwf gene was then cloned and put into the plasmid pMA5-sat. To investigate the effects of overexpressing the gene zwf on riboflavin production, the created plasmid pMA5-sat–zwf was subsequently inserted into a previously modified riboflavin-producing *B. subtilis* strain RF1. The findings showed that the recombinant *B. subtilis* strain RF1-pMA5-sat–zwf had 50 times more G6DP activity than the strain RF1, indicating that this enzyme had been successfully overexpressed. Furthermore, in a large-scale fermentation carried out in a 5 L. fermenter, a final riboflavin titer of 12.01 $g \cdot L^{-1}$ was obtained with this recombinant strain, which increased by 30.3% in comparison to that of the strain RF1. These findings suggested the possible uses of the recently created NTC-resistant plasmid pMA5-sat in modifying the genetic makeup of *B. subtilis* strains that produce riboflavin.

Furthermore, RibM from *Streptomyces davawensis* (*S. davawensis*) is a functional, energy-independent riboflavin transporter in *B. subtilis* that primarily catalyzes riboflavin export, according to a report by Hemberger et al. Through the integration of the *S. davawensis* gene ribM into a high-performance *B. subtilis* riboflavin production strain, the recombinant RibM overproducing strain was able to significantly increase riboflavin synthesis, contingent on the concentration of the inducer, isopropyl β-D-thiogalactoside (IPTG). This was the first instance of riboflavin production optimization in *B. subtilis* based on successful riboflavin export, indicating that increasing riboflavin excretion is another helpful tactic to boost riboflavin yield in *B. subtilis*.

To enhance riboflavin production in *B. subtilis*, fermentation condition optimization has also been suggested as a feasible approach, in addition to metabolic engineering.

One of the most crucial factors in the fermentation process is the amount of oxygen dissolved in the culture medium, which is directly linked to cell growth and product formation. Therefore, its availability is essential for *B. subtilis*'s riboflavin synthesis, which is readily increased by altering the agitation speed. Man et al. looked into how agitation speed affected the amount of riboflavin produced in fed-batch fermentation using recombinantly engineered *B. subtilis* RF1. Kinetic analysis results indicated that while a higher agitation speed of 900 $r \cdot min^{-1}$ was beneficial in the later phase of the fermentation process, a lower agitation speed (600 $r \cdot min^{-1}$) in the initial phase of the process facilitated cell growth and riboflavin production. Based on these findings, a two-stage control strategy was created, with the agitation speed being set at 600 r·min^{-1} for the first 26 hours and 900 r·min^{-1} for the remaining time, to maintain high cell growth and riboflavin production throughout the fermentation process. The sudden change in agitation speed in the two-stage control strategy, however, had a negative effect on cell growth and riboflavin production within a short period of time. As a result, a strategy of gradually increasing agitation speed from 600 to 900 r·min^{-1} was established, with the maximum riboflavin titer reaching 9.4 g·L^{-1} in a 48-hour fermentation and a production yield of 0.051 g per gram of glucose. There was a 20.5% increase in riboflavin titer and a 21.4% increase in production yield when compared to the best result from a fed-batch fermentation with a single-agitation speed (600 r·min^{-1}).

To find a riboflavin production accelerator, Wan et al. examined the effects of separately adding seven additives—calcium gluconate, citric acid, sodium citrate, calcium chloride, alanine, malic acid, and fructose-1,6-diphosphate (FDP)—on the ability of *B. subtilis* to produce riboflavin during shake-flask fermentation. To assess the effects, two indices were used: the riboflavin production yield and the bacterial density optical density $(OD)_{600}$. The findings showed that the addition of calcium gluconate, sodium citrate, and alanine significantly increased the production of riboflavin. Thus, these three additives were chosen for additional investigation using orthogonal analysis to examine their combined effects on riboflavin production. The findings indicated that the highest riboflavin titer of 6.46 g·L^{-1} was obtained under the ideal concentration ratios of calcium gluconate, sodium citrate, and alanine, which were 7.5, 5.0, and 1.5 g·L^{-1}, respectively. This represents an approximate increase of 40% over the titer obtained without the addition of additives.

In a recent work, Oraei et al. evaluated the effects of 13 different minerals ($CaCl_2$, CuCl, $FeCl_3$, $FeSO_4$, $AlCl_3$, Na_3MoO_4, $Co(NO_3)_2$, NaCl, KH_2PO_4, K_2HPO_4, $MgSO_4$, $ZnSO_4$, and $MnSO_4$) on riboflavin production by wild-type *B. subtilis* ATCC 6051 in shake-flask fermentation, to develop an appropriate fermentation medium for improving riboflavin production. Plackett–Burman (PB) design was first carried out to screen minerals with significant influence on riboflavin production, with the results revealing that the concentration of three minerals showed the greatest influence, that is, $MgSO_4$, K_2HPO_4, and $FeSO_4$. Subsequently, an optimization test was performed using response surface methodology (RSM) to determine the optimal concentrations of the five selected medium components, including two carbon sources used in all treatments, fructose and yeast extract, and three minerals, $MgSO_4$, K_2HPO_4, and $FeSO_4$. The highest riboflavin titer in a 72-h shake-flask fermentation was obtained (11.73 g·L^{-1}) when the concentration of these five components was 38.10, 4.37, 0.85, 2.27, and 0.02 g·L^{-1}, respectively.

8.2.2.2.4. Lactic Acid Bacteria

A diverse collection of Gram-positive, non-sporulating, non-respiring, and typically non-motile rods or cocci make up LAB. They are closely related to human life and are the primary or only byproduct of the fermentation of carbohydrates, producing lactic acid. The formation of antimicrobial compounds like bacteriocins by LAB as well as the acidic environment they create inhibit the growth of many pathogenic and spoilage bacteria in fermented foods. These microorganisms have been empirically employed as fermenting agents to enhance food bio-preservation since ancient times. These days, lactic starter cultures, or LABs, are widely used in the food industry to produce a wide range of fermented dairy products, including cheese and yogurt, pickles (fermented vegetables), and sausages.

Even though LAB typically require external sources for various vitamins due to their auxotrophic nature, it has become widely recognized that certain LAB species possess the capability to manufacture certain B-group vitamins, including folates (vitamin B_9), riboflavin (vitamin B_2), and cobalamin (vitamin B_{12}). The vitamin production capability of LAB can vary significantly among different strains, often being dependent on the specific LAB species or strain, which is often linked to disruptions (partial or complete) in genetic information necessary for vitamin synthesis. For example, Thakur et al conducted a PCR-based screening for riboflavin biosynthesis genes in 60 potential riboflavin-producing Lactobacilli strains. Out of these, only 14 strains were able to thrive in a commercial riboflavin-free medium. Moreover, the presence of the rib operon was found to be strain-specific among various Lactobacilli species. Additionally, a fragmented or unfinished ribosomal operon in LAB often correlates with the incapacity to produce riboflavin. According to research conducted by Kleerebezem et al. and Burgess et al., the *Lactobacillus plantarum* (*L. plantarum*) strain WCFS1 possesses an incomplete ribosomal operon, with the entire ribG gene and a portion of the ribB gene missing. As predicted, this strain failed to thrive in conditions devoid of riboflavin. Moreover, it was revealed that various wild variants of *L. plantarum* carry the entire ribosomal operon and are capable of synthesizing riboflavin, including *L. plantarum* strain RYG-GYY-9049 sourced from traditional Chinese pickle brines, *L. plantarum* strain NCDO 1752 derived from pickled cabbage, and *L. plantarum* strains UNIFG1 and UNIFG2 obtained from natural sourdoughs. Generally speaking, the preservation of the ribosomal operon in LAB is crucial for riboflavin production.

Riboflavin, a vitamin extensively researched due to its significance in human wellbeing and prevalent deficiency, is predominantly produced by LAB. Given its crucial role in human health and the common occurrence of its deficiency globally, mainly due to dietary inadequacies, riboflavin-fortified foods have garnered significant interest. Although riboflavin is abundant in various foods, a high incidence of deficiency is globally noted. As a result, riboflavin-fortified foods emerge as a more organic, consumer-friendly, and cost-effective substitute for chemically manufactured riboflavin, satisfying the growing consumer demand for natural products. The adaptability of LAB to industrial food fermentation and their capacity to produce riboflavin make them excellent prospects for riboflavin production in foods.

This enhanced functionality of LAB paves the way for developing innovative functional foods with heightened riboflavin content. Consequently, the examination of wild-type riboflavin-producing *Lactobacillus acidophilus* strains from various food sources has attracted the attention of numerous researchers. Previously, it was found that out of 179 strains of LAB isolated from various food items, only 42 of them were capable of growing in a commercially available riboflavin-free medium. On the basis of the concentrations of riboflavin determined by high performance liquid chromatography (HPLC), five of these strains with relatively high riboflavin producing capacity (extracellular concentrations of riboflavin in the range of 190–260 $ng \cdot mL^{-1}$) were selected and then inoculated into soymilk to evaluate both their growth and riboflavin production in this food matrix. After 12 h of fermentation at 37°C, only the strain *L. plantarum* CRL 725 showed a significant increase in the concentration of riboflavin from the initial 309 to 700 $ng \cdot mL^{-1}$. In another study, screening of wild-type riboflavin-producing LAB from raw goat milk and cheeses was performed, and a total of 179 strains of LAB were isolated. Among these strains, only eight were able to grow in a riboflavin-free medium, and produced riboflavin within a total concentration range of 173–532 $ng \cdot mL^{-1}$.

Nevertheless, since wild-type LAB strains screened from foods or other sources are not naturally optimized for maximal production rates of some specific bioactive compounds, like riboflavin, the riboflavin titer and productivity obtained from fermentations using these strains were relatively low. Therefore, increasing these strains' riboflavin-producing ability to attain riboflavin overproduction remains a challenge. Two concurrent approaches have been frequently employed for this purpose: the development of recombinant strains based on metabolic engineering and the screening of strains resistant to roseoflavin.

Roseoflavin is a naturally occurring toxic analog of riboflavin and FMN that can bind directly to the FMN riboswitch and inhibit rib operon transcription. This antibacterial substance can induce spontaneous mutations in LAB strains that produce riboflavin, leading to the development of riboflavin overproducers. This strategy has been successfully used to select natural riboflavin-overproducing mutants of food-grade strains of *B. subtilis*, *Lactococcus lactis* (*L. lactis*), *L. plantarum*, *Leuconostoc mesenteroides* and *Propionibacterium freudenreichii* (*P. freudenreichii*), and has proven to be reliable.

The wild-type *P. freudenreichii* NIZO B374, which generates more riboflavin than the parent strain, gives rise to the spontaneously roseoflavin-resistant mutant strain *P. freudenreichii* NIZO B2336. These two strains were added 24 hours before the Campina MUH306 yogurt starter culture was added in order to test them as adjunct cultures for yogurt fermentation. The findings demonstrated that the final riboflavin content of the yogurt made with *P. freudenreichii* NIZO B2336 as an adjunct culture was 19.7 $\mu g \cdot g^{-1}$, exceeding both the yogurt made with *P. freudenreichii* NIZO B374 (10.5 $\mu g \cdot g^{-1}$) and the yogurt made without the addition of an adjunct culture (12.9 $\mu g \cdot g^{-1}$). According to this study, *P. freudenreichii* NIZO B2336 may be utilized to create innovative fermented milk products with high riboflavin content, which could have further health advantages.

After screening a strain of *L. plantarum* CRL 725, Juarez del Valle et al. were able to double the initial riboflavin concentration in soymilk. Additional screening of these strains' mutants resistant to roseoflavin was done. One of the mutants, designated CRL 2130, was discovered to have increased the initial riboflavin concentration in soymilk six times (from 309 to 1860 $ng \cdot mL^{-1}$). Therefore, it was shown that strains resistant to roseoflavin could be used to create new soy products enhanced with riboflavin that have higher nutritional value.

Furthermore, screening of roseoflavin-resistant mutants of the wild-type *Lactobacillus fermentum* strain (designated *L. fermentum* PBCC11) was carried out after Russo et al. isolated the strain from sourdough. 15 mutants in all were obtained, with seven of them producing more riboflavin than 1 $mg \cdot L^{-1}$. To fortify bread, the best riboflavin-overproducing mutant, L. fermentum PBCC11.5, was chosen and co-inoculated with commercial yeast *Saccharomyces cerevisiae*, which produced the highest level of riboflavin, 1.203 $mg \cdot L^{-1}$. The bread fermented with co-inoculum of yeast and L. fermentum PBCC11.5 had a final amount of riboflavin of 6.66 $\mu g \cdot g^{-1}$, which was found to be approximately two times higher than that of conventional bread.

Our latest research effectively eliminated the native *L. plantarum* strain RYG-YYG-9049, which produced 0.734 $mg \cdot L^{-1}$ of riboflavin, from traditional Chinese pickle juices. This strain was then used to perform roseoflavin-induced spontaneous mutation, and the No. 10 mutant (designated as *L. plantarum* RYG-YYG-9049-M10) demonstrated the highest capacity to produce riboflavin. Under ideal circumstances, this mutant was utilized to ferment raw soymilk and increase its riboflavin content to 2.920 $mg \cdot L^{-1}$.

Previous reports state that single-base modifications and/or gene deletions in the upstream regulatory region of the rib operon were the most common forms of roseoflavin-induced spontaneous mutation. The upsurge in riboflavin production was likely caused by a novel mutation type that our recent study discovered, in which roseoflavin induced the insertion of a 1059 bp DNA fragment in the upstream regulatory region of the rib operon. Though more research is needed to determine the precise mechanisms, it is hypothesized that roseoflavin-mediated mutations were able to impact the stability of the terminator structure, resulting in a decreased formation of this terminator and allowing a continuous transcription of the rib operon.

Therefore, by carefully choosing LAB species and strains and optimizing fermentation conditions, it is possible to raise the riboflavin levels in fermented food products, such as yogurt, cheeses, and bread. Increased riboflavin levels in novel functional fermented foods would boost their nutritional content and benefit consumers' health.

When it comes to metabolic engineering, recombinant strains with higher riboflavin-producing capacities are typically created using this technique. To elucidate the impact of rib gene overexpression on riboflavin production, Burgess et al. conducted a genetic analysis of the rib operon in *L. lactis* ssp *cremoris* strain NZ9000. The expression vector pNZ8048 was cloned containing the four structural rib genes (ribG, ribB, ribA, and ribH) in various combinations, and the resultant strain was introduced into Lactis NZ9000. The outcomes demonstrated that simultaneous overexpression of all four rib genes (engineered *L. lactis* NZ9000 harboring the plasmid pNZBAH) was the only way to achieve a high level of riboflavin (24 $mg \cdot L^{-1}$). As a result, the riboflavin consumer strain *L. lactis* NZ9000 was effectively transformed into a riboflavin producer.

Sybesma and colleagues isolated a riboflavin-producing mutant from *L. lactis* strain NC9000 by exposing it to roseoflavin. Sequencing revealed that the mutant *L. lactis* CB010 only had a base change in the regulatory region upstream of the rib genes. The bifunctional enzyme amino-4-hydroxy-6-hydroxymethyldihydropteridine pyrophosphokinase and GTP cyclohydrolase I is encoded by the gene folKE, which was cloned into plasmid pZ8161. The resultant plasmid, pNZ7017, was then inserted into *L. lactis* CB010. It was discovered that the synthesis of folate was also increased by overexpressing the gene folKE in the riboflavin-producing mutant that was engineered (*L. lactis* CB010 harboring the plasmid pNZ7017). Thus, with the aid of metabolic engineering, it was possible to overproduce both folate and riboflavin in *L. lactis* simultaneously.

GTP cycohydrolase II (encoded by ribA) can use GTP as the common precursor for both riboflavin and folate biosynthesis, while GTP cycohydrolase I (encoded by folE) can use it for riboflavin biosynthesis. Since these two enzymes are in competition with one another for GTP molecules, inactivating the gene folE may increase the production of riboflavin due to the increased availability of GTP. The riboflavin-producing ability of *L. fermentum* MTCC 8711, a wild-type riboflavin-producing strain isolated from yogurt, was enhanced by the inactivation of the folE gene. In a chemically defined medium (CDM), the strain *Lb. fermentum* MTCC 8711 yielded 2.29 $mg \cdot L^{-1}$ of riboflavin in a 24-hour fermentation that was stable for 72 hours, while the mutant Lb. fermentum GKJFE, which was disrupted by folE, produced 3.49 $mg \cdot L^{-1}$ in a 72-hour fermentation, which was roughly 50% more than that of the parental strain.

Utilizing metabolic engineering is a dependable method for utilizing biotechnologically significant strains that have low or no riboflavin-producing capacity. However, due to consumer or regulatory concerns about genetic engineering, genetically engineered strains still cannot be used in the fermentation of food products intended for human consumption.

REFERENCES

1. Alam, A., Woo, J. S., Schmitz, J., Prinz, B., Root, K., Chen, F., Bloch, J. S., Zenobi, R., & Locher, K. P. (2016). Structural basis of transcobalamin recognition by human CD320 receptor. *Nature Communications, 7*, 12100.
2. Balabanova, L., Averianova, L., Marchenok, M., Son, O., & Tekutyeva, L. (2021). Microbial and genetic resources for cobalamin (vitamin B_{12}) biosynthesis: From ecosystems to industrial biotechnology. *International Journal of Molecular Sciences, 22*, 4522.
3. Bernhardt, C., Zhu, X., Schütz, D., Fischer, M., & Bisping, B. (2019). Cobalamin is produced by *Acetobacter pasteurianus* DSM 3509. *Applied Microbiology and Biotechnology, 103*, 3875–3885.
4. Biedendieck, R., Malten, M., Barg, H., Bunk, B., Martens, J. H., Deery, E., Leech, H., Warren, M. J., & Jahn, D. (2010). Metabolic engineering of cobalamin (vitamin B_{12}) production in *Bacillus megaterium*. *Microbial Biotechnology, 3*, 24–37.
5. Buey, R. M., Ledesma-Amaro, R., Balsera, M., de Pereda, J. M., & Revuelta, J. L. (2015). Increased riboflavin production by manipulation of inosine 5'-monophosphate dehydrogenase in *Ashbya gossypii*. *Applied Microbiology and Biotechnology, 99*(22), 9577–9589.
6. Burgess, C. M., Smid, E. J., Rutten, G., & van Sinderen, D. (2006). A general method for selection of riboflavin-overproducing food-grade microorganisms. *Microbial Cell Factories, 5*(1), 24.
7. Burgess, C., O'Connell-Motherway, M., Sybesma, W., Hugenholtz, J., & van Sinderen, D. (2004). Riboflavin production in *Lactococcus lactis*: Potential for *in situ* production of vitamin-enriched foods. *Applied and Environmental Microbiology, 70*(10), 5769–5777.
8. Calvillo, Á., Pellicer, T., Carnicer, M., & Planas, A. (2022). Bioprocess strategies for vitamin B_{12} production by microbial fermentation and its market applications. *Bioengineering, 9*(8), 365. https://doi.org/10.3390/bioengineering9080365.
9. Campbell, G. R. O., Taga, M. E., Mistry, K., Lloret, J., Anderson, P. J., Roth, J. R., & Walker, G. C. (2006). *Sinorhizobium meliloti* BluB is necessary for production of 5,6-dimethylbenzimidazole, the lower ligand of B_{12}. *Proceedings of the National Academy of Sciences of the United States of America, 103*, 4634–4639.
10. Chamlagain, B., Deptula, P., Edelmann, M., Kariluoto, S., Grattepanche, F., Lacroix, C., Varmanen, P., & Piironen, V. (2016). Effect of the lower ligand precursors on vitamin B_{12} production by food-grade propionibacteria. *LWT–Food Science and Technology, 72*, 117–124.

11. Chamlagain, B., Sugito, T. A., Deptula, P., Edelmann, M., Kariluoto, S., Varmanen, P., & Piironen, V. (2018). *In situ* production of active vitamin B_{12} in cereal matrices using *Propionibacterium freudenreichii. Food Science and Nutrition, 6*, 67–76.

12. Cheng, X., Chen, W., Peng, W., & Li, K. (2014). Improved vitamin B_{12} fermentation process by adding rotenone to regulate the metabolism of *Pseudomonas denitrificans. Applied Biochemistry and Biotechnology, 173*, 673–681.

13. Cheng, Y., Rao, Z., Yang, T., Man, Z., Xu, M., & Zhang, X. (2015). Construction of a new resistance plasmid capable of riboflavin production in *Bacillus subtilis* RF1. *Chinese Journal of Applied and Environmental Biology, 21*(3), 435–440. (in Chinese)

14. Debussche, L., Couder, M., Thibaut, D., Cameron, B., Crouzet, J., & Blanche, F. (1992). Assay, purification, and characterization of cobaltochelatase, a unique complex enzyme catalyzing cobalt insertion in hydrogenobyrinic acid a,c-diamide during coenzyme B_{12} biosynthesis in *Pseudomonas denitrificans*. *Journal of Bacteriology, 174*, 7445–7451.

15. Deptula, P., Chamlagain, B., Edelmann, M., Sangsuwan, P., Nyman, T. A., Savijoki, K., Piironen, V., & Varmanen, P. (2017). Food-like growth conditions support production of active vitamin B_{12} by *Propionibacterium freudenreichii* 2067 without DMBI, the lower ligand base, or cobalt supplementation. *Frontiers in Microbiology, 8*, 368.

16. Deptula, P., Kylli, P., Chamlagain, B., Holm, L., Kostiainen, R., Piironen, V., Savijoki, K., & Varmanen, P. (2015). BluB/CobT_2 fusion enzyme activity reveals mechanisms responsible for production of active form of vitamin B_{12} by *Propionibacterium freudenreichii. Microbial Cell Factories, 14*, 186.

17. Dmytruk, K., Lyzak, O., Yatsyshyn, V., Kluz, M., Sibirny, V., Puchalski, C., et al. (2014). Construction and fed-batch cultivation of *Candida famata* with enhanced riboflavin production. *Journal of Biotechnology, 172*, 11–17.

18. Fang, H., Kang, J., & Zhang, D. (2017). Microbial production of vitamin B_{12}: A review and future perspectives. *Microbial Cell Factories, 16*, 15.

19. Gray, M. J., & Escalante-Semerena, J. C. (2007). Single-enzyme conversion of FMNH2 to 5,6-dimethylbenzimidazole, the lower ligand of B_{12}. *Proceedings of the National Academy of Sciences of the United States of America, 104*, 2921–2926.

20. Gu, Q., Zhang, C., Song, D., Li, P., & Zhu, X. (2015). Enhancing vitamin B_{12} content in soy-yogurt by *Lactobacillus reuteri*. *International Journal of Food Microbiology, 206*, 56–59.

21. Hajfarajollah, H., Mokhtarani, B., Mortaheb, H., & Afaghi, A. (2015). Vitamin B_{12} biosynthesis over waste frying sunflower oil as a cost-effective and renewable substrate. *Journal of Food Science and Technology, 52*, 3273–3282.

22. Hardlei, T. F., Obeid, R., Herrmann, W., & Nexo, E. (2013). Cobalamin analogues in humans: A study on maternal and cord blood. *PLoS ONE, 8*, e61194.

23. Hazra, A. B., Han, A. W., Mehta, A. P., Mok, K. C., Osadchiy, V., Begley, T. P., & Taga, M. E. (2015). Anaerobic biosynthesis of the lower ligand of vitamin B_{12}. *Proceedings of the National Academy of Sciences of the United States of America, 112*, 10792–10797.

24. Hemberger, S., Pedrolli, D. B., Stolz, J., Vogl, C., Lehmann, M., & Mack, M. (2011). RibM from *Streptomyces davawensis* is a riboflavin/roseoflavin transporter and may be useful for the optimization of riboflavin production strains. *BMC Biotechnology, 11*(1), 119.

25. Hodgkin, D. C., Kamper, J., Lindsey, J., Mackay, M., Pickworth, J., Robertson, J. H., Shoemaker, C. B., & White, J. G. (1957). The structure of vitamin B_{12}. I. An outline of the crystallographic investigation of vitamin B_{12}. *Proceedings of the Royal Society of London. Series A, Mathematical and Physical Sciences, 242*, 228–263.

26. Hugenschmidt, S., Schwenninger, S. M., & Lacroix, C. (2011). Concurrent high production of natural folate and vitamin B_{12} using a co-culture process with *Lactobacillus plantarum* SM39 and *Propionibacterium freudenreichii* DF13. *Process Biochemistry, 46*, 1063–1070.

27. Jaffe, E. K. (2000). The porphobilinogen synthase family of metalloenzymes. *Acta Crystallographica Section D: Biological Crystallography, 56*, 115–128.

28. Jeong, B. Y., Wittmann, C., Kato, T., & Park, E. Y. (2015). Comparative metabolic flux analysis of an *Ashbya gossypii* wild-type strain and a high riboflavin-producing mutant strain. *Journal of Bioscience and Bioengineering, 119*(1), 101–106.

29. Jordan, P. M. (1991). The biosynthesis of 5-aminolaevulinic acid and its transformation into uroporphyrinogen III. In *New Comprehensive Biochemistry* (Vol. 19, pp. 1–66). Elsevier.

30. Juarez del Valle, M., Laiño, J. E., Savoy de Giori, G., & LeBlanc, J. G. (2014). Riboflavin producing lactic acid bacteria as a biotechnological strategy to obtain bioenriched soymilk. *Food Research International, 62*, 1015–1019.

31. Juzeniene, A., & Nizauskaite, Z. (2013). Photodegradation of cobalamins

in aqueous solutions and in human blood. *Journal of Photochemistry and Photobiology B: Biology, 122*, 7–14.

32. Kleerebezem, M., Boekhorst, J., van Kranenburg, R., Molenaar, D., Kuipers, O. P., Leer, R., et al. (2003). Complete genome sequence of *Lactobacillus plantarum* WCFS1. *Proceedings of the National Academy of Sciences of the United States of America, 100*(4), 1990–1995.
33. Kojima, I. (1985). Process for producing vitamin B_{12} by the fermentation technique, and vitamin B_{12}-producing microorganism. *U.S. Patent 4,544,633*.
34. Kolonne, S., Seviour, R. J., & McDougall, B. M. (1994). Effect of pH on exocellular riboflavin production by *Eremothecium ashbyii*. *Biotechnology Letters, 16*(1), 79–84.
35. Kośmider, A., Białas, W., Kubiak, P., Drozdzyńska, A., & Czaczyk, K. (2012). Vitamin B_{12} production from crude glycerol by *Propionibacterium freudenreichii* ssp. Shermanii: Optimization of medium composition through statistical experimental designs. *Bioresource Technology, 105*, 128–133.
36. Kusel, J., Fa, Y., & Demain, A. (1984). Betaine stimulation of vitamin B_{12} biosynthesis in *Pseudomonas denitrificans* may be mediated by an increase in activity of δ-aminolaevulinic acid synthase. *Journal of General Microbiology, 130*, 835–841.
37. Ledesma-Amaro, R., Serrano-Amatriain, C., Jiménez, A., & Revuelta, J. L. (2015). Metabolic engineering of riboflavin production in *Ashbya gossypii* through pathway optimization. *Microbial Cell Factories, 14*(1), 163.
38. Leeper, F. J. (1989). The biosynthesis of porphyrins, chlorophylls, and vitamin B_{12}. *Natural Product Reports, 6*, 171–203.
39. Li, K. T., Liu, D. H., Li, Y. L., Chu, J., Wang, Y. H., Zhuang, Y. P., & Zhang, S. L. (2008). An effective and simplified pH-stat control strategy for the industrial fermentation of vitamin B_{12} by *Pseudomonas denitrificans*. *Bioprocess and Biosystems Engineering, 31*, 605–610.
40. Li, K. T., Liu, D. H., Li, Y. L., Chu, J., Wang, Y. H., Zhuang, Y. P., & Zhang, S. L. (2008). Improved large-scale production of vitamin B_{12} by *Pseudomonas denitrificans* with betaine feeding. *Bioresource Technology, 99*, 8516–8520.
41. Li, K. T., Liu, D. H., Li, Y. L., Chu, J., Wang, Y. H., Zhuang, Y. P., & Zhang, S. L. (2008). Influence of Zn^2, Co^2, and dimethylbenzimidazole on vitamin B_{12} biosynthesis by *Pseudomonas denitrificans*. *World Journal of Microbiology and Biotechnology, 24*, 2525–2530.

42. Li, K. T., Zhou, J., Cheng, X., & Wei, S. J. (2012). Study on the dissolved oxygen control strategy in large-scale vitamin B_{12} fermentation by *Pseudomonas denitrificans*. *Journal of Chemical Technology and Biotechnology, 87*, 1648–1653.

43. Lin, Z., Xu, Z., Li, Y., Wang, Z., Chen, T., & Zhao, X. (2014). Metabolic engineering of *Escherichia coli* for the production of riboflavin. *Microbial Cell Factories, 13*(1), 104.

44. Liu, J., Liu, Y., Wu, J., Fang, H., Jin, Z., & Zhang, D. (2021). Metabolic profiling analysis of the vitamin B_{12} producer *Propionibacterium freudenreichii*. *MicrobiologyOpen, 10*, e1199.

45. Man, Z. W., Rao, Z. M., Cheng, Y. P., Yang, T. W., Zhang, X., Xu, M. J., et al. (2014). Enhanced riboflavin production by recombinant *Bacillus subtilis* RF1 through the optimization of agitation speed. *World Journal of Microbiology and Biotechnology, 30*(2), 661–667.

46. Martens, H., Barg, M., Warren, D., & Jah, J. H. (2002). Microbial production of vitamin B_{12}. *Applied Microbiology and Biotechnology, 58*, 275–285.

47. Mathur, Y., Sreyas, S., Datar, P. M., Sathian, M. B., & Hazra, A. B. (2020). CobT and BzaC catalyze the regiospecific activation and methylation of the 5-hydroxybenzimidazole lower ligand in anaerobic cobamide biosynthesis. *Journal of Biological Chemistry, 295*, 10522–10534.

48. Minot, G. R., & Murphy, W. P. (1926). Treatment of pernicious anemia by a special diet. *Journal of the American Medical Association, 87*, 1666.

49. Miyano, K. I., Ye, K., & Shimizu, K. (2000). Improvement of vitamin B_{12} fermentation by reducing the inhibitory metabolites by cell recycle system and a mixed culture. *Biochemical Engineering Journal, 6*, 207–214.

50. Mohammed, Y., Lee, B., Kang, Z., & Du, G. (2014). Development of a two-step cultivation strategy for the production of vitamin B_{12} by *Bacillus megaterium*. *Microbial Cell Factories, 13*, 102.

51. Moore, S. J., & Warren, M. J. (2012). The anaerobic biosynthesis of vitamin B_{12}. *Biochemical Society Transactions, 40*, 581–586.

52. Nguyen-Vo, T. P., Ainala, S. K., Kim, J. R., & Park, S. (2018). Analysis and characterization of coenzyme B_{12} biosynthetic gene clusters and improvement of B_{12} biosynthesis in *Pseudomonas denitrificans* ATCC 13867. *FEMS Microbiology Letters, 365*, fny211.

53. Oraei, M., Razavi, S. H., & Khodaiyan, F. (2018). Optimization of effective minerals on riboflavin production by *Bacillus subtilis* subsp. *subtilis* ATCC 6051 using statistical designs. *Avicenna Journal of Medical Biotechnology, 10*(1), 49–55.

54. Park, E. Y., Ito, Y., Nariyama, M., Sugimoto, T., Lies, D., & Kato, T. (2011). The improvement of riboflavin production in *Ashbya gossypii* via disparity mutagenesis and DNA microarray analysis. *Applied Microbiology and Biotechnology, 91*(5), 1315–1326.
55. Piao, Y., Kiatpapan, P., Yamashita, M., & Murooka, Y. (2004). Effects of expression of HemA and HemB genes on production of porphyrin in *Propionibacterium freudenreichii. Applied and Environmental Microbiology, 70*, 7561–7566.
56. Piao, Y., Yamashita, M., Kawaraichi, N., Asegawa, R., Ono, H., & Murooka, Y. (2004). Production of vitamin B_{12} in genetically engineered *Propionibacterium freudenreichii. Journal of Bioscience and Bioengineering, 98*, 167–173.
57. Pillai, V. V., Prakash, G., & Lali, A. M. (2018). Growth engineering of *Propionibacterium freudenreichii* Shermanii for organic acids and other value-added products formation. *Preparative Biochemistry & Biotechnology, 48*, 6–12.
58. Raux, E., Schubert, H. L., & Warren, M. J. (2000). Biosynthesis of cobalamin (vitamin B_{12}): A bacterial conundrum. *Cellular and Molecular Life Sciences, 57*, 1880–1893.
59. Roth, J. R., Lawrence, J. G., & Bobik, T. A. (1996). Cobalamin (coenzyme B_{12}): Synthesis and biological significance. *Annual Review of Microbiology, 50*, 137–181.
60. Russo, P., Capozzi, V., Arena, M. P., Spadaccino, G., Dueñas, M. T., López, P., et al. (2014). Riboflavin-overproducing strains of *Lactobacillus fermentum* for riboflavin-enriched bread. *Applied Microbiology and Biotechnology, 98*(8), 3691–3700.
61. Schwechheimer, S. K., Becker, J., Peyriga, L., Portais, J. C., & Wittmann, C. (2018). Metabolic flux analysis in *Ashbya gossypii* using ^{13}C-labeled yeast extract: Industrial riboflavin production under complex nutrient conditions. *Microbial Cell Factories, 17*(1), 162.
62. Schwechheimer, S. K., Becker, J., Peyriga, L., Portais, J. C., Sauer, D., Müller, R., et al. (2018). Improved riboflavin production with *Ashbya gossypii* from vegetable oil based on ^{13}C metabolic network analysis with combined labeling analysis by GC/MS, LC/MS, 1D, and 2D NMR. *Metabolic Engineering, 47*, 357–373.
63. Shi, T., Wang, Y., Wang, Z., Wang, G., Liu, D., Fu, J., et al. (2014). Deregulation of purine pathway in *Bacillus subtilis* and its use in riboflavin biosynthesis. *Microbial Cell Factories, 13*(1), 101.
64. Silva, R., Aguiar, T. Q., & Domingues, L. (2015). Blockage of the

pyrimidine biosynthetic pathway affects riboflavin production in *Ashbya gossypii*. *Journal of Biotechnology, 193*, 37–40.

65. Sobczyńska-Malefora, A., Delvin, E., McCaddon, A., Ahmadi, K. R., & Harrington, D. J. (2021). Vitamin B_{12} status in health and disease: A critical review. Diagnosis of deficiency and insufficiency—Clinical and laboratory pitfalls. *Critical Reviews in Clinical Laboratory Sciences, 58*, 399–429.
66. Stamford, N. P. J., Duggan, S., Li, Y., Alanine, A. I. D., Crouzet, J., & Battersby, A. R. (1997). Biosynthesis of vitamin B_{12}: The multi-enzyme synthesis of precorrin-4 and factor IV. *Chemistry & Biology, 4*, 445–451.
67. Sybesma, W., Burgess, C., Starrenburg, M., van Sinderen, D., & Hugenholtz, J. (2004). Multivitamin production in *Lactococcus lactis* using metabolic engineering. *Metabolic Engineering, 6*(2), 109–115.
68. Thakur, K., Tomar, S. K., Brahma, B., & De, S. (2016). Screening of riboflavin-producing lactobacilli by a polymerase-chain-reaction-based approach and microbiological assay. *Journal of Agricultural and Food Chemistry, 64*(9), 1950–1956.
69. Thirupathaiah, Y., Rani, C. S., Reddy, M. S., & Venkateswar Rao, L. (2012). Effect of chemical and microbial vitamin B_{12} analogues on production of vitamin B_{12}. *World Journal of Microbiology and Biotechnology, 28*, 2267–2271.
70. Vandamme, E. J., & Revuelta, J. L. (2016). *Industrial Biotechnology of Vitamins, Biopigments, and Antioxidants* (ISBN 9783527681754). Wiley-VCH.
71. Wan, H. G., Xue, C. L., Wang, Y. D., Huang, Z. X., & Peng, Y. C. (2014). Effect of exogenous supplement on fermentative production of riboflavin by *Bacillus subtilis*. *Food Fermentation Industries, 40*(4), 13–17. [In Chinese]
72. Wang, G., Bai, L., Wang, Z., Shi, T., Chen, T., & Zhao, X. (2014). Enhancement of riboflavin production by deregulating gluconeogenesis in *Bacillus subtilis*. *World Journal of Microbiology and Biotechnology, 30*(6), 1893–1900.
73. Wang, P., Jiao, Y., & Liu, S. (2014). Novel fermentation process strengthening strategy for production of propionic acid and vitamin B_{12} by *Propionibacterium freudenreichii*. *Journal of Industrial Microbiology and Biotechnology, 41*, 1811–1815.
74. Wang, P., Shen, C., Li, L., Guo, J., Cong, Q., & Lu, J. (2020). Simultaneous production of propionic acid and vitamin B_{12} from corn stalk hydrolysates by *Propionibacterium freudenreichii* in an expanded bed adsorption

bioreactor. *Preparative Biochemistry & Biotechnology, 50*, 763–767.

75. Wang, P., Wang, Y., & Su, Z. (2012). Improvement of adenosylcobalamin production by metabolic control strategy in *Propionibacterium freudenreichii*. *Applied Biochemistry and Biotechnology, 167*, 62–72.

76. Wang, P., Wang, Y., Liu, Y., Shi, H., & Su, Z. (2012). Novel *in situ* product removal technique for simultaneous production of propionic acid and vitamin B_{12} by expanded bed adsorption bioreactor. *Bioresource Technology, 104*, 652–659.

77. Wang, P., Zhang, Z., Jiao, Y., Liu, S., & Wang, Y. (2015). Improved propionic acid and 5,6-dimethylbenzimidazole control strategy for vitamin B_{12} fermentation by *Propionibacterium freudenreichii*. *Journal of Biotechnology, 193*, 123–129.

78. Wang, X., Wang, Q., & Qi, Q. (2015). Identification of riboflavin: Revealing different metabolic characteristics between *Escherichia coli* BL21(DE3) and MG1655. *FEMS Microbiology Letters, 362*(11), fnv071.

79. Wang, Z., Chen, T., Ma, X., Shen, Z., & Zhao, X. (2011). Enhancement of riboflavin production with *Bacillus subtilis* by expression and site-directed mutagenesis of zwf and gnd genes from *Corynebacterium glutamicum*. *Bioresource Technology, 102*(4), 3934–3940.

80. Wang, Z., Shi, H., & Wang, P. (2016). The online morphology control and dynamic studies on improving vitamin B_{12} production by *Pseudomonas denitrificans* with online capacitance and specific oxygen consumption rate. *Applied Biochemistry and Biotechnology, 179*, 1115–1127.

81. Wang, Z., Wang, H., Li, Y., Chu, J., Huang, M., Zhuang, Y., & Zhang, S. (2010). Improved vitamin B_{12} production by step-wise reduction of oxygen uptake rate under dissolved oxygen limiting level during fermentation process. *Bioresource Technology, 101*, 2845–2852.

82. Warren, M. J., Raux, E., Schubert, H. L., & Escalante-Semerena, J. C. (2002). The biosynthesis of adenosylcobalamin (vitamin B_{12}). *Natural Product Reports, 19*, 390–412.

83. Woodward, R. (1973). The total synthesis of vitamin B_{12}. *Pure and Applied Chemistry, 33*, 145–177.

84. Xia, W., & Chen, W. (2015). Industrial vitamin B_{12} production by *Pseudomonas denitrificans* using maltose syrup and corn steep liquor as the cost-effective fermentation substrates. *Bioprocess and Biosystems Engineering, 38*, 1065–1073.

85. Xia, W., Wei, P., & Kun, C. (2014). Interactive performances of betaine on the metabolic processes of *Pseudomonas denitrificans*. *Journal of Industrial Microbiology and Biotechnology, 42*, 273–278.

86. Xie, C., Coda, R., Chamlagain, B., Varmanen, P., Piironen, V., & Katina, K. (2019). Co-fermentation of *Propionibacterium freudenreichii* and *Lactobacillus brevis* in wheat bran for *in situ* production of vitamin B_{12}. *Frontiers in Microbiology, 10*, 1541.

87. Xu, D., Feng, Z., Hou, W. T., Jiang, Y. L., Wang, L., Sun, L., Zhou, C. Z., & Chen, Y. (2019). Cryo-EM structure of human lysosomal cobalamin exporter ABCD4. *Cell Research, 29*, 1039–1041.

88. Xu, Z., Lin, Z., Wang, Z., & Chen, T. (2015). Improvement of the riboflavin production by engineering the precursor biosynthesis pathways in *Escherichia coli. Chinese Journal of Chemical Engineering, 23*(11), 1834–1839.

89. Yatsyshyn, V. Y., Fedorovych, D. V., & Sibirny, A. A. (2014). Metabolic and bioprocess engineering of the yeast *Candida famata* for FAD production. *Journal of Industrial Microbiology and Biotechnology, 41*(5), 823–835.

90. Zhang, F., Rodriguez, S., & Keasling, J. D. (2011). Metabolic engineering of microbial pathways for advanced biofuels production. *Current Opinion in Biotechnology, 22*(6), 775–783.

91. Zhang, X., Ban, R., Liu, L., & Zhang, R. (2017). Riboflavin production by a genetically modified *Bacillus subtilis. Microbiology China, 44*(1), 59–67. [In Chinese]

92. Zhang, Y., Li, X., Wang, Z., Wang, Y., Ma, Y., & Su, Z. (2021). Metabolic flux analysis of simultaneous production of vitamin B_{12} and propionic acid in a coupled fermentation process by *Propionibacterium freudenreichii. Applied Biochemistry and Biotechnology, 193*, 3045–3061.

93. Zhang, Y., Liu, J. Z., Huang, J. S., & Mao, Z. W. (2010). Genome shuffling of *Propionibacterium shermanii* for improving vitamin B_{12} production and comparative proteome analysis. *Journal of Biotechnology, 148*, 139–143.

94. Zhao, X., Condruz, S., Chen, J., & Jolicoeur, M. (2016). A quantitative metabolomics study of high sodium response in *Clostridium acetobutylicum* ATCC 824 acetone–butanol–ethanol (ABE) fermentation. *Scientific Reports, 6*(1), 28307.

95. Zhao, X., Kasbi, M., Chen, J., Peres, S., & Jolicoeur, M. (2017). A dynamic metabolic flux analysis of ABE (acetone–butanol–ethanol) fermentation by *Clostridium acetobutylicum* ATCC 824, with riboflavin as a by-product. *Biotechnology and Bioengineering, 114*(12), 2907–2919.

Chapter 9

Polysaccharides

Contents

INTRODUCTION

Polysaccharides made extracellularly by microorganisms for food, medicine, and vaccination purposes have garnered increased attention during the past 20 years. The great diversity in their structural and functional properties, despite their construction from very similar building blocks (the carbohydrate ring structure of pyranose and furanose), accounts for their wide range of utility. The fungi Schizophyllan commune and *Sclerotium rolfsii* produce rod-shaped molecules called schizophyllan and scleroglucan, respectively; the bacteria Azotobacter vinelandii produces intermediate structures called alginates; and the fungus Aureobasidium pullulans produces linear random coil-type structures called pullulan. Neutral polysaccharides like dextrans from the bacteria *Leuconostoc mesenteroides* or highly electronegative or "polyanionic" molecules like xanthan secreted by *Xanthomonas campestris* are also possible.

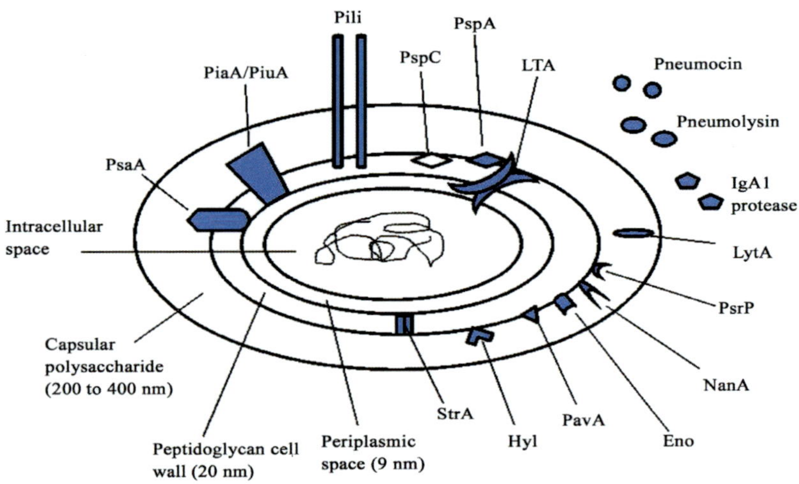

Figure 9.1. *Exocellular or 'capsular' polysaccharide layer from Streptococcus pneumoniae.*

Source: Safari, D., Rijkers, G., & Snippe, H. (2012). The Future of Synthetic Carbohydrate Vaccines: Immunological Studies on Streptococcus pneumoniae Type 14. InTech. doi: 10.5772/48326.

Microorganisms produce polysaccharides, which are secreted from the cell to form a layer covering their surface. This layer is frequently quite deep when compared to the dimensions of the individual cell (Figure 9.1). To separate them from any polysaccharides that may be present within the cell, they are referred to as exopolysaccharides due to their position. It is believed that the primary purposes are protective, acting as a physical barrier to block harmful substances from entering the area or, in a more focused manner, binding and neutralizing bacteriophages. They might stop dehydration in the right conditions. They might also stop immune system cells or other microbes from phagocytosing them. The unusual diversity of capsular polysaccharides (CPSs) may have evolved to prevent antibody responses, which makes them highly immunogenic. This characteristic can be used to develop vaccines. They contribute to the host's adhesion and penetration as well. In the case of plants, this entails potential specific interactions with the polysaccharide structures found in their cell walls. Plant defense against bacterial infection may involve the action of plant lectins, which are glycoproteins with unique binding characteristics to carbohydrate structures.

A small number of secreted polysaccharides may have a pathogenic role. *Pseudomonas aeruginosa* is frequently associated with respiratory tract infections. It is known to produce alginate, which exacerbates respiratory tract blockages and promotes secondary infections. Phloem blockages in plants have also been reported to occur. But the secreted polysaccharides themselves are attractive for biotechnological use because they almost never cause known toxicity issues and many can be harvested in large quantities for a reasonable price. Microbiological products play a central role in poly *Saccharomyces cerevisiae* aride biotechnology. They range from cheap bulk materials like xanthan, which is mostly used in food applications and costs about US$14 per kg, to expensive bulk materials like cyclic dextrans, which costs about US$50 per kg and is used in high-value pharmaceutical and research applications.

Nearly all of the major plant polysaccharides, including glucans, alginate-like substances, and even cellulose, are known to be produced by specific microbes in addition to complex materials unique to individual bacteria. Bacteria are a prime target for genetic manipulation because they have been studied for a longer time and are generally easier to manipulate genetically than higher organisms. This makes them ideal for modifying biosynthetic pathways and expressing heterologous genes to produce particularly desirable enzymes. Naturally, unlike enzymes, Polysaccharides cannot be directly controlled by genetic material; instead, they must be addressed indirectly by using the enzymes that are responsible for altering the biosynthetic or degradation pathways. This strategy has a lot of potential right now. In addition, biotechnology is responsible for all the advancements in this field because fermentation processes now involve the large-scale growth of bacteria. There are restrictions as well. For example, since processors currently use waste by-products from other processes as their raw material, it is unlikely that large-scale fermentation could ever compete with current processes for the manufacture of pectin. This is unrelated to the higher upfront costs associated with large-scale fermenters when compared to straightforward extraction techniques. Therefore, for bacterial products to be worth their likely higher cost, they need to possess especially beneficial qualities.

9.1. OVERVIEW OF POLYSACCHARIDES

In food, polysaccharides are the most prevalent type of carbohydrates. These are long-chain polymeric carbohydrates made up of glucose-sidic bonds securing monosaccharide units together. Using amylase enzymes as a catalyst, this carbohydrate can hydrolyze (react with water) to produce its constituent sugars, which can be either mono- or oligosaccharides. Their structures vary from being highly branched to being linear. Examples include structural Polysaccharides like chitin and cellulose and storage polysaccharides like starch, glycogen, and galactogen.

Polysaccharides frequently exhibit significant heterogeneity, with minor variations in the repeating unit. These macromolecules can have different properties from their monosaccharide building blocks depending on their structure. They could even be insoluble in water or amorphous. A polymer is referred to as a homopoly *Saccharomyces cerevisiae* aride (also known as a homoglycan) if every monosaccharide in it is of the same type; on the other hand, hetero polysaccharides (also known as heteroglycans) are polymers that contain multiple types of monosaccharidess.

Natural *Saccharomyces cerevisiae* arides are generally composed of simple carbohydrates called s with general formula $(CH_2O)_n$ where n is three or more. Examples of monosaccharides are glucose, fructose, and glyceraldehyde. Polysaccharides, meanwhile, have a general formula of $C_x(H_2O)_y$ where x and y are usually large numbers between 200 and 2500. When the repeating units in the polymer backbone are six-carbon monosaccharides, as is often the case, the general formula simplifies to $(C_6H_{10}O_5)_n$, where typically $40 \leq n \leq 3000$.

In general, oligosaccharides contain three to 10 monosaccharide units, while polysaccharides contain more than ten; however, the exact cutoff varies slightly depending on the convention. One significant class of biological polymers is polysaccharides. In living things, they typically serve structural or storage purposes. Plants use starch, a polymer of glucose, as a storage polysaccharide. Amylose and branched amylopectin are two forms of starch that are found in plants. The more densely branched glucose polymer found in animals, known as animal starch, is structurally similar to glucose. Because of its characteristics, glycogen can be metabolized more quickly, which is advantageous for animals with active lifestyles. They are crucial for bacterial multicellularity in bacteria.

Two examples of structural polysaccharides are cellulose and chitin. Cellulose is the most common organic molecule on Earth and is found in the cell walls of plants and other organisms. It is utilized in many different processes, including the production of rayon (through the viscose process), cellulose acetate, celluloid, and nitrocellulose. It also plays a big part in the paper and textile industries. Although chitin's structure is similar, its strength is increased by side branches that contain nitrogen. It can be discovered in the cell walls of some fungi and the exoskeletons of arthropods. It can be used for a variety of things, such as surgical threads. Callose or laminarin, Chrysolaminarin, Xylan, Abranoxylan, Mannan, Fucoidan, and Galactomannan are additional polysaccharides.

9.1.1. Classification of Polysaccharides

Glycans, or polysaccharides, are big molecules made up of over 20 monosaccharides or their derivatives connected in a linear or branching pattern by O-glycosidic bonds (poly- many; saccharide- sugar). They can form branched structures, which sets them apart from linear polymers like proteins and nucleic acids. This is because each monosaccharides consists of several -OH groups and one keto (aldehyde or ketone) group, which can react with other monosaccharides to form branched chain structures. Two classes exist for polysaccharides.

- **Homopolysaccharides:** Homoglycans, also referred to as homopolysaccharides, are Polysaccharides created by repeating the same monosaccharides. They can be further classified according to the identity of the constituent monosaccharides units. Glucans, for instance, are Polysaccharides made entirely of glucose. Similarly, fructans and galactans, respectively, are the names given to the polymers of fructose and galactose.
- **Heteropolysaccharides:** Various kinds of monosaccharidess are the building blocks of these Polysaccharides. Nonetheless, the majority of hetero *Saccharomyces cerevisiae* arides only contain two different types of monosaccharidess. Different configurations of monosaccharidess in homo and hetero Polysaccharides are depicted in Figure 9.2.

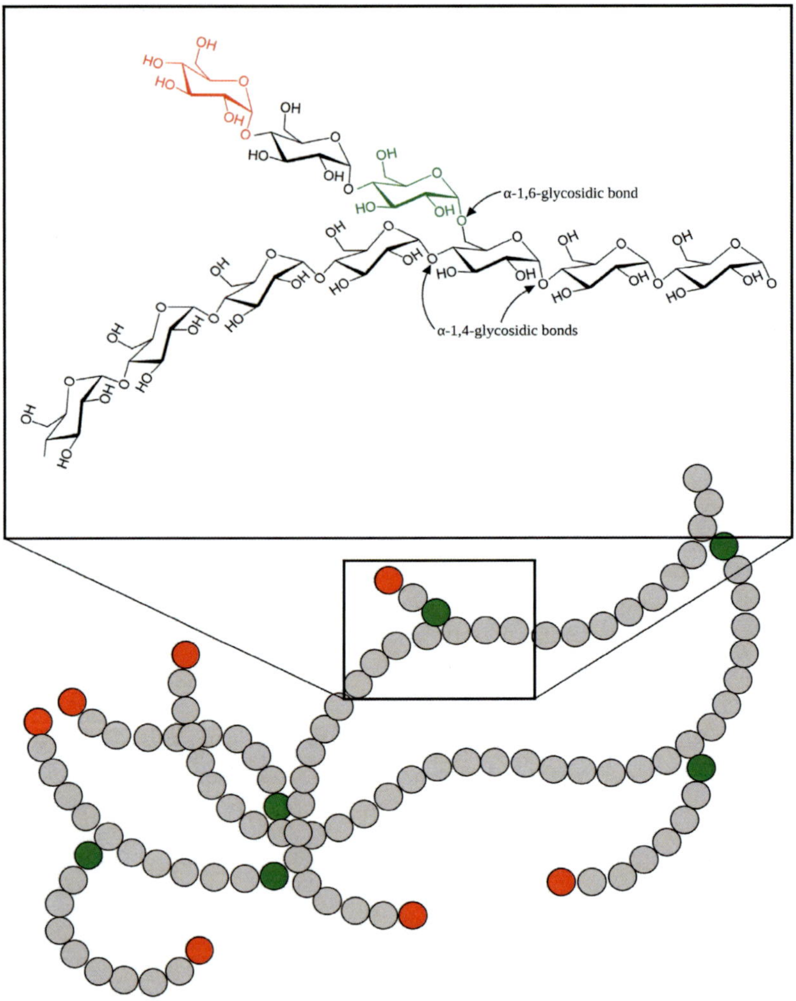

Figure 9.2. *Homo- and heteropolysaccharides.*

Source: Public Domain, https://commons.wikimedia.org/w/index.php?curid=611992.

9.1.2. Storage Polysaccharides

Because polysaccharides have a lower osmotic pressure than monosaccharides, organisms store energy in the form of polysaccharides. This group of functions includes those of glucose and glycogen. In plants, starch acts as a store of energy; in animals and microorganisms, glycogen acts as a poly*Saccharomyces cerevisiae* aride for storage.

9.1.2.1. Starch

It is stored as granules in the chloroplast of the plant cell and serves as the primary energy reserve in plants. Though practically all plant cells contain it, unripe fruits, tubers, and grain seeds are particularly abundant in it. Since starch is a homopolymer of α-D glucose, it is classified as a glucan. The two types of polymers that make up starch are amylopectin, which makes up the remaining 70–90% of the material, and α-amylose, which makes up 10% to 30%. Figure 9.3a shows that α-amylose is a linear polymer consisting of several thousand to half a million α-D glucose units connected by a (1→4) glycosidic bond. Because of the α(1→4) linkage in α-amylose, every glucose residue is angled in relation to the glucose before it, forming a coil or helical structure similar to that of a telephone wire. Similarly to α-D glucose units, amylopectin is composed of linear chains connected by (1→4) glycoside bonds. Nevertheless, it branches out every 24–30 glucose residues by forming (1→6) glycosidic bonds (Figure 9.3b), which inhibits the formation of coils.

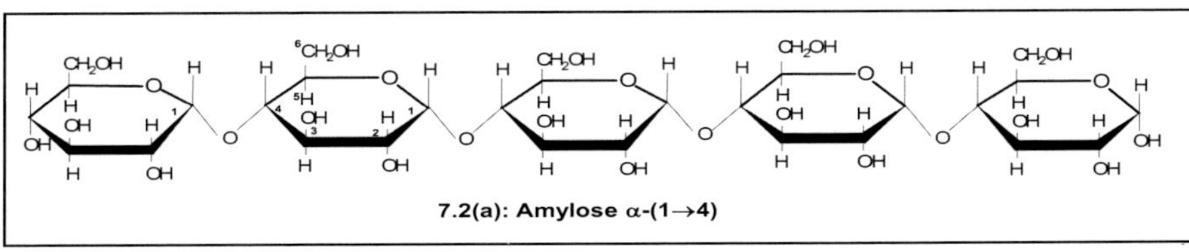

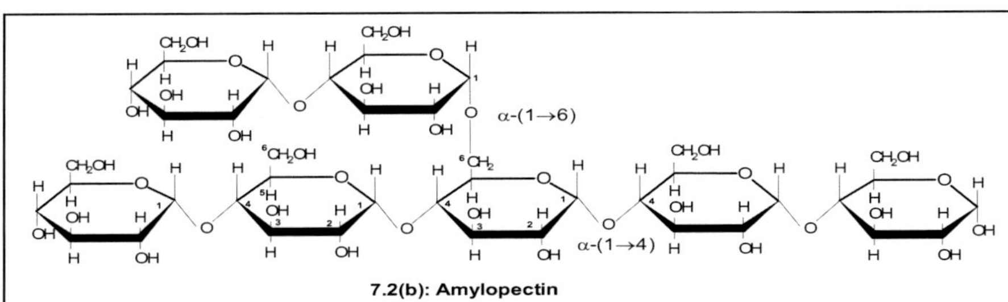

Figure 9.3. *Arrangement of α–D-glucose in amylose: a) and amylopectin; b) components of starch. (A). Linear structure of amylose having α – D-glucose joined by α (1→ 4) glycosidic bonds. (B). Branched chain structure of amylopectin; α – D-glucose joined by α (1→ 4) glycosidic bonds form linear strands which branch out by forming α (1→ 6) glycosidic bonds at intervals of 24-30 glucose molecules. Also note that one end of these strands is reducing while the other end is non- reducing.*

Source: https://egyankosh.ac.in/bitstream/123456789/68511/1/Unit-7.pdf.

The center of the coil created by amylose adsorbs iodine to form starchiodine complex, which gives an intense blue-black color when combined with potassium iodide or iodine in a solution of starch. This characteristic has been applied to the qualitative starch detection test. Many foods, including bread and potatoes, contain starch. The enzyme α-amylase, which is found in saliva, aids in the breakdown of food in the human body. When starch is partially broken down by heat, acids, or enzymes, a sticky mixture called dextrin is produced. This mixture is made up of smaller polysaccharides, monosaccharides, and oligosaccharides. The creation of dextrin is what gives bread its golden-brown crust when it is heated.

9.1.2.2. Glycogen

Animal starch is comparable to glycogen, which is a homopoly *Saccharomyces cerevisiae* aride of α-D glucose and is used by animals as a store for glucose. All cells contain it as granules, but the liver and skeletal muscles contain large amounts of it. Its structure is comparable to that of amylopectin, but it is more branched, occurring nearly every 8–14 glucose units. Additionally, the iodine test is positive. Glycogen phosphorylase, an enzyme found in the liver and muscles, hydrolyzes glycogen one residue at a time from its non-reducing end to produce glucose-1-phosphate when energy is required. Interestingly, most enzymes that hydrolyze starch and glycogen target the non-reducing ends, releasing one glucose at a time. You can see how the long polysaccharide chains would have continued to break down into tiny fragments and completely solubilize if these enzymes had been attacking anyplace in between these Polysaccharides. Rather, the branched-chain structures of both glycogen and amylopectin have been designed by nature so that the enzymes can attack multiple non-reducing ends at once, allowing for the quick mobilization of glucose when needed. Conversely, α-amylose possesses a single non-reducing end, making it primarily utilized for the extended storage of glucose.

9.1.3. Structural Polysaccharides

Microorganisms, animals, and plants all contain large amounts of structural polysaccharides. Although they behave very differently, structural and storage polysaccharides are substantially similar in composition. Their roles encompass a variety of duties, such as offering stiffness, durability, intercellular support, and lubrication.

9.1.3.1. Cellulose

Since it is present in the cell walls of almost all plants, cellulose is the most prevalent natural polysaccharide. Owing to its hard and fibrous texture, it is a major constituent of woody plant parts like tree bark, stems, and trunks. It is also a homopolysaccharide belonging to the glucan class. D-glucose is present in it as a linear polymer connected by β (1→4) glycoside bonding (Figure 9.4). As a result, you may have noticed that the type of glycoside bond I is the only way cellulose and α-amylose differ. E. Instead of α (1→4) linkage, it has β (1→ 4) linkage. But this little distinction gives cellulose an entirely different structure and set of properties.

Starch's α (1→ 4) linkages are bent by nature, giving it a coiled structure. In contrast, cellulose's β (1→ 4) linkages are more stable when alternating glucose units are flipped at 180°, resulting in fully extended cellulose chains. This conformation is referred to as extended ribbon. Cellulose chains create a lot of inter-chain hydrogen bonds when they are in this configuration. Consequently, cellulose is a linear, densely packed, rigid material that can withstand high pressures (Figure 9.4).

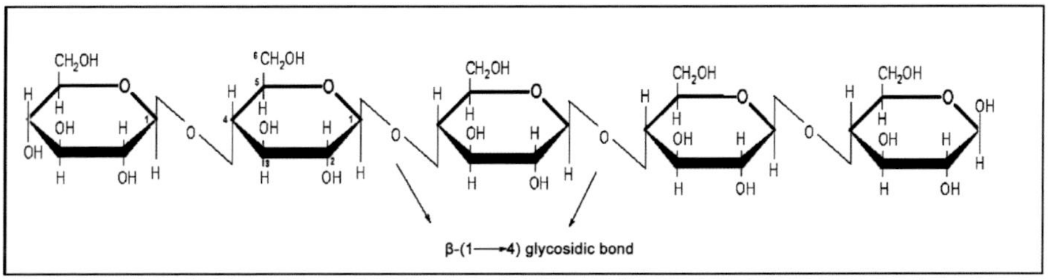

Figure 9.4. *Linkage of D-glucose residues by β (1→ 4) glycosidic bonds in cellulose.*

Source: https://egyankosh.ac.in/bitstream/123456789/68511/1/Unit-7.pdf.

The most common carbohydrate is cellulose, but vertebrates cannot eat cellulose because they lack the enzymes needed to hydrolyze the β (1→ 4) linkage. Consequently, fibrous food cannot be digested by humans. But these foods aid in excretion and bowel movement. The digestive tracts of herbivores, including deer, giraffes, and cows, are home to the enzyme cellulase, which aids in plant digestion by hydrolyzing β (1→ 4) linkages.

A comparison of starch, glycogen, and cellulose demonstrates how different types of glycosidic bonds (α or β) can join homopolymers of glucose when they branch to varying degrees or assume distinct three-dimensional structures with unique properties appropriate for their respective functions (Figure 9.5).

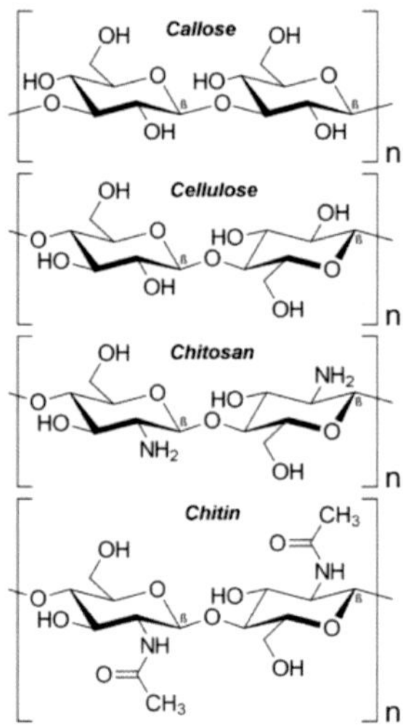

Figure 9.5. *Different polysaccharides.*

Source: By Qniemiec – Own work, CC BY-SA 4.0, https://commons.wikimedia.org/w/index.php?curid=94646819.

9.1.3.2. Chitin

Chitin is arguably the second most prevalent polysaccharide in nature, right behind cellulose. Figure 9.6 shows that it is a homopolysaccharide composed of N-acetyl-β-D-glucosamine connected by β (1→4) linkages. Chitin is a component of most fungi's cell walls as well as the hard exoskeleton of arthropods, including insects, crabs, and mollusks. Chitin provides the matrix for mineralization in a lot of these exoskeletons. Collagen serves as a matrix for mineral deposition in vertebrate bones, which is an example of this type of analogy.

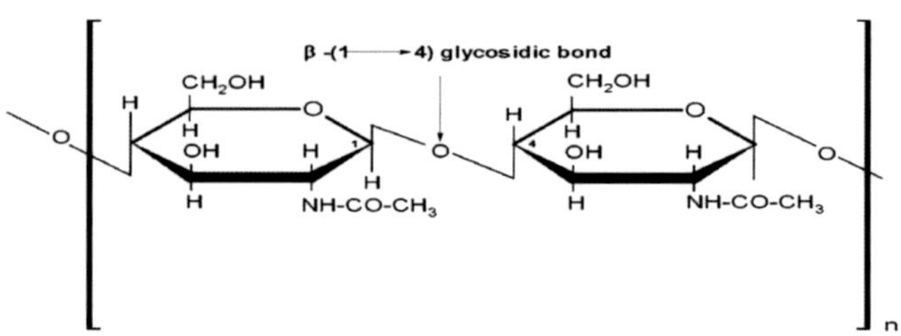

Figure 9.6. *N- acetyl glucosamine residues are linked by β (1→4) bonds.*

Source: https://egyankosh.ac.in/bitstream/123456789/68511/1/Unit-7.pdf.

9.1.3.3. Glycosaminoglycans (GAGs)

These are a family of linear heteropolysaccharides that make up a significant portion of the extracellular matrix (ECM) that envelops connective tissues like blood vessels, cartilage, skin, and tendons. GAGs are not present in plants; they are exclusively present in bacteria and animals. Repeating disaccharide units make up these unbranched polysaccharides. In most instances, one of the sugars in the repeating disaccharide is uronic acid, and the other sugar is an amino sugar like N-acetyl glucosamine or N-acetyl galactosamine. Esterified sulfate groups can also be present in some GAGs. GAGs are highly negatively charged due to the presence of sulfate and carboxylate groups, and in order to prevent repulsion between nearby groups, they adopt extended rod-like helix structures in solution.

Because GAGs can absorb large amounts of water and form a viscous jelly that reduces friction, they are also known as muco polysaccharides. They have the same capacity to withstand compression as a sponge, which will only compress so much before regaining its original shape.

Hyaluronic acid: It is composed of up to 50,000 D-glucoronic acid disaccharide units connected to N-acetyl-D glucosamine via a (β1→3) glycosidic bond (Figure 9.7). Due to its ability to form a highly viscous solution, it is a crucial component of both the vitrous humor of the eye and synovial fluid, which lubricates joints. It also contributes to the elasticity and high tensile strength of tendons and cartilage as part of their extracellular matrix (ECM).

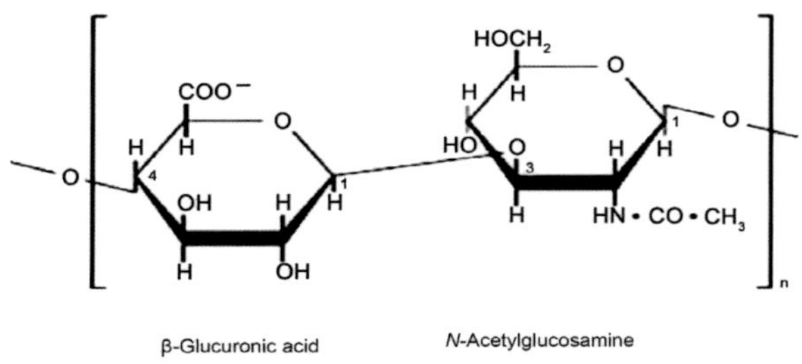

Figure 9.7. *Structure of repeating dimer in hyaluronic acid.*

Source: https://egyankosh.ac.in/bitstream/123456789/68511/1/Unit-7.pdf.

- **Heparin:** It is frequently added to blood samples for clinical analysis as a natural anticoagulant to stop clotting. It is also used as a medication to prevent blood clots in blood vessels. A combination of strongly sulfated heteropolysaccharides makes up heparin. Figure 9.8 depicts one of its frequently occurring repeating disaccharides,

L-iduronate-2-sulfate: (α1→ 4): N-sulfo-D-glucosamine-6-sulfate. Like other GAGs, heparin is not found extracellularly; rather, it is found in mast cells that line the arterial walls, particularly in the skin, liver, and lungs.

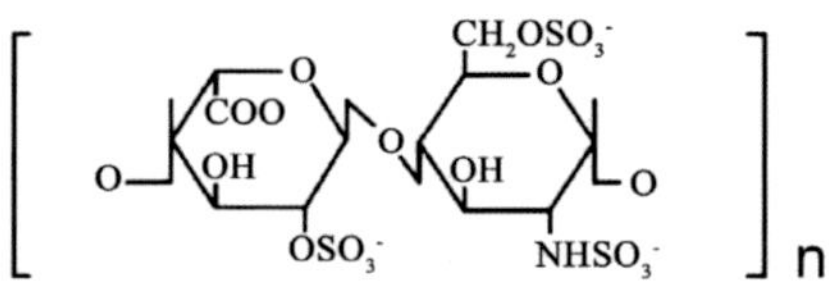

Figure 9.8. *Repeating disaccharide in heparin.*

Source: https://egyankosh.ac.in/bitstream/123456789/68511/1/Unit-7.pdf.

- **Chondroitin Sulfates:** There are two members in this category; chondroitin-4- sulfate which is polymer of disaccharide [D-glucoronic acid: (β1→3): N-acetyl-D galactosamine- 4-sulfate] (Figure 9.9a). Another is chondroitin-6- sulfate which has N-acetyl-D galactosamine- 6-sulfate in place of N-acetyl-D galactosamine-4-sulfate. Both provide tensile strenth to cartilage, tendons, ligaments and walls of aorta. Recently, these are being recommended as dietary supplement for osteoarthritis patients (Figure 9.9b).

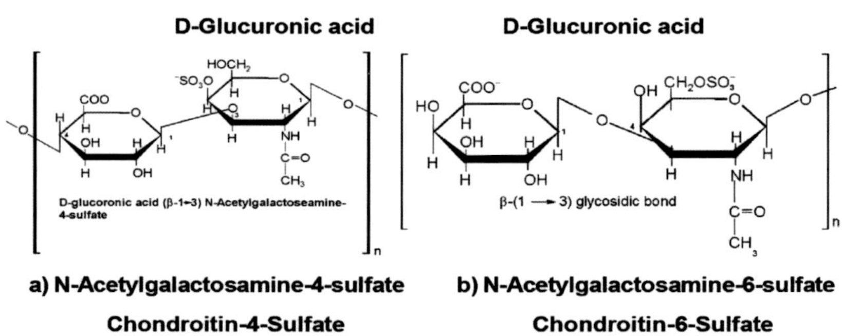

Figure 9.9. *Repeating disaccharide in (a) chondroitin-4- sulfate; and (b) chondroitin-6- sulfate.*

Source: https://egyankosh.ac.in/bitstream/123456789/68511/1/Unit-7.pdf.

- **Karatan Sulfate:** It can be found in bone, cartilage, the cornea, and horny tissues like the hoof, nails, and claws. No uronic acid is present. D-galactose: (β1→ 4): N-acetyl-D-glucosamine-6-sulfate are repeating disaccharides (Figure 9.10).

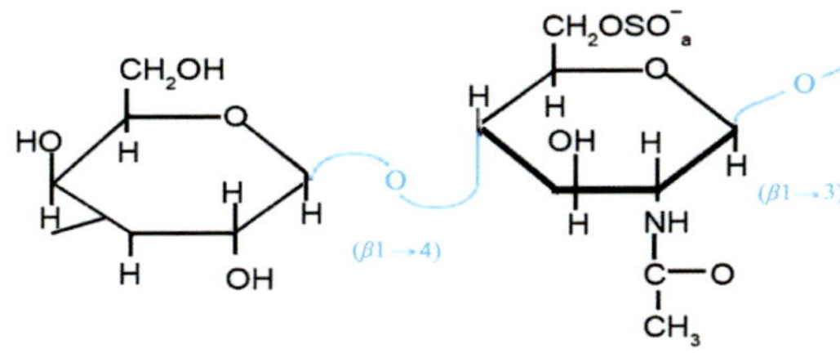

Figure 9.10. *Repeating disaccharide of keratin sulfate.*

Source: https://egyankosh.ac.in/bitstream/123456789/68511/1/Unit-7.pdf.

- **Dermatan Sulfate:** It is present in heart valves, blood vessels, and skin. The disaccharides N-acety-D-galactosamine-4-sulfate and L-iduronic acid are repeated, and the amounts of glucuronic acids vary (Figure 9.11).

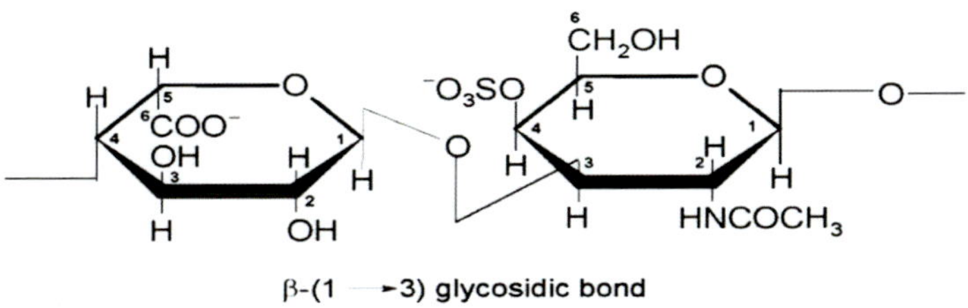

Figure 9.11. *Repeating disaccharide found in dermatan sulfate.*

Source: https://egyankosh.ac.in/bitstream/123456789/68511/1/Unit-7.pdf.

9.2. COMMERCIAL IMPORTANCE OF POLYSACCHARIDES

9.2.1. Market for Water-Soluble Gums

it's interesting, for instance, to compare gum consumption with import levels, as this provides some insight into the use of "home-grown material." It's also important to take notice of a drop in the import levels of some goods, especially gum Arabic, as this suggests a change in consumer behavior toward other water-soluble polymers. Gum quality, source, and the market in which they are sold all have an impact on market prices.

9.2.2. Problems Associated with the Production of Plant Polysaccharides

When trying to maintain guaranteed supplies of products with stable specifications, many higher plant and algal polysaccharide suppliers encounter several highly intractable challenges. The plant tissues themselves are the source of the initial issues. Polysaccharides are made by plants either as structural materials or as storage products, and their chemical composition can change in response to the needs of the plant's metabolism. These needs could result from things like seasonal variations in the environment or the plant's aging process. Brown seaweeds are plants that exhibit typical behavior of the type that has been described; these plants contain alginates, the content of which has been found to vary. Examining various plant sections has revealed differences that are somewhat correlated with the age of the tissues, as demonstrated by values for both new and mature fronds as well as for various stipe segments of *Laminaria hyperborea*. Manufacturers of alginate try to take advantage of this by separating stipe from frond in order to create products with different specifications. Laminaria digitata is a seaweed from which commercially useful alginates are collected. It has been shown that the stipe of this seaweed contains alginate with a higher guluronic acid content than that from the frond.

Development efforts on several common sources, including potatoes and corn, have improved starch products to a significant extent. A new frontier in starch technology has been opened up by the development of more recent amylopectin starches from agronomically modified corn varieties like waxy maize and waxy sorghum. These starches have produced a wide range of products with precise and repeatable specifications. With these compounds' improved optical and rheological qualities and unique gel-forming abilities, products that could not have been made with the mixed amylose-amylopectin starches that were previously on the market could be created. However, the manufacturer has little control over his raw material for many other industrial gums. For example, attempts to control the metabolism of seaweeds have yielded very little success, and the kind of development work needed to form starch products has not been feasible.

The production of gum Arabic, which is the exudate from the gum nodule that forms when the bark of Acacia Senegal is tapped, is another example of an uncontrolled situation. Nomadic tribesmen gather the gum from trees that grow naturally. After being exposed to the intense sun, the nodules are transferred to central collection locations. Government organizations grade the nodules before selling them to the gum supplier. Therefore, before or after harvesting, the supplier has no control over the product specification.

Processing-related product modification and degradation are other common challenges in the production of plant and algal gums. There are numerous difficult processing situations, such as alkali extraction, acid precipitation, hot-water leaching, and bleaching agent treatment. In order to extract the crude sodium alginate, the dry seaweed is first ground and shred, and then it is digested with sodium carbonate. This stage involves the removal of small amounts of cellulosic material, after which the polysaccharide precipitates as the calcium salt or, if suitable, as alginic acid using hydrochloric acid. It is necessary to leach the calcium alginate with hydrochloric acid to change it into the acid form, and then to titrate the sodium alginate. There are two different titration methods that can be used. The first is a heterogeneous reaction that stops the polysaccharides from dissolving by employing aqueous sodium hydroxide and aqueous sodium carbonate in the presence of isopropanol. After filtering out the aqueous isopropanol, the product is dried. The alternative method entails mixing the viscous dough in a powerful mixer and adding about a stoichiometric amount of dry sodium carbonate to the wet cake of alginic acid. After adding a small amount of sodium carbonate to correct the pH, the semisolid mass is dried by drum or roller drying. Hypochlorite may be used to bleach the polysaccharides prior to final drying (Booth, personal communication). These processing conditions can cause alkaline degradation by βelimination reactions, acid-catalysed hydrolysis and possible biological and physical degradation. It's also possible to add other undesired characteristics like color, odor, and chemical contaminants. In order to achieve a uniform product specification with the required viscosity or gel-forming property, the manufacturer frequently mixes various samples. However, when the user requires the polysaccharides to perform in a specific manner utilizing more than one physical property, e.g., the two rheological characteristics cannot be defined together if they are blended in order to have a particular viscosity and later to form a defined gel. Users might therefore be required to reformulate products.

A last issue could be a supply shortage. It can be challenging to supply any product made from plant and algae sources, both in the short and long terms. Due to weather patterns and disease, poor harvests of economically significant crops like corn, citrus fruit, and legumes are frequent and frequently result in brief shortages. Even more significant issues may arise from plant species that are not cultivated under strict control. Because of the defoliation and death of numerous acacia trees brought on by recent severe droughts in the Sudan, there is a shortage of gum Arabic. Changes in ocean currents and harsh weather have an impact on the availability of seaweed polysaccharides.

Growth zones shift over years, and more aggressive weed species may encroach on a weed species' native range. The latter condition is frequently brought on by pollution, as green seaweeds—which prefer to grow in environments richer in organic matter—invade areas inhabited by the more delicate red and brown varieties. Overfishing of some seaweed species, especially those of the Gelidium species, which is a common source of agar, can lead to shortages. It has been proposed that some algae species be grown in locations where they are not typically found to increase supplies. In this regard, a great deal of recent discussion has focused on one species of brown algae, *Macrocystis pyrifyra*. This weed is harvested off the coast of California for its alginate content. It grows naturally in several parts of the world. Since this weed cannot be found off the coast of Europe, it has been proposed that Macrocystis sp. be grown in order to increase the supply of weeds that contain alginate ought to be grown near the French region of Brittany. Many objections were raised due to the weed's 66-meter-long stipes, which can have a diameter of more than 2.5 cm, and its ability to grow in water that ranges in depth from 750 to 2,400 cm. It was mentioned that the presence of a seaweed with such enormous dimensions would most likely upset the ecological balance of the area and cause problems with coastal shipping due to the fouling of propellers and rudders.

9.2.3. Production of Polysaccharides of Commercial Significance via Chemical Modification

Chemical derivative formation is one method that has been used for many years to obtain polysaccharides with modified specifications. Successful chemical modification of polysaccharides has resulted in the development of significant commercial products like cellulose and starch ethers and esters. Two related ideas have formed the foundation of the strategy for chemical-derivative formation. The first idea stems from a desire to enhance a product's functionality for a specific use. The reaction of propylene oxide with partially neutralized alginic acid, which yields the propylene glycol ester derivative of this polysaccharide, is an illustration of derivative formation that falls into this category. Alginic acid's unsubstituted compound is less compatible with acids and multivalent metal cations than the product obtained from substituting several of the carboxyl groups in the acid. Soft drinks are one application of propylene glycol alginate, where the polymer serves to modify texture, stabilize foam, and control penetration. Unmodified alginate is not suited for this application because soft drinks' acidic content would make the polysaccharide insoluble.

The formation of entirely new polysaccharide products with properties entirely distinct from the parent polymer is the second overarching idea behind the chemical modification of polysaccharides. The synthesis of cellulose ethers is one application of this method. The complete insolubility of native celluloses in water limits their application. The water-soluble methyl and hydroxypropyl derivatives are designed for their non-ionic thickening, film-forming, gelling, and other characteristics. Whereas the propylene glycol alginates still have some characteristics of native alginate, they have developed into significant commercial products in their own right and have little in common with the compounds from which they originated.

However, there are restrictions on the range of products that can be produced through chemical modification that are both currently and potentially useful to industry. The native polysaccharide nature influences the chemically modified polysaccharide in part. The performance of the derivative may suffer if the latter possesses undesired traits, such as low molecular weight or acid-labile glycosidic bonds. Another thing to keep in mind is that chemical reactions are comparatively primitive compared to the enzyme-catalyzed reactions used in polysaccharide biosynthesis. Many modifications are not achievable with current technology; for example, it is currently not possible to specifically substitute hydroxyl groups in polysaccharide by chemical means, although one cell-free, enzymatic process is commercially viable for producing dextran.

Since the enzymatic systems mediate the production of polysaccharides from microorganisms, this method has garnered a lot of interest. It provides a more or less controlled synthetic method that is far more satisfying than many of the conventional processes, despite not being as well defined as many purely chemical reactions.

The aforementioned discussion makes clear that the manufacturing of natural and chemically modified polysaccharide gums is frequently a very lucrative and successful industry. But there are still problems, many of them unsolvable.

9.3. BIOPOLYMERS, POLYSACCHARIDES FROM MICROORGANISMS

To set the stage for this chapter on the synthesis of microbial gums, the conversation has up to this point been restricted to an examination of the current state of plant and algal gums. As previously mentioned, a great deal of research has been done on the development of chemical processes for the primary production of defined polysaccharide products; however, aside from starch, cellulose, and a few other gums, little real-world success has been attained. The current interest in the exploitation of micro-organisms stems from the failure of purely chemical reactions to significantly expand the range of industrial gums available. Numerous bacteria, fungi, and yeasts—many of which elaborate polysaccharides as exocellular capsules and slimes—are among the microorganisms involved. Numerous microorganisms secrete large amounts of slimy or gelatinous material into the aqueous medium in which they grow. Upon analysis, these materials are typically identified as carbohydrate polymers.

It will be helpful to examine microorganism polysaccharides by first providing a definition that allows a rational comparison with other polysaccharide types that are currently in use and of industrial significance. The carbohydrate polymers known as polysaccharides in question are those that are typically found outside of the cell walls of various bacteria, fungi, and yeasts. They possess unique rheological and gel-forming characteristics, making them interesting when compared to industrially produced Polysaccharides that are currently in use. It has been demonstrated that bacterial exopolysaccharides contain a range of monosaccharide residues, such as uronic acids, amino sugars, neutral hexoses and methyl-pentoses, and oxo sugars. Phosphate, pyruvate, succinate, and acetyl groups are additional elements.

9.3.1. Technical Advantages and Disadvantages of Industrial Polysaccharides Production by Fermentation

When compared to more conventional methods, the fermentation approach to polysaccharide production has several inherent advantages. They include: (i) Using well-characterized, affordable, and abundant raw materials, especially if it is possible to develop fully defined media; and (ii) Possessing the capacity to exert control over the synthetic process through continuous fermentation parameters, like temperature, pH, medium composition, inoculum composition, inoculum size, broth mixing and aeration, and fermentation time. It is possible to identify two distinct goals that are crucial for fermentation control. The first part involves using process control to maintain product specifications, i.e., the capacity to replicate fermentations within predetermined bounds to synthesize products. A manipulative control of the product type is the second aspect, for example, the capacity to create a variety of microbial gums with various molecular weights and structures by adjusting fermentation conditions in particular ways. To achieve this later goal, mutation may be crucial. Other factors that may help include: (iii) the potential to use gentle recovery techniques (like solvent precipitation) to prevent unintended product degradation and the resulting loss of quality; (iv) more flexibility in the selection of production sites; and (v) the use of continuous polymer production that yields high productivities.

However, there are many technical issues or drawbacks with fermentation compared to conventional techniques. These are primarily caused by the extremely viscous fermentation broths, which result in: (i) Low product concentrations—typically less than 5% w/v—which necessitate very large fermenters (50–200 m^3); (b) high water consumption; (c) heavy recovery solvent use; (ii) high power requirements—due to the high viscosities in cultures; (iii) challenging cell removal; and (iv) poor mixing and aeration, which results in oxygen limitation and decreased productivity.

9.3.2. Commercial Potential of Polysaccharides from Micro-Organisms

To understand the current state of development and the commercial standing of the most advanced microbial gums, it will be helpful to look at the potential for microbial polysaccharides. In contemplating the production of microbial gums, the potential manufacturer must answer a number of important questions, (i) What annual production capacity will the plant have? To answer this question the following information must be available: (a) the present and future market size for both products of specific interest, and for likely competing products: (b) the number of competing products and their manufacturers: (c) the sensitivity of direct manufacturing cost to plant size; (ii) If the product is a new one, whether the properties are sufficiently unique to merit a development program; (iii) If the product is: (a) either to be the same as a microbial gum already in production, e.g., dextran; or (b) the equivalent to a plant gum, e.g., microbial alginate, whether the new version can match or even surpass the properties of the material currently available; (iv) Can the new polysaccharides be produced at a cost and sold at a price that will yield a reasonable return on investment, considering the potential for price reductions by established manufacturers; and (v) The expected product life; and whether new, superior products are being considered. Due to the complexity of the global gum market and the lack of readily available market data, it will be very difficult, if not impossible, to address some of the above factors pertaining to market size and structure. Nevertheless, it is worthwhile for prospective microbial gum producers to make a sincere effort to find solutions.

Regarding the production of microbial gums, the great majority of them have not been used for commercial purposes. In the Northern Regional Research Laboratories (NRRL), a large portion of the early work that produced results with commercial significance was carried out.) in Peoria, Illinois, the home of the US Department of Agriculture in an endeavor aimed at making use of extra corn starch. This starch surplus was caused by an overabundance of corn in the Midwest of the United States, and it was thought to be a potentially useful fermentation substrate as a source of glucose.

9.3.3. Dextran

Dextran, a polysaccharide obtained from the condensation of glucose, is a complex branched glucan that was initially obtained from wine. Dextrans are defined by IUPAC as branched poly-α-d-glucosides originating from microorganisms, with glycosidic bonds primarily ranging from C-1 → C-6. The lengths of dextran chains range from 3 to 2,000 kilodaltons.

The main chain of the polymer is made up of branches from α-1,3 linkages and α-1,6 glycosidic linkages between glucose monomers. A dextran is distinguished from a dextrin, which is a straight-chain glucose polymer attached to α-1,4 or α-1,6 linkages, by its distinctive branching.

9.3.3.1. Microsurgery

Microsurgeons frequently use these drugs to reduce vascular thrombosis. By binding to erythrocytes, platelets, and vascular endothelium, dextran decreases erythrocyte aggregation and platelet adhesiveness while simultaneously boosting electronegativity. This mechanism underlies its antithrombotic action. Dextrans also lowers platelet function by reducing factor VIII-Ag Von Willebrand factor. Dextrans administration causes a change in the thrombus structure, which makes the clots easier to lyse (more evenly distributed platelets with coarser fibrin). Dextran plays the role of a plasminogen activator and has thrombolytic properties by blocking α-2 antiplasmin.

Outside of these characteristics, larger dextrans have been used urgently to treat hypovolemia because they are strong osmotic agents and do not exit the vessels. The hemodilution brought on by the volume expansion associated with the use of dextran enhances blood flow, thereby enhancing microanastomose patency and lowering thrombosis. Nevertheless, there is no discernible difference in the antithrombotic efficacy between intravenous and intra-arterial dextran administration.

There are several molecular weights of dextrans available, ranging from 3 kDa to 2 MDa. Larger dextrans (>60,000 Da) are poorly eliminated by the kidney and can stay in the blood for weeks before being broken down. As a result, they have colloidal and antithrombotic effects that last longer. Dextran-40 (MW: 40,000 Da) has been the most widely used member of this family in anticoagulation therapy. Within the first 24 hours of intravenous infusion, nearly 70% of the dextran-40 is eliminated through urine, with the remaining 30% being retained for a few more days.

9.3.3.2. Other Medical Uses

- Dextran is used in some eye drops as a lubricant. and in certain intravenous fluids to solubilize other factors, such as iron (in a solution known as Iron Dextran).

- Intravenous solutions with dextran function both as volume expanders and means of parenteral nutrition. Such a solution provides an osmotically neutral fluid that once in the body is digested by cells into glucose and free water. It is occasionally used to replace lost blood in emergency situations, when replacement blood is not available, but must be used with caution as it does not provide necessary electrolytes and can cause hyponatremia or other electrolyte disturbances.
- Dextran also increases blood sugar levels.
- Dextran can be used in an ATPS for PEGylation

9.3.3.3. Laboratory Uses

- Dextran is used in the osmotic stress technique for applying osmotic pressure to biological molecules.
- It is also used in some size-exclusion chromatography matrices; an example is Sephadex.
- Dextran has also been used in bead form to aid in bioreactor applications.
- Dextran has been used as an immobilization agent in biosensors.
- Dextran preferentially binds to early endosomes; fluorescent-labeled dextran can be used to visualize these endosomes under a microscope.
- Dextran can be used as a stabilizing coating to protect metal nanoparticles from oxidation and improve biocompatibility.
- Dextran coupled with a fluorescent molecule such as fluorescein isothiocyanate can be used to create concentration gradients of diffusible molecules for imaging and allow subsequent characterization of gradient slope.
- Solutions of fluorescently-labeled dextran can be perfused through engineered vessels to analyze vascular permeability.
- Dextran is used to make microcarriers for industrial cell culture
- Orally-administered dextran sodium sulfate is used to induce colitis in animal models of inflammatory bowel disease.
- Dextran is a common model compound to test the potential of drug formulations to facilitate intestinal absorption via the paracellular route.

9.3.4. 'Xanthan Gum'

Xanthan gum is a thickening, stabilizing, and emulsifying additive that is produced by feeding a glucose solution derived from corn, soy, dairy, or wheat to strains of the *Xanthomonas campestris* bacteria. The sugary solution is fermented by the bacteria, forming a sticky protective layer with a consistency that can be used to bind and thicken substances. These qualities make xanthan gum helpful in the food-making process; in particular, it helps starches combine and enhances the rising of gluten-free baked goods.

To enhance the texture, consistency, and shelf life of foods like salad dressings, soups, sauces, and baked goods, xanthan gum is used in food production. For people who need to follow a gluten-free diet due to coeliac disease or non-coeliac gluten sensitivity, it is especially helpful. This is because in order to produce a product that mimics the crumb and lightness of regular bakes, gluten-free alternatives to wheat flour require additives like xanthan gum. In order to replicate the elastic qualities of gluten, xanthan gum thickens and binds starches. Xanthan gum can be easily dissolved in water and is sold as a powder.

Additionally, xanthan gum is a common ingredient in many household goods, such as makeup and wallpaper paste. Products that include xanthan gum include toothpaste, face creams, shampoos, and body lotions. It facilitates the thickening and emulsification of these products, improving their appearance and ease of pouring or squeezing. In Peoria, fermenters ranging in capacity from 90 to 900 liters were used to scale up the laboratory culture of the bacteria and isolate the xanthan gum. The growth medium consisted of salts, distiller's solubles, and corn sugar (Rogovin et al., 1961). Centrifugation was used to get rid of the bacteria, then vacuum drying, potassium chloride-induced methanol precipitation, and finally, product recovery. In this early work, the highest conversion of substrate to polymer was obtained when the initial medium glucose concentration was limited to 1 percent; despite numerous attempts to modify the culture conditions, an increased glucose concentration resulted in maximal polysaccharide formation without full substrate utilization. The majority of the polymer was formed in the first 72 hours, during which the culture fluid's viscosity had reached nearly 4,000 cP. Although polysaccharide production continued for up to 8 days, this period saw the formation of most of the polymer. Higher viscosity resulted from longer fermentation (c. 7,000 cP) and difficulties getting rid of the bacteria.

9.3.4.1. Commercial Development

The work done in Peoria revealed the following: (i) high yield production of xanthan gum; (ii) high viscosity of gum solutions at low concentrations;

(iii) highly pseudoplastic aqueous solutions of the gum; (iv) temperature-independent viscosity of xanthan gum solutions for practical purposes; (v) stability of the gum over a wide pH range; and (vi) stable gel formation in aqueous solutions of the gum in the presence of locust bean gum. On the basis of the behavior described, and the high yields of gum, commercial development of xanthan gum was considered appropriate.

The U.S. Department of Agriculture contacted a number of industrial companies that they thought might be interested in commercializing the process, and this work on laboratory and pilot plant development continued at Peoria into the early 1960s. Initially, a number of groups showed interest. Exxon Production Research and General Mills have remained very interested in these Kelco. Following a pilot plant evaluation, Kelco obtained multiple US production patents and started commercial production in 1967. Local expansion of the plant has been announced, and a further plant to produce xanthan and other biopolymers is under construction at a cost of US \$35m. This plant is planned to go on-streamin late 1976. General Mills formed a joint venture group with the French chemical company, Rhone Poulenc, to produce xanthan gum in a plant at Meile in the departement of Deux-Sevres, reported to go on-stream during early 1975. General Mills has declared plans to build a plant in Iowa, which is expected to begin operations in 1978. Although there isn't much information on plant capacity, it is unlikely that plants with a capacity of much less than 2,000 tons annually will be profitable.

Exxon Production Research developed its own gum, focusing more on the user side. It was used as a viscosifier for oil-well drilling mud in primary applications and as a mobility control agent in conjunction with other chemicals, like surfactants, to enhance recovery from partially depleted reservoirs. The development of xanthan gum and potentially other biopolymers in the future will depend heavily on the latter application. Exxon have obtained a number of both user and producer patents, and now issue licences for the Kelco product to be used for oil drilling and flooding, and collect royalties reported to be U.S. 0.2\$/kilo. Although Exxon have conducted an extensive pilot plant program into the production of xanthan, commercial production of the gum has not yet been announced.

9.3.4.2. Applications for Xanthan Gum

Numerous non-food industries, including textile printing and dyeing, oil well drilling and flooding, ceramic glaze manufacturing, cleaners and polishes, latex emulsion paints, and rust-curing agents, have found use for the special physical qualities displayed by xanthan gum aqueous solutions. Crude oil production is the gum's most significant single use. It was discovered that in the bentonite muds used to drill oil wells, xanthan gum had a technical advantage over other polymer lubricants due to its pseudoplastic flow behavior and unique stability towards heat, acid, and alkali. Exxon and Kelco established Xanco, a subsidiary, in Houston, Texas, to pursue oil recovery applications.

The gum has recently had a chance thanks to research on enhanced oil recovery that is being conducted globally by numerous oil companies. The general method of extracting more oil from a well after natural pressure has stopped forcing the oil out is known as enhanced recovery for the purposes of this discussion. In reality, significant volumes of agents that were not previously present in the reservoir—either thermal or chemical—are injected. Steam flooding and fire flooding are examples of thermal approaches; carbon dioxide injection or a mixture of surfactants, hydrocarbons (typically alcohol), and a mobility control agent like a polymer are examples of chemical methods. The latter method, known as "surfactant flooding," has proven to be the most effective thus far, and among the polymers of current interest is xanthan gum (polyacrylamide being the other major polymer). It is theoretically possible to recover a significant amount of crude oil using improved methods. Of the 434 billion barrels of crude oil that have been found in the United States to date, according to economics and technology available at the end of 1974, only about 32% (140 billion barrels) have been produced or are considered a producible reserve (American Gas Association, 1975). Based on improved techniques, an estimated 40–60 billion barrels could be produced, and an additional 70–90 billion barrels could be produced. This means that 165–170 billion barrels cannot be recovered using current or available technology. Many U.S. companies are now actively interested in this area with an eye towards supplying the large volumes of materials which will be required, as it has been estimated that to obtain 100,000 barrels of oil per day would require 623 × 106 lbs of petroleum sulphonates per year, 29.7 × 106 liters of alcohols per year and 37.4 × 106 lbs of polymers such as xanthan gum or polyacrylamide (Sharp, 1975; Umland, 1974). When approval was released in the US, xanthan gum made significant strides in the human food industry (Federal Register, 1969). Since then, fruit-flavored beverages, canned and frozen foods, relishes, French dressings, instant meals, toppings, and whips have all been developed to use the gum. All of these products have had their mouth feel and texture improved by the addition of tiny amounts of gum, frequently in combination with locust bean gum, which together create a stable gel. The animal feed industry is another one with a keen interest in xanthan gum because it is in fierce competition with agar when using xanthan and locust bean gum as a jellying agent in canned pet foods or as a suspending agent in low-solids liquid animal feeds.

9.3.4.3. Fermentation Process

Commercial-grade D-glucose or starch that has been thinned using an acid-enzyme combination is used as the carbohydrate substrate in the traditional batch process used to produce xanthan gum. Corn-steep liquor (from the corn wet-milling industry), casein hydrolysate (from dairy processors), and distiller's dried solubles (from the alcoholic beverage production) are examples of nitrogen sources that have been used in the past but have mostly been replaced by yeast and soy extracts because of their somewhat erratic and unpredictable composition. Dipotassium hydrogen phosphate, for buffering and necessary mineral requirements, and magnesium chloride are the inorganic requirements of the fermentation.

The key factors that are carefully regulated in growth conditions are temperature, pH level, and fermentation duration. Because of the extremely high viscosities encountered, the oxygen-transfer rate is affected and will become limiting unless the fermenter baffle and impeller geometry are carefully designed for optimal gas transfer. For this reason, aeration and mixing of the fermentation broth are critical variables.

The broth has a pH of roughly 6 and a viscosity of more than 30,000 cP when the fermentation is stopped (measured on the Contraves viscometer at 25°C and a shear rate of one sec-1). Although recent patents have claimed the use of enzymes for cell digestion (Kelco, 1975a), there are no reports that the cells are removed from the fermentation broth in industrial fermentations. Normally, products are precipitated using methanol or isopropanol in the presence of potassium chloride. The product can then be shred before drying on a moving band drier. The wet cake can then be recovered by filtration or centrifugation. Typically, alcohol is extracted from the cake by passing hot gases from the drier through charcoal recovery columns, and it is recovered from the spent liquor by distillation. In a fermentation process like this, the product yield should be between 75 and 80%. However, due to the high broth viscosity encountered, the concentration of gum in the fermenter is limited to less than 5 percent for the previously mentioned reasons.

Although the continuous production of xanthan gum has been studied in the laboratory (Rogovin, 1969), no industrial fermentations using this culture mode have been found. Production costs would be significantly reduced if continuous xanthan culture could be achieved on an industrial scale. Batch processes take about 80 hours to complete, while a continuous process with a dilution rate of 0.05 h^{-1} would result in a fermentation that takes 20 hours. Therefore, a continuous process could reduce capital costs by increasing plant utilization efficiency.

9.3.4.4. Chemical Structure

When xanthan gum was first discovered, its chemical structure consisted of a 14 or 16 unit repeat structure made up of pyruvate and acetate substituents and the residues of glucose, mannose, and glucuronic acid Siddiqui, 1967. After a thorough revision of this initial structure, it is now known that the gum is made up of a cellulose backbone with a side chain that has mannose, glucuronic acid, and acetate residues on every second main-chain glucose unit. The pyruvate residues are non-stoichiometrically attached as a ketal to the terminal side-chain mannose unit.

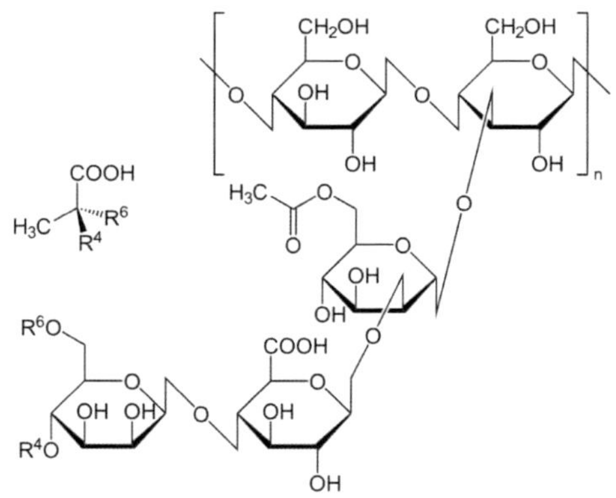

Source: By NEUROtiker – Own work, Public Domain, https://commons.wikimedia.org/w/index.php?curid=3744710.

Molecular weight was measured with light-scattering methods, yielding values between 13 × 10^6 and 50 × 10^6 Daltons. A polysaccharide dispersion heated in 4 M urea yielded a value of 2 × 106 daltons. These variations were attributed to variations in the conditions during fermentation and the development of microgels.

9.3.5. Pullulan

Pullulan is a polymer of maltotriose units, or α-1,4-;α-1,6-glucan. In maltotriose, a α-1,4 glycosidic bond connects three glucose units, while a α-1,6 glycosidic bond connects successive maltotriose units. The fungus *Aureobasidium pullulans* converts starch into pullulan. The cell uses pullulan primarily to fend off desiccation and predators. This presence also makes it easier for molecules to diffuse into and out of cells.

Pullulan is an edible, mostly tasteless polymer that is primarily used in the food industry to make edible films for use in mouthwash or breath fresheners like Johnson and Johnson's (USA) Listerine Cool Mint and Avery Bio-Tech

Private Ltd.'s Meltz Super Thin Mints. (India). In place of gelatine, pullulan and HPMC can be utilized as a vegetarian alternative for medication capsules. It is identified as food additive E 12004. Pullulan has also be explored as natural polymeric biomaterials to fabricated injectable scaffold for bone tissue engineering, cartilage tissue engineering, and intervertebral disc regeneration.

9.3.5.1. Applications for Pullulan

Patents asserting pullulan's exclusive use in a variety of food and industrial applications have been granted. However, there isn't much information available about the usefulness of the claims made because the products aren't yet for sale.

9.3.5.2. Chemical Structure

The gum is composed of maltotriose units or occasionally maltotetraose units linked together with $\alpha 1 \rightarrow 6$ bonds.

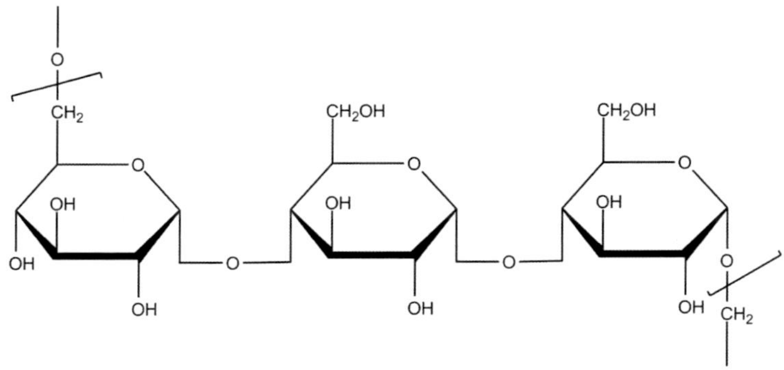

Source: By Klever – Own work, Public Domain, https://commons.wikimedia.org/w/index.php?curid=3825738.

9.3.6. Microbial Alginate

The food, textile, pharmaceutical, and paper industries use alginic acid, a commercially significant gum with gelling and Visco-elastic properties (Percival and McDowell, 1967). In the past, alginates were thought to be byproducts of Laminaria species and other closely related Phaeophyceae members, where they function as crucial structural polymers. The extracted polymer exhibits variations in the ratio of mannuronic acid to guluronic acid residues, and the alginate content varies significantly in these seaweeds. It was found that two species of Gram-negative bacteria, *Azotobacter vinelandii* and *Pseudomonas aeruginosa*, produced exopolysaccharides that were heteropolysaccharides with residues of D-mannuronic acid and L-guluronic acid, which were very similar to the products produced by algae (Gorin and Spencer, 1966; Linker and Jones, 1964, 1966). Mannuronic acid was the main component of *Azotobacter vinelandii*, and acetate and guluronic acid were also present. Similar periodate uptake was observed when compared to algal alginate, but there was also an additional presence of acetate and a higher ratio of mannuronate to guluronate residues. But keep in mind that algal material is typically extracted using weak alkali, which would eliminate any acetyl groups that may have been initially present. A later study of alginate from *Azotobacter vinelandii* (Larsen and Haug, 1971) indicated a similar ratio of mannuronate to guluronate residues in several preparations and an acetyl content to give one substituent group on every five monosaccharide residues. The product from *Pseudomonas aeruginosa* was found in strains isolated from cystic fibrosis patients and shown independently to resemble alginate (Linker and Jones, 1966; Carlson and Matthews, 1966).

Linker and Jones (1966) for a single preparation from *Pseudomonas aeruginosa* provided compelling evidence for a polymannuronate. The majority of the preparations' infrared spectra and range of uronic acid ratios were similar to those of seaweed alginate. Polysaccharides made from a variety of *Pseudomonas aeruginosa* types were the subject of an extension of this study (Evans and Linker, 1973). All of the polysaccharides shared a common characteristic: they were alginate-like polymers with acetyl groups and residues of D-mannuronic acid and L-guluronic acid, with the D-mannuronic acid residues being/3-linked. The uronic acid component ratios and acetyl contents differed. Additionally, the polysaccharides were compared to two enzyme preparations from a bacterium and ab alone, respectively, as substrates. There were some variations in the oligosaccharides generated by the various substrates, and the fragment pattern from bacterial enzyme digests was generally much simpler than that from abalone enzyme treatment.

Yet Polysaccharides derived from *Azotobacter vinelandii* in addition to Ps *Pseudomonas aeruginosa* have been recognized as alginates; however, the circumstances required for the synthesis of these polymers have not been extensively documented in research. When cultivated in nitrogen-free medium, isolates of Azotobacter are typically mucoid and appear to produce alginate. Like in the Enterobacteriaceae, there is presumably an effective nitrogen-limiting, carbohydrate-rich environment that favors polysaccharide synthesis (Wilkinson, 1958). The type of carbohydrate source—mannitol, sucrose, or another sugar—has no bearing on the polymer's composition.

In their investigation of *Pseudomonas aeruginosa* strains that synthesize alginate, Evans and Linker (1973) observed that the circumstances necessary for the highest level of polysaccharide synthesis also led to the least amount of reversion to the non-mucoid state. At 37°C, very little polysaccharide was produced; at 12°C, however, more polysaccharide was formed than at 25°C, although bacterial growth was substantially slower at the lower temperature. Maximum yields were achieved using supplemented MacConkey bile-salt agar. Additionally, scientists have observed that *Pseudomonas aeruginosa* forms more alginate on MacConkey agar than it does on a nitrogen-deficient, excess-carbohydrate medium, which was previously demonstrated to be effective in yielding significant amounts of the polysaccharide colanic acid from Enterobacteriaceae strains.

Some of the factors that promoted the production of alginate on MacConkey agar were the addition of 0.5 percent glycerol or 0.25% M sodium chloride. At 25–37°C, but not at 12°C, higher concentrations of sodium chloride or glycerol inhibited both bacterial growth and polysaccharide production. Shake flasks with a liquid medium can also produce polysaccharides well, but the ideal production conditions are still unknown. *Pseudomonas aeruginosa* was rejected for commercial study because of its association with pathogenic conditions in man, and *A. vinelandii* was selected for examination, the company concerned being Tate and Lyle Ltd. in the U.K. Research commenced in 1969, and by 1975 a process had been developed which was being operated on pilot plant (1 m^3)-scale fermenters.

The term 'alginates' is given to the group of acidic polysaccharides normally extracted from the brown seaweeds, alginates from different sources having been shown to vary within a structural framework based upon a linear block copolymeric arrangement of a-L-guluronic acid and β-D-mannuronic acid residues.

Early research using small-scale fermentation experiments demonstrated that cultures of *Azotobacter* sp. could yield several types of polysaccharides except for the bacterial product's partial acetylation, all falling within the previously described structural framework. Viscosity measurements also revealed that the molecular weight distributions appeared to be somewhat lower than those of the majority of algal alginates.

It was thought that a workable method for producing microbial alginate might be created if many goals could be met. The goals were to: (a) raise the product yield from the 5 percent found in the initial experiments to at least 50%; and (b) control and raise the molecular weight over the largest possible range. The viscosity that the material produces plays a significant role in determining the applications and selling prices of alginates. There are uses for low-viscosity materials, but higher viscosity products are typically more expensive; (c) to obtain control over the block structure and mannuronic: guluronic acid ratio. These criteria largely determine the gelling properties of the polysaccharide, and since the bacterial product must compete with a broad variety of algal products, this goal was deemed equally significant as accomplishing objective (b); (d) to create a continuous process.

The single most significant increase in both yield and molecular weight of microbial alginate was produced by lowering the concentration of phosphate in the medium to one-twentieth of that in the original growth medium (Imrie, 1973). This was among the many fermentation parameters examined in batch culture, such as medium composition, culture pH value, temperature, oxygenation of broth, inoculum age, and inoculum composition. The low-phosphate medium's reduced ability to function as a buffer meant that pH control was required to keep the amount of polysaccharide production at a constant level. As a result, the culture pH value dropped until cell growth stopped due to the production of carbon dioxide and alginate. Due to these enhancements, the product yield increased to the point where about 25% of the supplied sucrose was transformed into alginate. Higher viscosities were observed in bacterial alginate solutions as a secondary outcome of the medium composition alterations. This resulted from both an increase in the average molecular weight and a narrowing of the molecular weight distribution.

By employing continuous culture, the bacteria's ability to convert sucrose into alginate was further enhanced. The product concentrations stayed low, but the maximum efficiency was raised to 50% from the previously mentioned batch cultures' 25%. Continuous culture was also used to boost fermentation productivity, resulting in a five-fold increase in alginate concentration compared to batch culture (from 0.044 to 0.20 kg m^3h).

9.3.7. Scleroglucan

Scleroglucan, a ß-linked glucan with a covalent structure, is derived from the fungus *Sclerotium glucanicum*. The Pillsbury Company developed this polysaccharide for commercial use, and thcy sell it under the brand name Polytran. According to its claims, Polytran exhibits pseudoplasticity across a wide pH and temperature range, with the presence of different salts having little to no effect on the substance's functionality. In use, Polytran is said to improve ceramic glazes, latex paints, drilling muds, printing inks, and seed coatings when combined with acids and alkalis. It also stabilizes the rheology of bentonite clays during storage and exposure to high temperatures. However, not much is known about the applications of Polytran. Prices in 1975 were around £4,750 per ton, suggesting that in order for a sizable market to have grown, special end-use applications would have had to be created, ruling out the possibility of competing gums that were less expensive.

9.3.8. Curdlan

Curdlan is a high-molecular-weight, water-insoluble linear beta-1,3-glucan polymer of glucose. Curdlan is made up of glucose residues linked by β-(1,3) bonds, which when heated in an aqueous suspension, form elastic gels. It was claimed that *Alcaligenes faecalis* var. *myxogenes* produced it. This non-pathogenic curdlan-producing bacterium has since been reclassified as an Agrobacterium species in taxonomy.

Many pathogenic and soil-dwelling bacteria produce extracellular and capsular polysaccharides. Curdlan is an exopolysaccharide that is produced by soil bacteria belonging to the Rhizobiaceae family. It is a neutral β-(1,3)-glucan with possibly a few intra- or interchain 1,6-linkages. Four genes necessary for the synthesis of curdlan have been found in *Agrobacterium* sp. Two notable producers of curdlan are Agrobacterium tumefaciens and ATCC31749. CrdS (Q9X2V0, family GT2, Pfam PF13632), which codes for the catalytic subunit of β-(1,3)-glucan synthase, is a putative operon that is flanked by two other genes. A putative regulatory gene called crdR is located at a different locus. Maximum synthesis of curdlan with a large molecular mass also requires a membrane-bound phosphatidylserine synthase, which is encoded by pssAG. The curdlan operon is upregulated and the rate of curdlan synthesis is accelerated by nitrogen starvation.

Curdlan has been approved as a food additive by the U.S. Department of Agriculture and has many uses as a gelling agent in the food, construction, and pharmaceutical industries. Studies evaluating its use in fat replacement for foods like sausages, meat patties, and other meat products are currently underway.

9.4. DEVELOPMENT OF NEW MICROBIAL POLYSACCHARIDES

In order to produce exopolysaccharide, a laboratory may start a development program that goes like this: (i) screening material from different sources for appropriate microbial species; (ii) identifying mucoid (slime- or capsule-producing) micro-organisms; (iii) selecting many strains based on a qualitative assessment of the chemical and physicochemical properties of the polymers; and (iv) further developing a carefully chosen group of isolates to pilot-plant scale to gain a thorough understanding of the growth characteristics of the microorganisms, the prerequisites for producing polysaccharide, and the specifics of the polysaccharide.

9.4.1. Sources and Isolation

One of two theories may guide the choice to try isolating microorganisms that produce polysaccharides. Microorganisms known to produce large amounts of exopolysaccharides can be isolated and examined, or isolates from environments rich in carbohydrates or from other sources can be examined to see if any have the necessary culture characteristics. Extracellular Polysaccharides can be produced by a wide variety of bacteria and yeast species as slime or capsules. While the chemical makeup of these polysaccharides has been documented for numerous preparations, only a small number of microbial exopolysaccharides have undergone thorough structural investigations. They could be homopolysaccharides made up of a single type of monosaccharide residue, or they could be with up to four monosaccharidess in repeating unit structures ranging in size from di-hexa units. In at least one case, heteropolysaccharides without obvious ordered repeating structures, such as microbial alginates, could also be present. Any of these polymer kinds could have different kinds of acyl substituents in it. Yeasts and both gram-positive and gram-negative bacteria are among them. The characteristics and potential of the polysaccharides found on the surfaces of green and blue-green algae are not well understood, but it is likely that the slower growth rates of these microorganisms reduce the appeal of using their polymers for commercial production.

9.4.2. Carbohydrate-Rich Environments

One strategy is based on the observation that while relatively little nitrogenous substrate is available, many habitats contain significant amounts of waste products from industrial processes that are produced from carbohydrates. Numerous microorganisms produce more polysaccharides under these circumstances. Thus, studying such environments may yield valuable exopolysaccharide -producing microorganisms that can be screened, identified, and assessed after being isolated in pure culture. Acquiring microbial species from these types of environments has the added benefit of potentially enabling them to transform the waste (i.e., inexpensive) carbohydrates into an extracellular polymer that could have commercial value. Possible locales and their carbohydrate sources are:

- Sugar (beet or cane) refineries—sucrose
- Paper and pulp mills—cellulose
- Potato crisp or chip-processing plants—starch
- Brewery or distillery effluents—various
- Corn or vegetable-processing plants—starch, cellulose
- Activated sludge—various

Although enhanced treatment techniques should guarantee that the water released from these processes has a low Biological Oxygen Demand and low carbohydrate content, the liquors before treatment or in settling tanks or lagoons may serve as a breeding ground for microorganisms that produce polysaccharide. Examining samples from machinery or the buildings that house it is also necessary when processing plants are thought to be a potential source of microorganisms.

9.4.3. Identification

Unless the microorganisms are part of a well-characterized group, like the plant-pathogenic genus Xanthomonas or the nitrogen-fixing bacterial genus Azotobacter, identification may prove difficult. Many of the isolated bacteria have characteristics common to the family Pseudomonadaceae, so Hendrie and Shewan's (1966) method of identifying such cultures is a good one to use. To ascertain the Gram reaction, cell morphology, and whether the bacteria are motile, microscopic analysis is required. If flagella are found, they are examined under an electron microscope. Any pigmentation is recorded, and cultures' ability to fluoresce when exposed to UV light is examined. The process of glucose breakdown is investigated, along with oxidase activity and additional features. After examining the Gram-staining reaction and cell morphology initially, additional tests will be necessary for other bacterial groups. The reports by Gibbs and Skinner (1966) and Gibbs and Shapton (1968) contain these for several genera.

The identification process varies from that used for bacteria if the isolates that produce polysaccharides are yeasts. Tests for the utilization of nitrate, culture pigmentation, and sugars and related compounds can be very useful. Beech et al. (1968) provide comprehensive methods for identification using these techniques. According to Campbell (1970, 1971), the use of serological techniques, along with other conventional tests, has become more significant in the identification of yeasts. These techniques are used in the numerical taxonomy and identification of different yeast genera.

9.4.4. Selection and Further Testing

The isolates that produce comparatively large amounts of polysaccharides with intriguing viscosity characteristics will be selected for additional analysis. Strains that grow poorly under all observed culture conditions or that only produce small amounts of polymer ought to be discarded. This portion of the development program is the most labor- and equipment-intensive since it will likely require testing strains for a variety of traits in batch and continuous cultures in addition to preparing and characterizing the amounts of exopolysaccharides generated.

A simple growth medium should be used to test each of several variables. The variables include the media's buffering, mineral concentrations, aeration, pH value, and the concentrations of carbon and inorganic nitrogen sources. The reader is directed to Calam (1969) for general aspects of laboratory culture media formulation and Blakebrough (1969) for fermenter design. Other relevant information is to be found in 'Methods in Microbiology' edited by J. R. Norris and D. W. Ribbons (1969 onwards). The utilization of waste carbohydrate products as carbon and energy sources, as well as the degree of substrate conversion to polymer, are potentially significant from an economic perspective in the production of polysaccharides. Organisms that produce polysaccharides at yields of less than 40% and in batch times greater than 90 hours are unlikely to be very successful commercially.

9.5. THE PRODUCTS

9.5.1. Chemical and Physicochemical Properties

Identifying the chemical and physical properties of polysaccharides at an early stage will be relevant. The reader is referred to the numerous texts on for specifics, but it will be challenging, if not impossible, to determine the impact of different fermentation manipulations on polymer structure until a fairly clear understanding of certain basic structural features is obtained (nature of sugars, sugar ratios and acyl substituents, positions of linkage, and molecular weight). Also, having as much chemical and physical data as possible is crucial for writing patents and applying to food authorities. Physically, it is necessary to ascertain whether and how the polysaccharides form gels, as well as the flow behavior of aqueous solutions.

9.5.1.1. Flow Behavior

A significant number of separated polysaccharides will exhibit non-Newtonian characteristics, meaning that the apparent viscosity of the polymer solutions will not be a constant function of the shear rate. Therefore, it will be crucial to determine the rate of shear, perhaps using the "Power Law" method. In actual use, capillary viscometers and rotational viscometers with vague shear boundaries are laborious to operate and yield somewhat inaccurate findings. It is strongly recommended that the reader give serious consideration to cone and plate instruments, such as the Wells Brookfield, Shirley Ferranti, Haake Rotovisco, Contraves Rheomat, and Hercules models, which function at specific rates of shear. The fact that many viscometers require several 100 milliliters to operate and that the volumes of culture available for resting in laboratory fermentations are frequently small (2–10 ml) is another practical consideration. The following characteristics need to be ascertained after a suitable amount of dry polysaccharides was obtained for assessment. (i) relationships between shear stress and shear rate at standard concentrations and temperatures; (ii) temperature and apparent viscosity at defined shear rates; (iii) pH and apparent viscosity at defined shear rates; (iv) salt and apparent viscosity at defined shear rates; (v) concentration and apparent viscosity at defined shear rates; (vi) shear stability, (the impact of elevated shear on the structure of polymers); (viii) stability during freeze-thaw cycles, which refers to the influence of repeated cycles on the polymer structure.

9.5.1.2. Gel Properties

It will be necessary to conduct a number of empirical tests based on the gel behavior of known polysaccharide systems in order to determine whether and how the polysaccharide forms stable gels. Polysaccharide aqueous solutions should be exposed to various conditions, such as (i) extremes in pH; (ii) temperature swings; (iii) the addition of polyvalent cations; and (iv) heating and cooling in conjunction with other gums, for example, quaternary ammonium compounds, carob, and guar. It is sufficient to evaluate gel or precipitate formation visually. However, using tools like the Instron gel tester will be necessary for a more accurate assessment of the gel strength in terms of elasticity and rigidity. Biological testing will eventually be necessary to guarantee that the product is free of toxicity and undesirable side effects. A decision to continue developing products cannot be made without considering all of this and additional pertinent data, as well as the required costings.

9.5.2. Biological Criteria

There are two issues that could arise when isolating exopoly polysaccharide-forming microorganisms and limit the number of truly viable strains. The potential risk to workers, particularly during large-scale production, and the ethical requirements for the finished product—all pathogenic microorganisms must be effectively removed from the product—require that certain species that produce polysaccharides be disregarded. These pathogens may be human or animal. The yeast Cryptococcus neoformans and the opportunistic bacterial pathogen Pseudomonas aeruginosa (a possible source of microbial alginate) are two examples of species that are unusable for these reasons. Plant pathogen use presents another possible risk. Numerous pathogenic bacterial species for plants generate exopolysaccharides. Members of the genera *Corynbacterium, Pseudomonas, Xanthomonas, and Agrobacterium* are examples of such species. It goes without saying that the cultivation of these bacteria on a large scale can only be considered if the system for their growth and harvesting is totally contained, preventing environmental contamination, and if the final product is treated in a way that ensures the bacteria are completely destroyed. For example, laws governing the use of plant pathogens are in place in Britain. Certain species are considered legally non-native to the nation; therefore, before cultivating these species, a license from the Ministry of Agriculture, Fisheries, and Food must be obtained. Regarding the use or importation of microbial cultures, other nations have their own laws.

The is produced needs to be free of harmful microbial components, and there should be limitations on the use of microorganisms that have the potential to be harmful to humans, animals, or plants. This is particularly true if the polysaccharide is meant for use in the food, cosmetic, or pharmaceutical industries. Microorganisms can produce extracellular or intracellular toxic products. The latter are typically mold-derived aflatoxin-type compounds or proteins, if the species is bacterial. When polysaccharide isolation is carried out after precipitation using organic solvents, certain toxin types may be eliminated through solvent dissolution or destruction, respectively. Cell-bound substances that are challenging to fully extract from the viscous polysaccharide solutions are probably a bigger concern. Therefore, the cell wall of gram-negative bacteria contains lipopolysaccharide. While the bacteria per se are not harmful to humans or animals, the lipopolysaccharide has endotoxic qualities. When consumed in large enough quantities, it can cause symptoms like nausea, vomiting, and diarrhea that are similar to salmonella food poisoning. It goes without saying that a comparatively high amount must be consumed, and purified exopolysaccharide preparations are unlikely to meet this requirement. Nevertheless, there may be a chance that workers who are exposed to the material on a regular basis will become sensitized.

The Ministry of Agriculture, Fisheries, and Food currently sets regulations for the use of stabilizers and emulsifiers in food in Great Britain. On June 18, 1974, the Council of Ministers adopted a directive (74/329/EEC) on emulsifiers and stabilizers for food use, which also applies to member countries of the European Economic Community. Due to this directive, a statutory list of approved stabilizers and emulsifiers will be created. As a result, no food that is sold, consigned, delivered, or imported into the Economic Community may contain any additional emulsifiers that are not allowed. The Food Chemicals Codex (1972) lists alginic acid, sodium, potassium, ammonium, and calcium alginates, as well as propylene glycol alginates, in the new list. This contains xanthan gum but does not currently contain bacterial alginates. For this microbial product, there are two more requirements: the nitrogen content must not exceed 1 percent and there must be no living bacteria present. Additionally, all emulsifying agents must meet general purity standards for heavy-metal ions like arsenic, lead, copper, and zinc. Other non-EU nations have comparable laws governing food additives. The xanthan gum specifications are particularly interesting because they might reveal the purity requirements for any future microbial products added to the approved list. Although there is no need for such strict standards when preparing materials for non-food uses, it is clearly preferable to have no live microorganisms present. The nitrogen content permitted in the specification for xanthan gum is considerably higher than that obtained in laboratory preparations of the polysaccharides from *Xanthomonas campestris* (Jeanes et al., 1961) or of the polymer from the yeast *Cryptococcus laurentii var. flavescens* (Jeanes et al., 1964). The amounts of nitrogen in the preparations were 0.08% and 0.03%, respectively. It is not good to have relatively high levels of nitrogenous impurities present because they indicate the presence of dead bacterial cells and because they provide nutrients that encourage the growth of microbial contaminants. The producers of xanthan gum advise adding preservatives to polymer solutions before storing them, and it appears likely that the growth of contaminating microorganisms will affect the nitrogenous material rather than the polymer itself. These concerns would apply to any microbial exopolysaccharides held in solution; the issue is more acute for cruder grades used in applications other than food or medicine, where the polysaccharide solutions must be kept at room temperature or higher for extended periods of time.

In the United States of America, unless an ingredient is generally recognized as safe (GRAS) for the purposes for which it is intended, it is considered a food additive and is subject to regulation. Therefore, the so-called GRAS list acts as the prerequisite for the acceptance of microbial (and other) polysaccharides added to food. The Food and Drug Administration is in charge of keeping an eye on this list, and as of right now, xanthan gum is allowed under their approval. Foods can use this a thickener, stabilizer, emulsifier, suspending and bodying agent, and foam enhancer. Rocks (1971) has covered a number of potential applications as well as an overview of the key characteristics of xanthan gum solutions. Almost 12 years were spent on the xanthan gum test program, which involved feeding animals. Dogs received daily intakes of up to two grams per kilogram of body weight (Robbins et al., 1964). Animals fed the highest levels of polysaccharide did not exhibit any toxic effects, other than persistent diarrhea. Additionally, the gastrointestinal tract did not exhibit any polysaccharide degradation (Booth et al., 1963). When it comes to the development and introduction of other microbial polysaccharides, the acceptance of xanthan gum for use in food and other applications could provide a helpful precedent.

9.5.3. Modification of Products

For all possible uses, the polysaccharide in its isolated and purified form from culture might not be the best option. Its potential applications can most likely be expanded through chemical or enzymatic modification.

9.5.3.1. Chemical Alteration of Polysaccharides

In their natural state, many microbial exopolysaccharides, including xanthan gum and bacterial alginates, are frequently acetylated. The polymer can be chemically modified most easily by treating it with a mild alkali, even though the presence of acetyl groups makes the polymer more lipophilic than it would otherwise be. The solubility in water and other solvents will probably change if acetyl groups are removed, but the general rheological properties of polysaccharide solutions might not be significantly impacted.

By starting with material that has a molecular weight significantly higher than necessary, dextran is modified through carefully regulated acid hydrolysis. Sodium sulfite is used as an antioxidant, calcium carbonate is used to maintain neutrality in thermal degradation, and acid with a pH of 2 is used in mild-acid hydrolysis. Another method for producing dextran sulphate from dextran is to sulphate it with chlorsulphonic acid in pyridine. The well-known sephadex line of adsorbents is made from various modifications of the neutral dextran polymer. They include materials with functional groups added and glucose units connected to the functional groups via ether linkages, as well as crosslinked polymers of different sizes. This modification of the original polysaccharides yields anion-exchange groups that can act as either cation exchangers (CM-Sephadex) or anion exchangers (DEAE-Sephadex). By adding hydroxypropyl groups, lipophilic activities can also be introduced.

Although changing all polysaccharides in this way isn't feasible or even required, the variety of dextran products shows how many different products are available. Alginates can also be chemically modified when certain characteristics are required. Triethanolamine alginate forms soft films and is soluble in both water and 75% ethanol. Despite being soluble in hot water, alginic acid amide gels when it cools. Propylene glycol alginate is used in products like beer and soft drinks where the acidity would cause the previously mentioned gel of alginic acid. In this case, the carboxyl groups are esterified in the free acid. This specific alginate derivative also finds additional uses in the creation of photographic emulsions.

9.5.3.2. Enzymic Modification of Polymers

Microorganisms that produce an enzyme that hydrolyzes their own product can occasionally be found to modify preformed polymers. A case in point are certain bacteria that produce hyaluronic acid and, at a later stage of their growth, hyaluronidases. Aside from producing enzymes that break down the polymer into smaller molecular weight fragments, certain microorganisms that form alginate can also modify the monosaccharides that make up the polymer through a process called epimerization. If the activities of the enzymes cannot be satisfactorily controlled, then obviously such modifications are not desirable. Dextran is the only microbial exopolysaccharides that is presently heavily dependent on enzymatic modification. To create material in a specific molecular-weight range, carefully treat dextran with the appropriate enzymes. Chemical modification or alteration of molecular weights with enzymes will likely find an increasingly important role as more biopolymers are introduced and an increasing variety of uses developed for them.

9.6. POTENTIAL FOR STRAIN IMPROVEMENT BY MUTAGENESIS

It is possible to proceed with further development once a strain of bacteria or yeast that produces a polysaccharides with commercially useful properties has been isolated and preliminary cultural studies have been finished. One of five objectives can be achieved by mutagenesis of the microbial strain: (i) increased production of polysaccharides; (ii) modification of the polysaccharides produced; (iii) modification of the microbial cells' surface properties; (iv) removal of polysaccharide synthesis; and (v) change of species.

9.6.1. Increased Polysaccharide Production

This necessitates a deep understanding of the relevant biosynthetic pathways and, if applicable, the genetic regulatory network. As is the case for a homopolysaccharide like dextran, the formation of the polysaccharide may be a relatively straightforward process in terms of the number of enzymes involved. As complex as the kind produced by *Xanthomonas campestris*, if the polymer is a heteropolysaccharide (Jeanes et al., 1961) or Salmonella species and *Escherichia coli* (Lawson et al., 1969; Sutherland, 1969), a large number of additional enzymes and genetic information are involved. We will use bacterial alginate as an example of a relatively simple system and compare it with the much more complex polysaccharides from *Xanthomonas campestris*. It appears that D-mannuronic acid residues are first converted into a linear homopolymer to produce bacterial alginate. An extracellular epimerase that was isolated from an Azotobacter species transforms the homopolymer into a heteropolymer (Haug and Larsen, 1971). D-mannuronic acid and L-guluronic acid residues are dispersed along the polymer molecule in an apparently non-uniform manner to form the product, which is a heteropolymer.

Assuming that the basic mechanism of synthesis is similar to that found in other bacterial exopolysaccharides, the formation of the homopolymer requires the production of GDP-mannose and its oxidation to GDP-mannuronic acid (Pindar and Bucke, 1975). Thereafter several enzymes assemble mannuronosyl residues on a carrier lipid and extrude the finished polymer from the bacterial surface. Although relatively little research on these fundamental processes involved in alginate synthesis has been performed, it is probable that, in two alginate-producing bacterial species, namely *Azotobacter vinelandii* (C. Bucke, unpublished results) and *Pseudomonas aeruginosa* (I. W. Sutherland, unpublished results), some at the least of the enzymic processes. Thus the complete process is probably analogous to that involved in production of the capsular or slime polysaccharides of *Klebsiella aerogenes* (Sutherland and Norval, 1970; Troy et al., 1971). In addition there is the unusual role of an extracellular epimerase.

The path of alginate formation that has been postulated suggests that there are three possible ways to increase the production of polysaccharides: by increasing the specific activities of the enzymes involved, by changing control mechanisms like feedback inhibition, or by increasing the amounts of precursors or cofactors that are available. Regretfully, there are currently no particular selection techniques available to produce mutants with elevated amounts of the enzymes necessary for alginate synthesis. Following the steps of mutagenesis, mutation expression, and plating on an appropriate medium, colonies should be observed. Notable morphological changes should be noted in particular, as should colonies that seem to produce more polysaccharides (or possibly also those producing a mucoid colony with a different viscosity from the wild type). Increased GDP-mannose pyrophosphorylase production may result in an increase in alginate since pyrophosphorylases are generally recognized as important enzymes in the regulation of nucleoside diphosphate sugars and their derivatives (e.g., Bernstein and Robbins, 1965). This effect might not be produced as frequently by other enzymes. Or, enhanced carrier lipid synthesis C_5 isoprenoid alcohol phosphate; syn. increased production of polysaccharides by the bacteria may result from the use of bactoprenyl phosphate. Other polymers found outside the membrane of gram-negative bacteria like *Azotobacter* or *Pseudomonas* species are also produced using the same carrier lipid. These are lipopolysaccharides and peptidoglycans. There is no constraint on this as there would be for the two structural polymers of the bacterial wall, as one must take into consideration potential spatial effects if significantly more than normal amounts of lipopolysaccharides or peptidoglycan were formed. However, in the presence of increased amounts of carrier lipid, a more likely outcome is the synthesis of greater quantities of exopolysaccharide. Unlike mutations that directly impact GDP mannose or polysaccharide formation-related enzymes, which are involved in polysaccharide synthesis, there is likely a way to isolate bacteria that can produce higher quantities of bactoprenol through selective means.

By binding with prenyl phosphate in the presence of divalent cations, bacitracin is known to prevent the dephosphorylation of the isoprenoid alcohol pyrophosphate (Stone and Strominger, 1971). In order to avoid the effects of the bacitracin, mutants with higher-than-normal cellular levels of carrier lipid should be isolated through selection for bacitracin resistance. Initial investigations involving *Klebsiella aerogenes* have demonstrated the possibility of obtaining mutants with significantly elevated bacitracin resistance. Additionally, testing these mutants revealed a higher rate of hexose 1-phosphate transfer from UDP-glucose to lipid phosphate compared to the original strains. The mutants also produced more polysaccharides in washed suspensions, or non-growing cells. This implies that the degree of polysaccharide synthesis may be influenced by the amount of isoprenoid lipid present in the cell.

9.6.2. Modification of the Polysaccharide

The enzymes that are impacted when microorganisms generate a modified exopolysaccharides are almost completely unknown. The majority of mutations influencing the formation of extracellular polysaccharides have an "all or none" relationship with the product. Since the majority of the steps in the final polymer's biosynthesis are sequential, the absence of one enzyme in the sequence prevents the formation of any polysaccharides. This may not always be the case, however, with acylating enzymes. The Enterobacteriaceae family of bacteria includes several species that produce polysaccharides with a common carbohydrate structure. Colanic acid is the name given to this polymer (Goebel, 1963). Later research revealed that the amount of acyl groups in the polysaccharides that was isolated from various species varied significantly. One strain generated a non-acetylated polymer despite the side chain's terminal galactosyl residue always retaining a carboxyethylidene or comparable substituent, presumably due to a shortage of the proper particular polysaccharide acetylase. It remains unclear how common this phenomenon is or how restricted it is. Given that different acyl residues are present on the exopolysaccharides of serologically identical capsulate bacteria in the genus Klebsiella (Sutherland, 1972), it appears likely that acyl groups can be lost or altered without compromising the ability to synthesize polysaccharides.

9.6.3. Alteration of Surface Properties

The ease of extracting extracellular polysaccharides can be significantly impacted by changes to the microbial cell's surface characteristics. For proper sedimentation, most bacteria with sizes between 2 and 6 μηι × 1–2 μηι would need a centrifugal force of 4 to 6,000 g. The polysaccharides cause the culture fluid's viscosity to rise to the point where much stronger centrifugal forces are needed. Harvesting yeasts and larger bacteria is not too tough. The culture has a propensity to "auto-agglutinate" if the bacteria's surface is changed, especially if the cells become "rough" due to the loss of some of the surface polymeric material (such as lipopolysaccharides from Gram-negative bacteria). This indicates that flocculation of the cells happens naturally in unstirred cultures; as a result, significantly less centrifugation is needed to recover bacteria from culture fluids. Naturally, this also holds true for the more viscous solutions created during the formation of exopolysaccharides. A disadvantage of producing mutants devoid of some of the polymeric material on their surface is that some of these mutants are "leaky," meaning that they leak cell components—particularly proteins that are typically found in the periplasmic space—into the growth medium, where they run the risk of contaminating the final product. Additionally, lysis may happen more easily than it would in organisms of the wild type, either in the culture or when the polysaccharide is being harvested by centrifugation. The polysaccharide product is subsequently contaminated by the cell components released during lysis, some of which—such as nucleic acids—may be very challenging to remove or destroy.

Microorganisms that primarily produce exopolysaccharides as tightly bound capsules on the cell surface can employ a different strategy. One way to get rid of the issues with extracting capsules from a lot of cells is to look for mutants that form slime (SI). These mutants are stable and only produce slime, not distinct capsules. The polysaccharides that have been studied from SI mutants and the corresponding capsular material of the parent strains have the same compositions, as far as can be ascertained by chemical and enzymic techniques.

Under some circumstances, the properties of a particular type of mutant found in the Klebsiella aero genes may be useful for the production of polysaccharides. According to Norval and Sutherland (1969), this particular set of conditional mutants exhibited the ability to only produce polysaccharides at the restricted temperatures once the bacteria had stopped growing or had been removed from the nutrient medium and placed in a buffer that contained salts, a source of carbohydrates, but no nitrogen. A two-stage procedure could be built around this. In nutrient-rich environments, the primary stage would be used to generate a lot of cells. The cultivated bacteria would be moved to a secondary fermenter so that polysaccharide synthesis could occur there without the need for additional growth.

There is always a chance that strains of bacteria that are used as a source of polysaccharides and are heterotrophic species could contract bacteriophages. As a result, there is no production of exopolysaccharides and viability is lost. Normal aseptic procedures and filtered air supplies are necessary to prevent bacteriophage issues, but stock cultures are always susceptible to infection. Therefore, keeping recognized phage-resistant strains on hand is advised. Although isolating them is not too difficult, it is important to make sure that all phages have been removed from the resistant derivatives, that the strain has not undergone lysogenization, and that the production of polysaccharides has not been impacted. Phage resistance is linked to receptor loss, as many phage receptors are fragments of the lipopolysaccharides found in the walls of Gram-negative bacteria (Lindberg, 1977). Therefore, it's possible that the phage-resistant mutants lost the enzymes necessary for the metabolism of intermediary carbohydrates, which also resulted in the loss of their ability to synthesize exopolysaccharides. The necessary kind of resistant mutant is one that affects only receptor synthesis and forms exopolysaccharides in the same way as the wild type.

9.6.4. Elimination of Polysaccharases

To date, it has not been demonstrated that many microorganisms that synthesize exopolysaccharides can also break them down. There are several exclusions. Certain microorganisms release exopolysaccharides during a particular phase of their growth cycle, which is subsequently broken down by the release of hydrolytic enzymes at a later stage. Thus, *Azotobacter vinelandii* synthesizes alginate and small amounts of alginase, *Streptococcus mutans* produces both insoluble dextran-type polymers and "mutanase," which degrades them, and Streptococcus species are capable of producing both hyaluronic acid and hyaluronidases. If a microorganism is employed to make a polysaccharides on a commercial basis and it subsequently ruins or degrades the product, this is obviously unacceptable. It is possible to carefully regulate the cultural environment to inhibit the production of polysaccharides both during growth and during product isolation. This is feasible in strains where the release of the hydrolytic enzymes and the synthesis of polysaccharides occur at distinctly different times. A far better option would be to employ mutants devoid of polysaccharides. After mutagenesis, microbial colonies can be examined to achieve this. In contrast, wild-type clones would eventually lose their mucoid appearance and resemble non-mucoid variants if exposed to extended incubation. Colonies lacking the enzymes are likely to remain mucoid under all incubation conditions.

9.6.5. Inter-Strain Transfer of Genetic Material

Strains belonging to the same species can share the genetic determinants responsible for polysaccharide synthesis. Transduction in *Escherichia coli* and transformation in *Streptococcus pneumoniae* have both been used to observe this (Markovitz et al., 1968). It might be able to transfer the relevant genetic material in a fully functional form into another bacterium by mobilizing the genetic material from a bacterial species that generates an exopolysaccharides with desired properties. By transferring genetic material from *Klebsiella pneumoniae*, nitrogen-fixing strains of *Escherichia coli* have been successfully created through a variety of related functions (Dixon and Postgate, 1972). The ability to form alginate from the pathogen *Pseudomonas aeruginosa* to other completely harmless Pseudomonas species could be one of the future uses for similar transfers involving polysaccharide synthesis. Alternatively, another species of this group of bacteria may be able to synthesize "xanthan gum" through introduction from the plant pathogenic *Xanthomonas campestris*; this would not interfere with national regulations for the control of plant diseases. Transfer of alginate synthesis from the slow-growing *Azotobacter vinelandii* to a different, faster-growing bacterial species is another potential scenario.

The genetic systems of the bacteria under investigation would require extensive research, and success rates cannot currently be predicted. There could be issues with the recipient bacteria's genetic information's stability. Similar to the nif genes in *Klebsiella pneumoniae*, which regulate nitrogen fixation, the original genes in *E. coli* were presumably chromosomal, but after being transferred from the donor, they were typically found on plasmids (extrachromosomal DNA), making them less stable.

REFERENCES

1. American Gas Association, American Petroleum Institute, & Canadian Petroleum Association. (1975). *Reserves of Crude Oil, Natural Gas Liquids, and Natural Gas in the United States and Canada, and United States Production Capacity as of December 31, 1974.*
2. Austrian, R., Bernheimer, H. P., Smith, E. E. B., & Mills, G. T. (1959). *Journal of Experimental Medicine*, *110*, 585.
3. Baird, J. K., Longyear, V. M. C., & Ellwood, D. C. (1973). *Microbios*, *8*, 143.
4. Beech, F. W., Davenport, R. R., Goswell, R. W., & Burnett, J. K. (1968). In B. M. Gibbs & D. A. Shapton (Eds.), *Identification Methods for Microbiologists* (part B, pp. 151–176).
5. Bernstein, R. L., & Robbins, T. W. (1965). *Journal of Biological Chemistry*, *240*, 391.
6. Blakebrough, N. (1969). In J. R. Norris & D. W. Ribbons (Eds.), *Methods in Microbiology* (Vol. 1, pp. 473–504).
7. Campbell, I. (1971). *Journal of General Microbiology*, *67*, 223.
8. Cannon, F. C., Dixon, R. A., Postgate, J. R., & Primrose, S. B. (1974). *Journal of General Microbiology*, *80*, 227.
9. Carlson, D. M., & Matthews, L. W. (1966). *Biochemistry*, New York, *5*, 2817.
10. Devlin, T. M. (n.d.). *Textbook of Biochemistry*. John Wiley & Sons.
11. Dixon, R. A., & Postgate, J. R. (1972). *Nature*, London, *237*, 102.
12. Federal Register. (1969). Xanthan gum. *Section 121. 1224*, 5376–5377.
13. Fischer, F. G., & Dörfel, H. (1955). Zeitschrift für Physiologische Chemie, *301*, 224.
14. *Food Chemicals Codex* (1972). (2nd ed.). National Academy of Sciences.
15. Goebel, W. F. (1963). Proceedings of the National Academy of Sciences of the United States of America, *49*, 464.
16. Harada, T., Fujimori, K., Hirose, S., & Masada, M. (1966). Growth and glucan (10C3K) production by a mutant of *Alcaligenes faecalis var. myxogenes* in defined medium. *Agricultural and Biological Chemistry, 30*, 764–769.
17. Haug, A. (1964). *Composition and Properties of Alginates* (Report No. 30). Norwegian Institute of Seaweed Research.
18. Jain, J. L. (n.d.). *Fundamentals of Biochemistry*. S. Chand & Company Ltd.

19. Jansson, P. E., Kenne, L., & Lindberg, B. (1975). *Carbohydrate Research*, *45*, 275.
20. Jeanes, A., Haynes, W. C., Wilham, C. A., Rankin, J. C., Melvin, E. H., Austin, M. J., Cluskey, J. E., Fisher, B. E., Tsuchiya, H. M., & Rist, C. E. (1954). *Journal of the American Chemical Society*, *76*, 5041.
21. Jeanes, A., Wilham, C. A., Tsuchiya, H. M., & Haynes, W. (1957). *Archives of Biochemistry and Biophysics*, *71*, 293.
22. Lawrence, A. A. (1973). *Edible Gums and Related Substances*. Food Technology Review No. 9. Noyes Data Corporation.
23. Lawson, C. J., & Symes, K. C. (1977). In P. A. Sandford & A. Laskin (Eds.), *Extracellular Microbial Polysaccharides* (Symposium no. 45, pp. 183–191). American Chemical Society.
24. Lawson, C. J., McCleary, C. W., Nakada, H. I., Rees, D. A., Sutherland, I. W., & Wilkinson, J. F. (1969). Biochemical Journal, *115*, 947.
25. Lehninger, A. L. (1984). *Principles of Biochemistry*. Worth Publishers.
26. Lindberg, A. A. (1977). In I. W. Sutherland (Ed.), *Surface Carbohydrates of Eukaryotes* (pp. 289–356).
27. Meltzer, Y. L. (1972). *Water Soluble Polymers*. Chemical Process Review No. 64. Noyes Data Corporation.
28. Murray, R. K., Bender, D. A., Botham, K. M., Kennelly, P. J., Rodwell, V. W., & Weil, P. A. (2022). *Harper's Illustrated Biochemistry* (29th ed.). McGraw-Hill Education.
29. Norris, J. R., & Ribbons, D. W. (1969). *Methods in Microbiology* (Vol. 3A). Academic Press.
30. Penman, A., & Sanderson, G. R. (1972). *Carbohydrate Research*, *25*, 273.
31. Pindar, D. F., & Bücke, C. (1975). *Biochemical Journal*, *152*, 617.
32. Satyanarayana, U., & Chakrapani, U. (n.d.). *Biochemistry*. UBS Publishers Distributors Pvt Ltd.
33. Sawai, T., Toriyama, K., & Yano, K. (1973). *Journal of Biochemistry*, Tokyo, *75*, 105.
34. Sutherland, I. W. (1972). In A. H. Rose & D. W. Tempest (Eds.), *Advances in Microbial Physiology* (Vol. 8, p. 143).
35. Sutherland, I. W. (1977). In P. A. Sandford & A. Laskin (Eds.), *Extracellular Microbial Polysaccharides* (Symposium no. 45, pp. 40–57). American Chemical Society.
36. Umland, C. W. (1974). *Presentation for the Federal Energy Administration: Enhanced Oil and Gas Recovery Symposium*. Washington, D.C.

37. Voet, D. J. (n.d.). *Principles of Biochemistry*. John Wiley & Sons.

38. Whistler, R. L., & BeMiller, J. N. (1965). In R. L. Whistler & M. L. Wolfrom (Eds.), *Methods in Carbohydrate Chemistry* (Vol. 5). Academic Press.

39. Zevenhuizen, L. P. T. M. (1968). *Carbohydrate Research*, *6*, 310.

40. Zhan, X. B., Lin, C. C., & Zhang, H. T. (2012). Recent advances in curdlan biosynthesis, biotechnological production, and applications. *Applied Microbiology and Biotechnology*, *93*(2), 525–531.

Chapter **10**

Production of Polyhydroxy Alcohols by Osmotolerant Yeasts

Contents

INTRODUCTION

Glycerol, erythritol, arabinitol, and mannitol are the polyhydroxy alcohols produced by microorganisms with the highest potential for use. These are all made from sugars in comparatively high yields by different yeast species. While glycerol can be obtained as a byproduct of the petrochemical industry and mannitol can be obtained from other sources, erythritol and arabinitol cannot be produced chemically in a way that is as simple and inexpensive as the microbial process. A few lesser-known polyhydroxy alcohols can also be produced by different microbial synthesis processes than the ones described above. For example, if D-xylose is used as the original substrate, a strain of *Pichia farinosa*, originally identified as *Pichia miso*, forms xylitol; two heptitols are formed as minor byproducts of this process. If suitable microbial species are found, a few other polyhydroxy alcohols could likely be formed by analogous processes.

Many microorganisms, particularly yeasts and organisms that resemble yeast and are salt- and sugar-tolerant, produce significant amounts of polyhydroxy alcohols. These fungi are found in nature close to flowers, fruits, and insects like bumblebees and domestic bees. They also act as spoilage agents for confections, jams, jellies, honey, and some kinds of preserved fruits. A few can be found in similar habitats and in salt meats. Certain species are used to produce a variety of salt-heavy fermented foods that are popular in Indonesia, Japan, and other countries in that region of the world.

All of the known metabolic pathways that are involved in the formation of the carbon skeletons of the different polyhydroxy alcohols have been identified. It can be assumed that combinations and modifications of the pentose phosphate and Embden-Meyerhof pathways will lead to the biosynthesis of all polyols. The nature of the control systems, which identify which of the various products are formed by different species and under different cultural conditions, is the source of the remaining issues.

It has now been proposed that the various polyols found in osmotolerant yeasts serve a purpose. They seem to function as compatible solutes and are crucial in keeping enzymes from being inhibited or inactivated when there is little water activity, such as when yeast is growing in artificial or natural media with high salt or sugar concentrations.

There are countless real-world and prospective applications for polyhydroxy alcohols. Erythritol and arabinitol have no known applications as of yet, however, the latter substance might be used in certain situations as a substitute for sweeteners. Although arabinitol has a nice, "cool" taste, there have been reports of digestive issues; these would need to be resolved before arabinitol could be used in food. The commercial applications of mannitol, and particularly glycerol, are too numerous to list here. Similar to ethanol, glycerol has long been a byproduct of the petrochemical sector. The price difference between the petrochemical and the fermentation product, however, may narrow due to the recent rise in crude oil prices. As a result, fermentation-produced glycerol could become economically viable and account for a sizeable portion of glycerol used in industry.

10.1. HISTORY

10.1.1. Organisms

While a diverse range of microorganisms can produce polyhydroxy alcohols, this chapter will focus solely on those produced by yeasts for discussion purposes. The majority of yeasts have been found to produce traces of one or more of these compounds, but the species that exhibit the strongest tolerance to high salt and sugar concentrations have produced the highest yields. Numerous osmotolerant yeast strains have been identified from diverse origins and given diverse appellations; however, the majority of them have subsequently been recategorized into a limited number of species. The majority of the highly osmotolerant species are found in *Saccharomyces rouxii*, *Saccharomyces cerevisiae* aromyces mellis, *Pichia farinosa*, *Pichia etchellsii*, and Schizo*Saccharomyces cerevisiae* aromyces octosporus.

The majority of the sugar-tolerant Saccharomyces species have been isolated from environments such as fermenting honey, spoiled canned fruits, fermented soft-centered candies, dried fruits, and byproducts of sugar refineries. Salt-tolerant strains of *Saccharomyces rouxii* and *Pichia farinosa* isolated from soy sauce and miso paste, as well as other strains of Pichia isolated from pickles, make up a second significant group of species and strains that have been isolated from salty environments. Onishi reported that several species of *Pichia, Hansenula, Debaryomyces, Candida*, and *Torulopsis* produce glycerol and arabinitol. A few of these species produced erythritol in trace amounts. A number of the yeasts that Onishi isolated were linked to multiple new species, while some were previously recognized species.

Because of their function as spoilage agents, the yeasts isolated from honey, jams and jellies, confections, pickles, and related foods were studied. However, the strains and species of *Saccharomyces rouxii*, in particular, that have been isolated from miso paste, soy sauce, and other similar products are crucial to the production of these foods. Currently, it is unknown if the production of polyhydroxy alcohols has a major impact on the end product's quality.

10.1.2. Process Development

Pasteur noted that between 2.5 and 3.0% of the sugar fermented during the brewing and winemaking processes was turned into glycerol. Research on the phenomenon was initially restricted to the creation of procedures for recovering glycerol from distillation residues. However, after it was found that altering the fermentation conditions could significantly boost the yield of glycerol, numerous large-scale procedures for producing this polyhydroxy alcohol by fermenting different sugars using yeasts for baking or brewing were developed. These procedures were created by Cocking and Lilly in England, Connstein and Ludecke in Germany, and Eoff et al. were founded in the United States on the use of *Saccharomyces cerevisiae* strains to disrupt the natural alcoholic fermentation process. This was achieved by adding "steering" agents, such as bisulfites or alkalis, which significantly reduced the amount of ethanol produced and increased the amount of glycerol. Byproducts of the process included ethanol and acetate or acetaldehyde (as the bisulfite complex), and approximately 25% of the initial sugar was transformed into glycerol. The recovery of the glycerol was hampered by the presence of relatively high concentrations of sulfites, bisulfites, or alkalis; for these and other reasons, the process was never profitable (Eoff et al., 1919).

The focus of later studies on the microbiological synthesis of glycerol was primarily on small adjustments and other modifications to the alkali and bisulphite processes, such as substituting different reagents for bisulphites in the complex formation process with acetaldehyde. The majority of these are merely of minor historical interest because none of them produced a commercially viable process or disclosed any novel fundamental ideas. A brief discussion of these investigations was included in a previous review. Nonetheless, a variation on the bisulphite method, created at the Forest Products Laboratory of the United States Department of Agriculture in Madison, Wisconsin, merits consideration. Centrifugation was used to recover the yeast from the spent mash and return it to the culture vessel, while automatic titration was used to continuously monitor the concentration of bisulphite in the culture. Additionally, sulfur dioxide was extracted and reused. The yields of acetaldehyde (11%), ethanol (17%), and glycerol (25%), based on the amount of sugar metabolized, could be achieved by running the process continuously. Glycerol reached a final concentration of 5% in the culture. Regretfully, no fresh approaches to the glycerol recovery conundrums were created.

Nickerson and Carroll proposed a potential remedy for the issues brought about by the steering agents' presence in the final culture, including slower and less predictable fermentations, a higher culture's susceptibility to contamination, and challenges in recovering the glycerol. They discovered that, in the absence of bisulfites, alkalis, or other steering agents, a yeast strain that was subsequently identified as *Zygosaccharomyces* acidifaciens appeared to convert roughly 22% of the sugar that was metabolized into glycerol. After the Dutch taxonomists abandoned the term *Zygosaccharomyces*, the yeast used by Nickerson and Carroll was eventually reclassified in the genus *Saccharomyces*, where it is currently known as *Saccharomyces bailii*. Other employees' investigations revealed that arabinitol was most likely present in some amount of the product. The fermentation process took roughly ten days under the essentially microaerophilic conditions of Nickerson and Carroll's work, and the utilization of sugar was not completed. Figard returned to using sulphites as steering agents after attempting to increase the yields of glycerol from starch hydrolysates using the same yeast strain. His yields showed little improvement, and the fermentation proceeded slowly and incompletely.

Studies on sugar-tolerant yeast strains were initiated in 1951 at the National Research Council of Canada's Prairie Regional Laboratory in Saskatoon, Saskatchewan. The yeasts used were mostly *Saccharomyces rouxii* and *Torulopsis magnoliae* strains that were isolated from flowers, fermenting honey, and dried fruits. The former species' strains yielded glycerol and arabinitol in varying ratios based on the culture conditions, while the latter strains produced glycerol and erythritol. Afterwards, it was discovered that strains of *Trichosporonoides oedocephalis* and *Endomycopsis capsularis,* a newly discovered species, produced polyhydroxy alcohols. *Trichosporonoides oedocephalis* generated comparable amounts of erythritol, while *Endomycopsis capsularis* yielded good amounts of arabinitol.

The process differed significantly from the' steered' fermentations in that it used aerobic growth conditions instead of micro-aerophilic or anaerobic conditions, had a higher concentration of sugar and a relatively high nitrogenous compound content in the medium, had a better rate of converting sugar to product, and produced a significantly higher yield of polyhydroxy alcohols. Polyhydroxy alcohol formation peaked and ethanol production was eliminated under ideal aeration conditions.

Researchers at the Forest Products Laboratory of the United States Department of Agriculture in Madison, Wisconsin, confirmed the findings from the Prairie Regional Laboratory. Numerous strains of "Zygo saccharomyces," the majority of which have since been reclassified as *Saccharomyces rouxii*, were tested, and the results they obtained were nearly identical to those obtained from the Prairie Regional Laboratory. They also isolated different species and strains of yeasts and organisms that resembled yeasts that produced different kinds of polyhydroxy alcohols. They discovered a strain of *Endomycopsis chodatii* that produced arabinitol, a strain of *Torulopsis magnoliae* (T2b) that produced good yields of glycerol, and an unidentified fungus that produced erythritol. The final species might have been the same or comparable to the one that researchers at Prairie Regional Laboratory isolated from honeycomb in California.

It has been reported that *Candida lipolytica* and *Candida zeylanoides* can produce mannitol and erythritol from n-alkanes, respectively. Many species of Aspergillus can also produce good yields of mannitol. Furthermore, Onishi and Suzuki have shown that the yeast *Trigonopsis variabilis* can produce erythritol from glucose or glycerol.

10.2. THE PROCESS

The normal growth processes of the various yeasts that produce polyhydroxy alcohols include the formation of these molecules, and the growth requirements dictate the requirements for polyol production. However, the growth and production conditions can affect the yields of the various polyhydroxy alcohols. The medium's composition—particularly the selection of carbon and nitrogen sources—aeration level, and temperature have the biggest effects on fermentation rates and yields.

10.2.1. Composition of the Medium

10.2.1.1. Carbon Sources

a. Sugars: When it comes to osmotolerant yeasts producing polyols, the type of carbon source relies on the type of yeast and the sugars it typically consumes. During growth on glucose, maltose, and occasionally galactose, trehalose, and sucrose, *Saccharomyces rouxii* produces glycerol and arabinitol. Maltose and trehalose are not assimilated by *Torulopsis magnoliae*, but glucose, galactose, raffinose, and sucrose are. On the other hand, *Pichia farinosa* utilizes a broader variety of common sugars, such as xylose, ribose, sorbose, cellobiose, and occasionally lactose and L-arabinose. All of the sugars used in the current identification scheme are used by *Endomycopsis burtonii* (Echodatii), except lactose, melibiose, Z)-arabinose, and L-rhamnose. Galactose, glucose, maltose, and sucrose are fermented and assimilated by *Trichosporonoides oedocephalis*. If the right circumstances are met, the majority of these sugars ought to be useful for the synthesis of polyhydroxy alcohols.

Practically speaking, very few sugars have been investigated as carbon sources for the synthesis of polyhydroxy alcohols. *S. acchrouxii* can convert maltose to glycerol and arabinitol, although the yields are not as high as when glucose is utilized. Sucrose is converted to glycerol by *Hansenula subpelliculosa* and *Torulopsis magnoliae*, arabinitol by *E. burtonii*, and erythritol by the yeast-like fungus that Hajny and his colleagues investigated. *Pichia farinosa*, sometimes known as *P. imiso*, exhibited behavior akin to *Saccharomyces rouxii* in that it could not form polyols from galactose or sucrose but could form them from glucose, fructose, and mannose. Onishi claims that although the latter two sugars were partially metabolized, a sizable amount of product was not produced. It should be noted that Onishi's strain of *Pi. farinosa* is a variant that uses sucrose; the type strain does not, according to the standard description of *Pi. farinosa.*

The yields and ratios of the products formed are influenced by the concentrations of salts and sugars. *Saccharomyces rouxii* yielded more glycerol when the glucose concentration was increased by approximately 50%, but the yield of arabinitol was largely unaffected. The rate of glucose utilization increased when the sugar concentration was above 30%, but the percentage of conversion to polyhydroxy alcohols decreased. The Wisconsin workers discovered that the same theories also explained their findings regarding *T. magnolia* glycerol production. As the glucose concentration was raised at first, yields rose; however, once an optimal concentration was reached, yields fell. The ideal glucose concentration was not set; rather, it was correlated with the rate of aeration, allowing for a significant increase in the optimum sugar concentration for maximum glycerol yield at the optimal oxygen availability. Hajny along with others achieved a 42.5% conversion rate of glucose to glycerol and a 7.9 percent final glycerol concentration. Similarly, Onishi discovered that his strain of *Pi. farinosa* could produce arabinitol and glycerol at a sugar concentration of about 29%. Only about 30% of the sugar was metabolized when he increased the sugar concentration above 48%, which resulted in a significant decrease in yield.

Furthermore, he discovered that *Saccharomyces rouxii's* could metabolize different sugars was impacted by the medium's high concentration of sodium chloride (18%). His strains used glucose in the presence of sodium chloride, but not galactose, maltose, or sucrose, which they typically used to varying degrees. Similarly, when sodium chloride was added to the medium, *Torulopsis etchelssii* and *Hansenula subpelliculosa* did not use galactose or maltose, respectively. Onishi tested a number of other yeast species, though, and found that high sodium chloride concentrations had no such effect.

B. Hydrocarbons: The most recent developments in yeast production of polyhydroxy alcohols have resulted from an investigation into species that utilize n-alkanes. Mannitol is generated from n-alkanes by *Candida lipolytica*, a species that is typically employed in the transformation of hydrocarbons into single-cell protein. What's more intriguing is how *Candida zeylanoides* produces erythritol. Hattori and Suzuki employed strain KY6166, an auxotroph that requires glycerol and has the ability to synthesize citric acid from n-alkanes. Erythritol formation started and citric acid production was inhibited if the pH was dropped to 3.0 or lower. Mannitol was produced instead of erythritol if the phosphate concentration was raised to 0.1–0.2 percent (as KH_2 $P0_4$). Hattori and Suzuki employed an aerated culture medium containing mineral salts and n-alkanes in the C_{12}–C_{15} range to achieve yields of 180 mg of erythritol per milliliter after 160 hours. After about 20 hours, or when cell growth had stopped, production got underway (Hattori, and Suzuki, 1974)

Typically, yeasts that use hydrocarbons, such as *Candida zeylanoides* and Clipolytica, do not convert carbohydrates into polyhydroxy alcohols. Both species exhibit strong aerobic metabolism, with either minimal or nonexistent sugar fermentation.

10.2.1.2. Nitrogen Sources

The majority of yeasts that are frequently used to produce polyhydroxy alcohols only use ammonium salts as a source of nitrogen; the only two that can assimilate nitrate as a source of inorganic nitrogen are *Trichosporon oedocephalis* and *Torulopsis magnoliae*. However, most yeasts are able to use a range of amino acids and other organic nitrogen compounds; in fact, complex nitrogen sources are typically employed in media that are used to produce polyhydroxy alcohol. Both yeast extract and commercial steep liquor provide enough nitrogen and a sufficient amount of growth factors.

Numerous studies have employed yeast extract in combination with different supplements to produce a range of polyols. Depending on the circumstances of the studies, yeast extract concentrations have varied from 0.5 to 4 percent, and the researchers haven't always agreed on the ideal concentration. The researchers at the Prairie Regional Laboratory obtained maximum yields at concentrations of roughly 1 percent, whereas the Wisconsin workers generally found decreased yields of all three polyols as the yeast extract concentration was increased beyond 0.5 percent. Lower polyol yields may have resulted from the conversion of a sizable portion of the sugar to ethanol under conditions of reduced oxygen supply, meaning that the increased cell populations brought about by an increased nitrogen concentration depleted the amount of dissolved oxygen in the medium that was actually available to the cells. In certain cases, the effects of increased nitrogen concentration and increased cell growth were combined. The data from Onishi et al. demonstrated the significance of this effect where yeast extract at 0.1 and 4.0 percent was utilized. The percentage of glucose converted to ethanol rose from 0.6 to 64%, while the percentage of glucose converted to polyols dropped from 49% to 6%. The aeration rate was typically adjusted in work at the Prairie Regional Laboratory to maximize the conversion of sugar to polyhydroxy alcohols and suppress the production of ethanol.

The majority of researchers supplemented the yeast extract with tiny amounts of urea, which led to marginally higher yields. An elevated urea concentration was necessary when yeast extract was replaced with corn-steep liquor, a more cost-effective source of nitrogen. Corn-steep liquor was insufficient as a source of nitrogen on its own. Although the pH of the fermentation was much lower than it was when yeast extract was used, adding calcium carbonate to the pH did not increase the yield. However, yields of polyols equivalent to those obtained with yeast extract as a nitrogen source were obtained when the concentration of corn-steep liquor was reduced to 0.75% and the concentration of urea was increased from 0.15% to 0.35%.

Several ammonium salts were attempted as additional sources of nitrogen, but they were not very successful. While ammonium lactate or tartrate produced yields roughly as good as those obtained with urea, the presence of ammonium sulphate, chloride, or phosphate in the medium typically resulted in lowered yields. By using buffered medium containing ammonium sulphate or chloride, Onishi was able to increase yields to some extent. However, the decreased yields obtained with these compounds may have been partially due to increased acid formation as the ammonium ion was used up. Nevertheless, the full rationale behind urea's advantageous action remains unclear.

It is possible to use other nitrogen sources with success. Hajny obtained moderately good conversions of sugar to product by using a medium containing blackstrap molasses, glucose, and urea for the production of arabinitol by *E. burtonii*. As additional sources of nitrogen, he and his colleagues were able to employ distiller's dried solubles, malt, and malt sprouts. To produce a mixture of polyhydroxy alcohols by Pi. farinosa, Onishi employed polypeptone, beef extract, yeast extract, and corn-steep liquor, with positive outcomes. Particularly, polypeptone effectively converted glucose into polyols. Additionally, casamino acids provided an adequate supply of nitrogen (Hajny et al., 1964).

Only inorganic nitrogen sources were utilized in the process by which *C. zeylanoides* and *C. lipolytica* produced mannitol and erythritol. As the maximum cell yield was reached after roughly 20 hours, growth was deemed satisfactory.

10.2.1.3. Sodium and Potassium Chlorides

Rather than being directly involved in the organism's metabolism, these salts are significant because of their impact on the osmotic tension of the medium. Concentrations of sodium chloride up to 18% are typically used in the production of soy sauce and other fermented foods, where *Saccharomyces rouxii* plays a key role, and the yeast strains are acclimated to it. Nonetheless, excessive salt concentrations can occasionally have a negative impact on the synthesis of polyhydroxy alcohols. When potassium chloride concentration was raised from zero to three milligrams (mM), Hajny et al. observed a significant decrease in glucose utilization along with a decrease in erythritol yield and an increase in yield of glycerol, which did not form in the potassium chloride-free medium. Onishi also discovered a change in the composition of the products that *Pi. farinosa* produced from arabinitol when 18% NaCl was present. The yeast produced both polyols in the absence of sodium chloride, but as the concentration of salt increased, the concentration of arabinitol decreased until it was almost completely absent at 6% NaCl, or 1 M. The only product that was formed was glycerol. The same change in the nature of the products formed as the concentration of salt was increased was observed in other yeast species that were tested.

Onishi also noticed that the yeasts were able to grow at much higher temperatures when there was a high concentration of salt present. This phenomenon is comparable to what Lochhead and Heron observed, where they discovered that the yeasts were not necessitating high sugar concentrations at 30°C, but were obligately osmophilic at 37°C. Researchers at Prairie Regional Laboratory saw the same effect of elevated sugar concentrations providing protection against high temperatures.

10.2.1.4. Phosphate and Other Mineral Nutrients; Growth Factors and Amino Acids

It now seems that the primary impact of these nutrients is to promote cell proliferation. Early research suggested that phosphate concentration variations might directly affect ethanol and glycerol yields; however, subsequent research showed that the effect was caused by increased cell growth and the ensuing depletion of oxygen from the under-aerated medium. As a result, there is a drop in glycerol and an increase in ethanol yield due to an increase in fermentation and a decrease in oxidative metabolism. With the exception of glycerol, changes in phosphate concentration had no discernible impact on the yields of polyhydroxy alcohols. For normal cell growth and maximum product yields, the amounts of other nutrient salts in the media typically used for polyol production are sufficient.

The strains and species of yeast that are used to produce polyhydroxy alcohol are known to need a variety of growth factors, such as thiamin and biotin. Nevertheless, glycerol or arabinitol yields were unaffected by the addition of growth factors or amino acids to the medium, suggesting that the organisms' needs for these substances are met adequately without the addition of these substances. Many amino acids can be used by the majority of the yeasts employed as their only source of nitrogen for growth, and they may also be used to produce polyhydroxy alcohol. Most strains can use some of the L-isomers and some of the D-isomers.

10.2.2. Aeration

Like *Saccharomyces cerevisiae*, *Saccharomyces rouxii* exhibits the Pasteur Effect when the level of aeration is changed. This means that the amount of glucose (or other fermentable sugar) that is converted to ethanol increases at low levels of aeration and decreases as the level of aeration is increased. In addition, the Prairie Regional Laboratory reported early on that the yield of glycerol rose as the ethanol yield dropped. Initial results suggested that the amount of aeration had no effect on arabinitol yield, but more research revealed some evidence that raising the aeration levels also increased arabinitol yields (Figure 10.1).

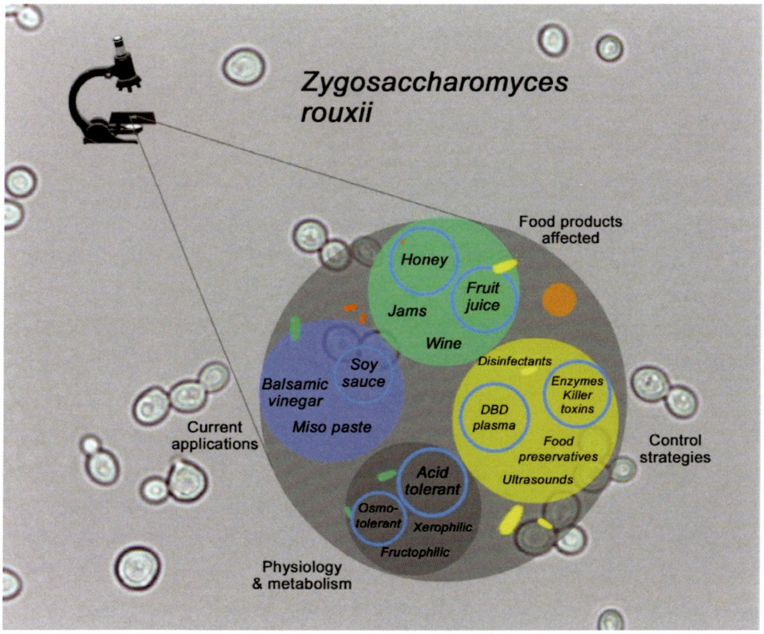

Figure 10.1. *Saccharomyces rouxii in a nutshell.*

Source: Escott C, Del Fresno JM, Loira I, Morata A, Suárez-Lepe JA. (2018). Zygosaccharomyces rouxii: Control strategies and applications in food and winemaking. Fermentation, 4(3), 69. https://doi.org/10.3390/fermentation4030069.

Overall, the findings from the Prairie Regional Laboratory were supported by the data collected by the Wisconsin employees. The yield of ethanol obtained in some of the early work was high enough to indicate that the levels of dissolved oxygen and aeration rates were insufficient for maximum yields of polyols. Although they used aeration rates at the other end of the optimal range for their later results, the yields were roughly the same as those at the Prairie Regional Laboratory. Hajny along with others additionally produced yields of glycerol, with an ideal aeration level falling between 42 and 200 mmoles 0 2 /l/h. When they studied the production of arabinitol by *E. burtonii*, they came to similar conclusions.

The density of the yeast population in relation to the rate of oxygen solution in the medium determines the rate of oxygen supply per unit of biomass, which is a critical factor determining the percentage conversion of glucose to polyhydroxy alcohols. This in turn depends on other nutritional aspects of the culture, particularly the kind and quantity of the nitrogen source and the medium's carbon to nitrogen ratio. A similar effect was noted by Campbell and Hegbom in their investigations into the variables influencing *Saccharomyces fragilis* cell growth.

10.2.3. Temperature

Early studies on the impact of temperature on *Saccharomyces rouxii*-based production of polyols showed that raising the fermentation temperature from 30° to 37°C significantly increased the concentration of glycerol to arabinitol. Later experiments did not show these results. Conversely, higher levels of arabinitol production led to higher yields of total polyols. As would be expected, there was a notable increase in the rate of glucose utilization at the higher temperatures. Peterson's group at the University of Wisconsin obtained similar results. *E. burtonii*, *T. magnoliae*, and different strains of "Zygosaccharomyces" produce polyol best at temperatures between 30°C and 35°C. At these temperatures, yields are slightly higher and rates of sugar utilization are faster. The organism that Hajny employed to produce erythritol reduced its yield at 35°C, but it increased its rate of sugar utilization and produced more glycerol at the expense of erythritol. It is unknown how these events relate to Lochhead and Heron's finding that these yeasts turned into obligate osmophiles at 37°C.

10.2.4. Product Recovery

There are no technical challenges in recovering polyols from aqueous solutions, which is basically the issue with extracting them from fermentation media. The process basically involves acidifying the broth, then settling or centrifuging the yeast cells out of the mixture as a rough primary clarification step. The majority of the water is then eliminated through evaporation, ideally in a vacuum. After filtering the gums out and adding hot ethanol, activated charcoal is added to the solution to decolorize it. In cold glycerol and ethanol mixtures, erythritol and arabinitol are only weakly soluble and separate from the solution as it cools. Following the filtration of the solids, the liquid is then subjected to atmospheric pressure distillation to recover the alcohol, while conventional vacuum distillation is used to recover the glycerol with a respectable yield. Due to arabinitol's solubility in glycerol-alcohol mixtures, which rises noticeably when even minute amounts of water are present, some of it is lost. Making sure that as much water is eliminated as possible prior to the crystallization stage is essential.

Recovery of these products essentially comes from alcohol-water mixtures if an organism is used that only produces erythritol or arabinitol, with little to no glycerol formed. Naturally, losses are significantly reduced if the water is removed entirely or almost entirely first. After dissolving the crude crystals in a small amount of water, recrystallization from ethanol usually yields satisfactory results for further purification of the product, regardless of the method used.

10.3. PHYSIOLOGY AND BIOCHEMISTRY OF POLYHYDROXY ALCOHOL FORMATION

Yeasts that are resistant to both sugar and salt belong to multiple genera and this trait is not species-specific. The primary interest of the investigators in these microorganisms was their role as spoilage agents. Early work on the physiology of these microorganisms was done on isolates from spoiled foods, especially honey. The finding that the moisture content of the substrate was the main factor limiting the growth of the yeasts in honey was particularly intriguing. The moisture content of properly ripened honey was too low, and as a result, the sugar content was too high, to allow yeast growth. On the other hand, honey that was exposed to a damp environment absorbed between 26 and 33% of the water. The critical water content required for the yeasts to grow was about 21%, which is significantly less than the concentration attained when honey is left to absorb moisture from the atmosphere.

The majority of *Saccharomyces* strains are likely the ones that were isolated by previous researchers, and the metabolic characteristics that have been reported are essentially those of this species: fermentation of only glucose and sucrose, the production of roughly equimolar amounts of carbon dioxide and ethanol, the formation of trace amounts of volatile acids, and enhanced fermentation in the presence of oxygen. *Saccharomyces rouxii* can tolerate a wide range of concentrations of sugar, salt, and acidity.

The growth factors that the osmotolerant yeasts that Lochhead and his colleagues studied required in different amounts. While each strain had different needs for pantothenic acid, inositol, and thiamin, they were all in need of biotin. Nickerson and Thimann discovered that in these haploid *Saccharomyces cerevisiae* species, a factor from *Aspergillus Niger* stimulated conjugation. Similar stimulatory effects were produced by riboflavin and α-oxoglutaric acid, but the *A. Niger* factor was not conclusively linked to them. Three of the leading researchers in this field have noted relationships between temperature and osmotic pressure. At 37°C, Lochhead and Heron's strains needed high sugar concentrations to grow; however, at 30°C, growth was observed at all sugar concentrations tested. In his research on salt-tolerant yeasts from soy sauce mashes, Onishi noticed the same thing: yeast growth at high temperatures required a high (18%) concentration of sodium chloride. Researchers at the Prairie Regional Laboratory discovered that, as long as the sugar content was higher than 30%, growth and the synthesis of polyhydroxy alcohol could occur at temperatures above 40°C, albeit at somewhat lower rates and yields.

At these temperatures, growth and the production of polyol ceased when the sugar concentration dropped below 30%. Nevertheless, S.E. Windisch isolated several *Saccharomyces aromyces* strains that did not exhibit a temperature-concentration effect. It is unknown why Windisch's findings differed from those of the other researchers. Furthermore, the nature of the effect is unclear. The behavior of temperature-sensitive osmotic remedial mutants of *Saccharomyces cerevisiae*, which use higher concentrations of potassium chloride in the medium to bypass a metabolic block at the restrictive temperature and permit growth, is analogous to this. In this instance, the mutant most likely produced a changed protein that isn't functional at the restrictive temperature unless higher salt concentrations are present to stabilize it. The same mechanisms that cause an enzyme or enzymes to become inactive in diluted media to be stabilized by high concentrations of sugar or salts in *Saccharomyces cerevisiae* mutants may also account for the osmotolerant yeasts.

The majority of important knowledge about the characteristics of salt tolerance in yeasts isolated from soy sauce and miso fermentations has been provided by Onishi. He discovered that, particularly at high temperatures, freshly isolated yeasts from soy mashes occasionally grew more slowly on regular media. In media containing sodium chloride as opposed to sugars, the limiting osmotic pressure was significantly lower, and the toxicity of different salts decreased as they went through the series: LiCl, $CaCl_2$, Na_2SO_4, $MgSO_4$, NaCl, and KCl. This series differs slightly from the findings at the Prairie Regional Laboratory, which suggested that magnesium salts were not as toxic as sodium chloride. The yeasts used by the latter investigators, however, were isolated from honeycomb and dried fruits and similar sources, and had not been previously acclimatized to high concentrations of salts.

Onishi discovered that cells cultured in sodium chloride-containing media grew more permeable, and that these cells needed a metabolizable substrate to survive in a phosphate buffer that contained 18% sodium chloride. Additionally, he looked for enzymes that could withstand or require high salt concentrations, but none could be found. Since high concentrations of salt inhibited the formation of glycerol in cell-free extracts, some or all of the enzymes needed for glycerol production were sensitive to high sodium chloride concentrations. Onishi proposed that osmotolerance in yeasts relied on an exclusion mechanism because there didn't seem to be any halo-tolerant enzymes in the cell-free extracts, but he provided no additional support for this theory.

The nature of osmotolerance in yeasts is now better understood thanks to Brown's recent work. The evidence presented previously appears to rule out the presence of halotolerant enzymes as one of the two possible explanations for the phenomenon. Additionally, if an exclusion mechanism is to blame, it cannot be one that causes the cell contents to be more diluted than the surrounding medium because there is no mechanism in place to stop water loss from the cell to the more concentrated solution outside the cell membrane. In addition to demonstrating that *Saccharomyces rouxii* cells have high concentrations of arabinitol while *Saccharomyces cerevisiae* cells do not, Brown has confirmed that the NAD+-specific isocitrate dehydrogenase from *Saccharomyces rouxii* has the same kinetics, water relationships, and electrophoretic properties as the same enzyme from *Saccharomyces cerevisiae*. According to Brown, at physiologically low levels of water activity, arabinitol acts as a compatible solute, shielding intracellular enzymes from inhibition or inactivation. More research may reveal whether, as Onishi proposed, the high intracellular arabinitol concentration is essentially a component of an active exclusion mechanism that enables the cell to actively exclude salts like sodium chloride and others that inhibit or inactivate any sensitive enzymes without also losing water.

10.3.1. Enzymic Processes and Regulatory Mechanisms in Polyhydroxy Alcohol Formation

10.3.1.1. The Pasteur Effect

Perhaps the most well-known and poorly understood phenomenon in yeast metabolism is the Pasteur Effect. Its role in controlling the ratio of ethanol to glycerol production accounts for its significance in the formation of polyhydroxy alcohols. Glycerol yield rises while ethanol yield falls under increasingly aerobic cultural conditions. Many theories have been put forth to explain the Pasteur Effect, but the majority of them have been overly simplistic. For a while, it was preferred that the fermentative and respiratory pathways compete with one another for inorganic phosphate, ADP, or both, but this theory is probably not the whole picture. According to a mechanism proposed by Sols et al., allosteric feedback mechanisms affecting multiple enzymes are responsible for regulating the balance between the fermentative pathways.

- NAD^+ -dependent isocitrate dehydrogenase, which catalyzes the first irreversible step on the pathway of citrate oxidation through the tricarboxylic acid cycle;
- phosphofructokinase, catalyzing the first irreversible step of the common glycolytic pathway from glucose 1-phosphate. The action of this enzyme is controlled by the content of ATP following anaerobiosis, and of ATP and citrate during aerobiosis,
- enzymes involved in glucose transport across the cell membrane.

The scheme as it is now has a lot of appealing aspects. It has not been investigated in osmotolerant yeasts, and initially it does not appear to explain the simultaneous rise in glycerol and other polyol production and fall in ethanol formation with increasing aeration. The other half of the phenomenon, as it is not observed in *Saccharomyces cerevisiae* to any significant degree, has not been attempted to be explained by Pasteur Effect investigators. Research on the Pasteur Effect in *Saccharomyces rouxii* may provide more encouraging results than *Saccharomyces cerevisiae* has so far produced.

10.3.1.2. Polyol Dehydrogenases

Numerous polyol dehydrogenases have been found in yeasts, some of which are required in the last stages of the formation of the different polyhydroxy alcohols in osmotolerant yeasts. The final reaction or reactions on the pathways leading to the formation of the various polyols are catalyzed by polyol dehydrogenases, which typically reduce one or both of the corresponding ketoses.

This scheme only explains half of the NADPH that is produced. The distribution of radioactivity in arabinitol formed from C-l- and C-2-labelled glucose is thought to be explained by transketolase-transaldolase catalyzed reactions, which can account for the remaining NADPH.

While other researchers have observed somewhat different reactions, Weimberg's scheme proposed a polyol dehydrogenase that catalyzed a reversible oxidation-reduction reaction producing arabinitol from glucose. For example, a reductase from *Saccharomyces rougeuxii* reduced xylulose instead of ribulose to arabinitol. Nonetheless, xylulose and ribulose were both reduced by crude cell-free extract, resulting in the presence of both enzymes in the intact yeast. Dehydrogenases found in brewer's yeast cause mannitol, sorbitol, dulcitol, and erythritol to oxidize.

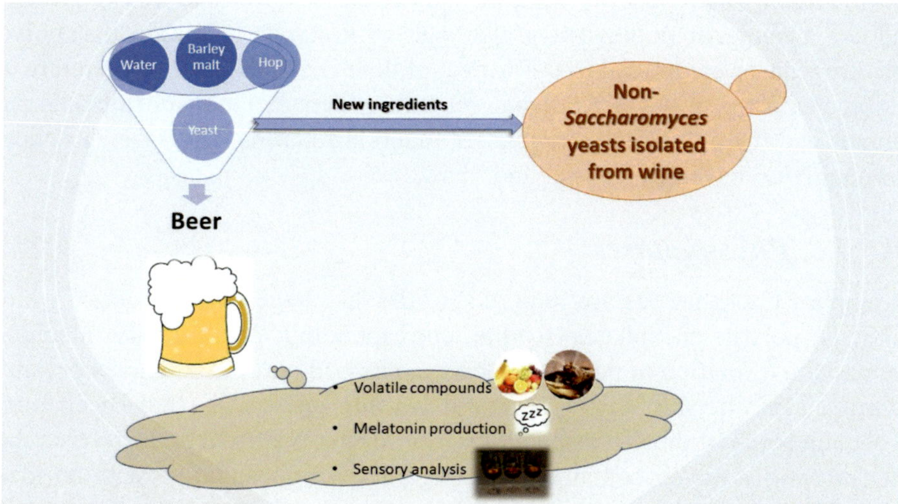

Source: Postigo V, Sánchez A, Cabellos JM, Arroyo T. (2022). New approaches for the fermentation of beer: Non-saccharomyces yeasts from wine. Fermentation, 8(6):280. https://doi.org/10.3390/fermentation8060280.

The enzyme facilitated the conversion of xylulose to arabinitol or arabinitol to xylulose. When it came to catalyzing the oxidation of various polyalcohols, such as mannitol and sorbitol to fructose, dulcitol to (possibly) tagatose, ribitol to ribulose, xylitol to xylulose, erythritol to erythrulose, and glycerol to dihydroxyacetone, Onishi's preparation was different from Blakley and Spencer's. This preparation needed Fe^{2+} ions and had an ideal pH of roughly 7.0, which was higher than the enzyme that Blakley and Spencer used. It might have included a variety of enzymes (Blakley and Spencer, 1962). Numerous other microorganisms have been reported to contain polyol dehydrogenases. An enzyme system found in *Penicillium chryosgenum* changes L-arabinose into L-arabinitol, which is then converted to L-xylulose. Small amounts of L-ribulose were also produced, most likely through a different enzyme-catalyzed reaction. Similarly, L-arabinitol is metabolized by *Aerobacter aerogenes* via a pathway that includes phosphorylated intermediates.

D-xylulose was directly oxidized from xylitol, and subsequently D-xylulose 5-phosphate was produced. Additionally, ribitol dehydrogenase, the synthesis of which is stimulated by ribitol and arabinitol, is present in Aerobacter aerogenes and only oxidizes ribitol. The presence of hydrogen acceptors restricted the amount of ribitol that could be oxidized in anaerobic environments. Two

NAD+-linked polyol dehydrogenases are present in Cellvibrio polyoltrophicus. These enzymes are induced by mannitol, sorbitol, and arabinitol. Solitol is converted to fructose by one enzyme, while arabinitol is converted to xylulose and mannitol to fructose by another. Numerous bacteria and yeasts produce and use a range of polyhydroxy alcohols, so it stands to reason that polyol dehydrogenases are crucial to this aspect of their metabolism. The literature on the biochemistry of acyclic polyols, including their metabolism in plants and animals and the polyol dehydrogenases found in mammalian tissues, has been thoroughly reviewed by Touster and Shaw.

10.3.1.3. Phosphatases

A range of phosphatases are required for the last stage of the process, which converts polyol phosphates to the corresponding polyhydroxy alcohols, because the formation of polyols appears to proceed via several phosphorylated intermediates. *Saccharomyces mellis* and *Saccharomyces rouxii* have been found to contain some of these phosphatases. The former researchers discovered that *Saccharomyces mellis* contained an inorganic pyrophosphatase with an ideal pH of 7.5, as well as an acid phosphomonoesterase with broad specificity. The presence of phosphate in the culture medium inhibited the formation of the former enzymes but not the latter. It is unknown how these enzymes contribute to the synthesis of polyhydroxy alcohols. Similarly, Ingram and Wood found that *Saccharomyces cerevisiae* possessed an acid phosphatase, but they were unable to identify a particular D-ribulose 5-phosphatase. However, the enzymes these researchers discovered are adequate to meet the requirements of the final reaction on the pathways leading to the formation of polyols.

10.3.2. Metabolic Pathways in Biosynthesis of Polyhydroxy Alcohols

10.3.2.1. Glycerol

The Embden-Meyerhof pathway produces glycerol in *Saccharomyces rouxii* and other osmotolerant yeasts, brewer's and baker's yeasts, and other fungi and organisms. In short, aldolase scission of fructose 1,6-diphosphate produces dihydroxyacetone phosphate, which is reduced and dephosphorylated to form glycerol.

The mechanism that regulates the relative yields of glycerol and ethanol is a more intriguing aspect of glycerol formation than the well-known pathway. In *Saccharomyces cerevisiae*, the majority of glucose is converted to ethanol during normal growth conditions. NAD+ is first reduced to NADH during the process, and then it is reoxidized when acetaldehyde is reduced in the last step of the ethanol formation process. Reduction of dihydroxyacetone phosphate to glycerol phosphate, which is subsequently dephosphorylated to yield glycerol, uses up a small amount of NADH. The challenge in fermentation-based glycerol production processes is to manipulate the Embden-Meyerhof pathway's reactions to suppress ethanol formation and direct more of the NADH generated earlier toward the reduction of dihydroxyacetone phosphate to glycerol phosphate (Figure 10.2).

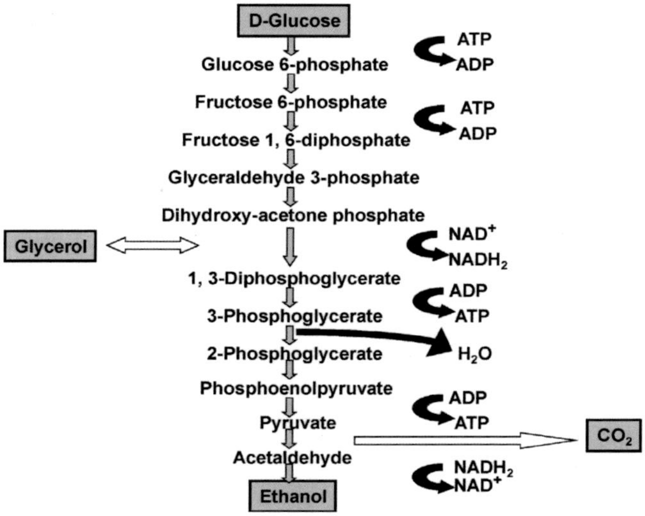

Figure 10.2. *Formation of ethanol, glycerol, and CO_2 from glucose by the Embden-Meyerhof-Parnas (EMP) pathway.*

Source: Stewart GG. (2016). Saccharomyces species in the production of beer. Beverages, 2(4), 34. https://doi.org/10.3390/beverages2040034.

This is accomplished in "steered" fermentation processes through the use of alkalis, sulfites, bisulfites, or other acceptors. When acetaldehyde binds to bisulfite, a complex is created that prevents NADH from reacting with it. NADH then reacts with dihydroxyacetone phosphate to produce glycerol. Similar actions are taken by other reagents, such as dimedone, which binds acetaldehyde before NADH can reduce it. A much higher percentage of the metabolized glucose is converted to glycerol, indicating that the control is not entirely achieved. However, ethanol is still formed in lower yield.

The Canizzaro reaction, in which one molecule of acetaldehyde is oxidized to acetate and another is concurrently reduced to ethanol, is what drives the alkaline process' control mechanism. This reaction was long thought to be nonenzymic, but since it happens even in the absence of live yeast cells, it is actually an alteration brought on by alkalis in the course of regular metabolism. Thus, one acetaldehyde molecule acts as a hydrogen donor, causing another molecule to be removed from the regular reaction system. This results in an excess of NADH once more, which can participate in the reduction of dihydroxyacetone phosphate, increasing the yields of glycerol.

In procedures for producing polyhydroxy alcohol with osmotolerant yeasts, steering agents are not required, and the factors that regulate the process are unknown, although they seem to be connected to respiratory metabolism. Maximum product yields require an adequate but not excessive oxygen supply. Growth of the organisms under suboptimal aeration conditions results in increased amounts of ethanol being formed and a significant decrease in the yield of glycerol. Similar to baker's yeast, the metabolic pathway is the same. Glycerol is labeled in the —CH_2OH carbons when it is formed from [l- ^{14}C]-glucose and in the —CHOH carbons when it is formed from [2–^{14}C]-glucose. The labeling in the ethanol produced also corresponds to the Embden-Meyerhof pathway's metabolism of glucose to produce these products.

10.3.2.2. Other Polyhydroxy Alcohols

The synthesis of D-arabinitol involves at least two processes. Initially, it was thought that the Warburg-Dickens pathway, which involves decarboxylation of 6-phosphogluconate, would be the primary pathway. This is supported by the fact that the specific activity of $^{14}CO_2$ formed from [1-^{14}C] -glucose was roughly four times that of $^{14}CO_2$ from [2-^{14}C] -glucose. But if the hexose monophosphate pathway is the only one involved in D-arabinitol's formation, then why is it labeled in C-l and C-5 when it is formed from [l-^{14}C] -glucose? Likewise, C-l, C-2, and C-4 are used to label D-arabinitol, which is produced from [2-^{14}C]-glucose. While the label in positions C–l is expected, the label in the other two cannot be explained by the straightforward decarboxylation of 6-phosphogluconate.

Nonetheless, these labeling patterns can be understood by presuming that the transketolase enzyme in this system is active. Two-carbon units from ketoses or ketose phosphates are transferred to a range of acceptors by transketolase. Here, fructose 1,6-diphosphate is the substrate and aldolase forms triose phosphate fragments, which label carbons 1 and 5 from [1- ^{14}C]-glucose and carbons 2 and 4 from [2- ^{14}C]-glucose. This results in the formation of 5-carbon chains. The label in C-5 arises from that in the triose phosphate which originates from C-l of the glucose, and is probably less than that in C-l of the arabinitol because some of the 2-carbon units are transferred by transketolase to unlabeled triose phosphate arising from carbons 4, 5 and 6 of glucose. Since the hexose monophosphate pathway is the only one that directly leads to the labeling of C-l in arabinitol formed from [2-^{14}C]-glucose, the significant excess of ^{14}C in C-l of arabinitol formed from [2-^{14}C]-glucose likely provides an approximation of the amount of arabinitol formed via this pathway. The label is further rearranged from [1- ^{14}C]- or [2- ^{14}C] -glucose by a number of subsequent reactions in the pentose phosphate pathway of the hexose monophosphate pathway, but these likely have little to no effect on the synthesis of D-arabinitol.

Similar to this, fructose 6-phosphate is likely primarily converted to erythritol by transketolase, leaving behind a 4-carbon fragment that is likely reduced and dephosphorylated. When [6–^{14}C] glucose is utilized as the substrate, the majority of the label is found in erythritol's C–4 chain. Other erythritol carbon atoms can be labeled by the other reactions in the hexose monophosphate pathway, but the transketolase-catalyzed reaction is the main source of this polyol.

In conclusion, Onishi discovered that *Pi-farinosa* produced heptitols from D-xylose, which are most likely the result of transketolase's transfer of 2-carbon units to pentose or pentose phosphate. The simultaneous high yields of xylitol were most likely the result of direct reduction of pentose phosphate or D-xylose. Thus, transketolase appears to be involved in the process by which osmophilic yeasts produce polyhydroxy alcohols.

To the best of our knowledge, no studies have been conducted on the metabolic pathways involved in the conversion of hydrocarbons into polyhydroxy alcohols. Nonetheless, enough information about the mechanisms underlying microbial hydrocarbon catabolism is available to propose potential routes. An n-alkane can oxidize one or both of its terminal carbons to produce the corresponding carboxylic acid. The remaining carbon chains can then be reduced in length by undergoing 3-oxidation, which produces 2-carbon units that can be added to the organism's other metabolic pathways for growth or the synthesis of other metabolites. For example, the condensation of two or three acetyl-CoA units could yield erythritol or mannitol. There is clearly a diversion of 2-carbon units from citrate synthesis to erythritol at the lower pH value in *Candida zeylanoides*, where citric acid production is increased by a decrease in pH value to below 3.0, but the mechanism for this is unknown.

The primary unresolved issue in the formation of polyhydric alcohols is the poorly understood metabolic control. The Pasteur Effect is linked to the regulation of glycerol formation in unsteered fermentations utilizing osmotolerant yeasts, which likely calls into question the viability of the currently proposed mechanisms for this effect. The mechanisms by which a given yeast or other fungal species will produce a particular polyhydroxy alcohol remain unknown.

Regardless of the species used, the culture medium typically contains traces of most possible polyols; however, significant yields of more than one or, at most, two are rare. The switch in Candida zeylanoides' production from citric acid to erythritol at pH values below 3 has not yet been explained. Though it doesn't address why any particular polyhydroxy alcohol should form, Brown's work has provided a reasonable explanation for the osmotolerant yeasts' ability to produce them.

Table 10.1. *Examples of S. cerevisiae Metabolic Engineering Studies for Bioethanol Production*

Purpose		Modification	Improvement
Lowering ATP yield	Introduction of futile cycle	Expression of *E. coli* FBPase insensitive to fructose-2,6-bisphosphate inhibition	8.8% higher bioethanol yield
	Introduction of ATPase enzyme	Heterologous expression of the F1 part of *E. coli* H^+-ATPase enzyme	10% higher bioethanol yield
	Increasing the unspecific alkaline phosphatase activity	Overexpression of *PHO8* gene	13% higher bioethanol yield
	Introduction of Entner-Doudoroff (ED) pathway	Expression of KDPG (2-keto-3-deoxy-6-phosphogluconate) aldolase from *E. coli*	–
	Relocation of sucrose hydrolysis from the extracellular space to the cytosol	Engineering the promoter and 5′ coding sequences of *SUC2* gene	4% higher bioethanol yield
Sustainable reduction of glycerol formation	Fine-tuning of glycerol 3-phosphate dehydrogenases (GPDH)	Introducing lower-strength TEF1 promoters to *GPD1* and *GPD2* genes	5% higher bioethanol yield

	Fine-tuning of glycerol 3-phosphate dehydrogenase	Engineering the promoter of *GPD1* in a *gpd2*Δ background	7% higher bioethanol yield
	Introduction of direct cofactor regulation strategies	Expressing *B. cereus gapN* gene, *E. coli frdA* gene and *mhpF* gene independently	Increased bioethanol yield
	Engineering of ammonium assimilation	Overexpression of *GLN1* and *GLT1* genes, deletion of *GDH1* gene	10% higher bioethanol yield
	Introduction of Calvin-cycle enzymes	Deletion of *GPD2* and heterologous expression of PRK and RuBisCO	15% higher bioethanol yield
Prevention of bacterial contamination	Introduction of phage endolysin on the cell surface	Anchoring recombinant peptidoglycan hydrolase, endolysin LysKB317, by using cell surface display	83.8% decrease in bacterial cell counts
Introduction of xylose catabolism	Construction of xylose reductase-xylitol dehydrogenase (XR-XDH) pathway	Genomic integration of *XYL1* and *XYL2* genes from *S. stipitis*, *XKS1* from *S. cerevisiae*, *BGL1* from *A. aculeatus*, and *GXS1* from *C. intermedia*	Xylose consumption rate of 6.62 g/L/h and an ethanol yield of 0.394
	Construction of xylose reductase-xylitol dehydrogenase (XR-XDH) pathway	Expression of *XYL1.2* from *S. passalidarum* and *XYL2* from *S. stipitis*	0.40 g g^{-1} CDW ethanol yield and 0.33 g g^{-1} CDW h^{-1} productivity
	Construction of xylose isomerase (XI) pathway	Expression of XI gene (xylA) from the bacterium *B. cenocepacia*	5-fold increase in xylose consumption and over 1.5-fold increase in ethanol production

	Construction of xylose isomerase (XI) pathway	Expression of XI from the bacterium *P. acidipropionici* with the co-expression of GroEL-GroES chaperonin complex from *E. coli*	Yield of 0.44 g ethanol/g xylose
Improving xylose catabolism	Lowering xylitol production	Deletion of *GRE3* gene in *S. cerevisiae*	Increased ethanol yield of 0.47 g/g of total sugars during fermentation of corn-cob hydrolysate
	Introduction of a synthetic reductive PPP for carbon dioxide recycling	Expression of RuBisCO and PRK enzymes in a *S. cerevisiae* strain harboring the XR/XDH pathway	Increased ethanol yield and reduced release of carbon dioxide
Increasing accessibility of lignocellulosic biomass	Conversion of cellulose into glucose	Expression of cassette carrying a cellulase gene from *A. gigas* Spix	37.7-fold higher ethanol yield
Increasing stress tolerance	Increasing osmotolerance	Overexpression of *TPS1* and *TPS2* genes in *S. cerevisiae*	Increased ethanol yield and osmotolerance, decreased glycerol production
	Increasing thermotolerance	Overexpression of *RSP5* gene	Thermotolerance at 41°C and ability to tolerate higher temperatures.

	Increasing ethanol tolerance	Overexpression of a truncated version of the *MSN2* gene in an industrial fuel ethanol strain	Increased ethanol tolerance (up to 14%)
	Increasing acetate tolerance	Overexpression of *HAA1* gene	The addition of acetate at 0.5% (*w*/*v*, pH 4.5) does not inhibit ethanol production
	Increasing HMF tolerance	Overexpression of *ADH1* and *ADH6* genes	Higher specific ethanol productivity in the presence of HMF
	Increasing acetic acid tolerance	Deletion of *ADY2* gene	14.7% increase in ethanol yield, in the presence of 3.6 g/L acetic acid
	Increasing coniferyl aldehyde tolerance	Overexpression of *ATR1* and *FLR1* genes	Increased coniferyl aldehyde tolerance

Source: Topaloğlu A, Esen Ö, Turanlı-Yıldız B, Arslan M, Çakar ZP. (2023). From Saccharomyces cerevisiae to ethanol: unlocking the power of evolutionary engineering in metabolic engineering applications. Journal of Fungi, 9(10), 984. https://doi.org/10.3390/jof9100984.

REFERENCES

1. Adler, L., Blomberg, A., & Nilsson, A. (1985). Glycerol metabolism and osmoregulation in the salt-tolerant yeast *Debaryomyces hansenii*. *Journal of Bacteriology, 162*, 300–306.
2. Blakley, E. R., & Spencer, J. F. T. (1962). *Canadian Journal of Biochemistry and Physiology, 40*, 1737.
3. Brown, A. D. (1974). *Journal of Bacteriology, 118*, 769.
4. Connstein, W., & Ludecke, K. (1919). *Chemische Berichte, 52*, 1385.
5. Eden, L., Van Nedervelde, M., Drukker, N., Benvenisty, N., & Debourg, A. (2001). Involvement of branched-chain amino acid aminotransferases in the production of fusel alcohols during fermentation in yeast. *Applied Microbiology and Biotechnology, 55*(3), 296–300.
6. Eoff, J. R., Lindner, W. V., & Beyer, G. F. (1919). *Industrial and Engineering Chemistry, 11*, 842.
7. Fossit, D., Mortlock, R. P., Anderson, R. L., & Wood, W. A. (1964). *Journal of Biological Chemistry, 239*, 2110.
8. Groleau, D., Chevalier, P., Tse Hing Yuen, T. L. S. (1995). Production of polyols and ethanol by the osmophilic yeast *ZygoSaccharomyces cerevisiae aromyces rouxii*. *Biotechnology Letters, 17*, 315–320.
9. Hajny, G. J., Smith, J. H., & Garver, J. C. (1964). *Applied Microbiology, 12*, 240.
10. Hattori, K., & Suzuki, T. (1974a). *Agricultural and Biological Chemistry, 38*, 581.
11. Huh, G. H., Damsz, B., Matsumoto, T. K., Reddy, M. P., Rus, A. M., Ibeas, J. I., et al. (2002). Salt causes ion disequilibrium-induced programmed cell death in yeast and plants. *Plant Journal, 29*, 649–659.
12. Kim, S. Y., Lee, K. H., Kim, J. H., & Oh, D. K. (1997). Erythritol production by controlling osmotic pressure in *Trigonopsis variabilis*. *Biotechnology Letters, 19*, 727–729.
13. Loray, M. A., Spencer, J. F. T., Spencer, D. M., & Figueroa, L. I. C. de. (1995). Hybrids obtained by protoplast fusion with a salt-tolerant yeast. *Journal of Industrial Microbiology, 14*, 508–513.
14. Lucca, M. E., Loray, M. A., Figueroa, L. I. C. de., & Callieri, D. A. S. (1999). Characterization of osmotolerant hybrids obtained by fusion between protoplasts of *Saccharomyces cerevisiae aromyces cerevisiae* and heat-treated protoplasts of *Torulaspora delbrueckii*. *Biotechnology Letters, 21*, 343–348.

15. Marvin, G. E., Peterson, W. H., Fred, E. B., & Wilson, H. F. (1931). *Journal of Agricultural Research*, *43*, 121.
16. Onishi, H. (1960). *Bulletin of the Agricultural Society of Japan*, *24*, 131, 226.
17. Peterson, W. H., Hendershot, W. F., & Hajny, G. J. (1958). *Applied Microbiology*, *6*, 349.
18. Prior, B. A., & Hohmann, S. (1997). Glycerol production and osmoregulation. In F. K. Zimmermann & K. D. Entian (Eds.), *Yeast Sugar Metabolism* (pp. 313–337). Technomic Publishing.
19. Rapin, J. D., Marison, I. W., von Stockar, U., & Reilly, P. J. (1994). Glycerol production by yeast fermentation of whey permeate. *Enzyme and Microbial Technology*, *16*, 143–150.
20. Roxburgh, J. M., Spencer, J. F. T., & Sallans, H. R. (1956). *Canadian Journal of Technology*, *34*, 248.
21. Spencer, J. F. T., & Gorin, J. F. T. (1960). The biosynthesis of erythritol and glycerol by *Torulopsis magnoliae*: Studies with C14-labeled glucose. *Canadian Journal of Biochemistry and Physiology*, *38*, 157–164.
22. Spencer, J. F. T., & Gorin, P. A. J. (1960). *Canadian Journal of Biochemistry and Physiology*, *38*, 157.
23. Spencer, J. F. T., & Spencer, D. M. (1978). Production of polyhydroxy alcohols by osmotolerant yeasts. In A. H. Rose (Ed.), *Primary Products of Metabolism* (pp. 393–425). Academic Press.
24. Spencer, J. F. T., Neish, A. C., Blackwood, A. C., & Sallans, H. R. (1956). *Canadian Journal of Biochemistry and Physiology*, *34*, 495.
25. Valero, E., Moyano, L., Millan, M. C., Medina, M., & Ortega, J. M. (2002). Higher alcohols and esters production by *Saccharomyces cerevisiae*: Influence of the initial oxygenation of the grape must. *Food Chemistry*, *78*(1), 57–61.
26. Van Eck, J. H., Prior, B. A., & Brandt, E. V. (1993). The water relations of growth and polyhydroxy alcohol production by ascomycetous yeasts. *Journal of General Microbiology*, *139*, 1047–1054.

INDEX

I

L

M

MIX

Papier | Fördert gute Waldnutzung

FSC® C083411

Kolibri 360 GmbH
Ferdinand-Jühlke-Straße 7
99095 Erfurt, Deutschland
produktsicherheit@kolibri360.de